WATER RESOURCES MANAGEMENT

WATER RESOURCES MANAGEMENT

Editor

Isidor Seeger

scitus
academics

Water Resources Management

Edited by **Isidor Seeger**

This Book is Printed on Acid-Free Paper. ♾

ISBN: 9781681179742

Library of Congress Control Number: 2017949112

SCITUS Academics LLC,

616, Corporate Way, Suite 2, 4766,

Valley Cottage, NY 10989

www.scitusacademics.com

Notice

Printed in United States of America.

Preface

Water, once an abundant natural resource, is becoming a more valuable commodity due to droughts and overuse. Security and sustainable development of our water resources is one of the key problems of the 21st century. Improved water management can make a significant contribution to achieve the Sustainable Development Goals related to availability and sustainable management of water resources. As with other resource management, this is rarely possible in practice. Water is an essential resource for all life on the planet. Of the water resources on Earth only three percent of it is fresh and two-thirds of the freshwater is locked up in ice caps and glaciers. Of the remaining one percent, a fifth is in remote, inaccessible areas and much seasonal rainfall in monsoonal deluges and floods cannot easily be used. At present only about 0.08 percent of all the world's fresh water is exploited by mankind in ever increasing demand for sanitation, drinking, manufacturing, leisure and agriculture. Effective and sustainable management of water resources is vital for ensuring sustainable development. In view of the vital importance of water for human and animal life, for maintaining ecological balance and for economic and developmental activities of all kinds, and considering its increasing scarcity, the planning and management of water resource and its optimal, economical and equitable use has become a matter of the utmost urgency.

The aim of this book is to focus attention on the management of surface water and groundwater resources. The contributions from outstanding scientists and experts provides detailed information about different topics and gives a general overview of the current issues in water resources assessment, development, conservation and control, emphasizing policies and strategies. It examines planning and design of water resource systems, and operation, maintenance and administration of water resource systems. This book will be of invaluable for the practicing professionals and students, mathematical modelers, hydrogeologists and water resources specialists.

Table of Contents

CHAPTER 1

Water Quality Assessment, Trophic Classification and Water Resources Management

Arkadi Parparov[1]*, Gideon Gal[1], David Hamilton[2], Peter Kasprzak[3], Alexandr Ostapenia[4]

[1]*Israel Oceanographic and Limnological Research, Kinneret Limnological Laboratory, Migdal, Israel*

[2]*Centre for Biodiversity and Ecology Research, University of Waikato, Waikato, New Zealand*

[3]*Leibniz-Institute of Freshwater Ecology & Inland Fisheries, Department of Stratified Lakes, Neuglobsow, Germany*

[4]*Belarus State University, Laboratory of Hydroecology, Minsk, Belarus*

ABSTRACT

Quantification of water quality (WQ) is an integral part of scientifically based water resources management. The main objective of this study was comparative analysis of two approaches applied for quantitative assessment of WQ: the trophic level index (TLI) and the Delphi method (DM). We analyzed the following features of these conceptually different approaches: A. similarity of estimates of lake WQ; B. sensitivity to indicating disturbances in the aquatic ecosystem structure and functioning; C. capacity to reflect the impact of major management measures on the quality of water resources. We compared the DM and TLI based on results from a series of lakes covering varying productivity levels, mixing regimes and climatic zones. We assumed that the conservation of aquatic ecosystem in some predefined, "reference", state is a major objective of sustainable water resources management in the study lakes. The comparison between the two approaches was quantified as a relationship between the DM ranks and respective TLI values. We show that being a classification system, the TLI does not account for specific characteristics of aquatic ecosystems and the array of different potential uses of the water resource. It indirectly assumes that oligotrophication is identical to WQ improvement, and reduction of economic activity within the lake catchment area is the most effective way to improve WQ. WQ assessed with the TLI is more suitable for needs of natural water resources management if eutrophication is a major threat. The DM allows accounting for several water resource uses and therefore it may serve as a more robust and comprehensive tool for WQ quantification and thus for sustainable water resources management.

Keywords: Water Quality, Trophic Level Index, Delphi Method, Sustainable Management, Lakes

1. INTRODUCTION

The most challenging problem of modern theoretical and applied hydroecology is to understand the fundamental principles of ecology for its application in effective management of water resources for both hydrological availability and water quality [1]. Quality is not absolute; the terms "good" or "poor" water quality have meaning only in relation to the use of water and the assessment of the user. Quantification of water quality (WQ) aims at describing the condition of a water body with reference to human needs. Investigations of the eutrophication phenomenon in the 1960s and 1970s, resulted in quantification of a trophic classification system [2,3]. WQ has been considered synonymous with 'trophic status' in many cases [4-6]. Further progress in WQ assessment is associated with implementation of optimization approaches to establishment of the natural resource sustainable management policies contributing to conservation of aquatic ecosystems within some desired reference condition [1,7]. We analyze and compare two key approaches to WQ quantification in relation to the increasing need for natural water resources management. The approaches include:

1) Quantitative modification of the expert panel method developed by Horton [8] and Ott [9] and integrated into the Delphi method (DM [10]). In the DM, the WQ ranking is completely defined by the needs of the water

resources uses and management [11]. DM has been implemented to WQ assessment for several natural waterbodies in the USA [12,13], New Zealand [14], Belarus [15] and Israel [16];

2) The trophic classification method [2] and its modification, the Trophic Level Index [4] (hereafter termed 'TLI'). This approach involves ranking lakes according to their productivity and nutrient regimes and thereby evaluating WQ changes based solely on trophic state responses to various forcing factors (e.g., nutrient loading), the TLI in many cases serves as a monitoring tool [17-19].

Comparative studies of the Delphi and TLI approaches to WQ assessment for water resources management are not available in the literature. In this study, we compare the DM and TLI using a quantitative WQ assessment based on results from a series of lakes covering varying productivity levels, mixing regimes and climatic zones. The lakes include Kinneret (Israel), Stechlin (Germany), Naroch Lakes (Belarus) and Rotorua Lakes (New Zealand), all of which have extensive databases, are major focal points for regional lake management. We will search for a reply to the following questions: Do these conceptually different approaches provide similar estimates of lake WQ? Do these approaches successfully indicate disturbances in the aquatic ecosystem structure and functioning? How well do these approaches reflect the impact of major management measures on the quality of water resources?

2. MATERIAL AND METHODS

2.1. Water Quality Assessment for Water Resources Management Purposes

Quantitative assessment of WQ should be based on the following principles [11]:

1) WQ system consists of water quality indices (WQI), and their permissible ranges corresponding to acceptable WQ.
2) The following functional correspondences should be established: between the ecological values of the WQI (e.g. nutrient concentration or Secchi depth) and some numeric rating value (R): R = f(WQI); between WQI and the intensity of the management measures (MM. e.g., economic activity in the watershed and water supply): WQI = Φ(MM). These two relationships allow establishment of a direct relationship between WQ and the management measures: R = F(MM)

3) Any system of WQ is subjective, reflecting a compromise between different stake holders and partners in water resources use and management.

4) An established WQI system should serve as a common language for all partners using a water resource, and in this sense, it is a tool for water resources management. Quantified WQ should also serve as a management target.

2.2. Brief Description of Study Lakes

Lake Kinneret (Israel) is a subtropical meso-eutrophic lake located at about an altitude of −210 m (i.e., below mean sea level). The lake water level depends on climatic inputs and withdrawal for water supply. Limnology of the lake is well documented [20]. The WQ system for conservation of L. Kinneret was assessed with the Delphi method [16]. The major environmental threat to the conservation of the lake ecosystem is the progressive lowering of its water level. The most critical issue for lake ecosystem stability and WQ is the progressively increasing proportion of cyanobacteria amongst the algal assemblage [21] and increase of water salinity above an assigned acceptable level of 240 mg $Cl\ l^{-1}$.

The Naroch Lakes (Belarus) consist of three connected lakes: Batorino (eutrophic), Miastro (meso-eutrophic) and Naroch (oligo-mesotrophic). The Naroch Lakes has been intensely studied [22]. The WQ system for the Naroch Lakes was also established based on the Delphi method [15].

Lakes Rotorua and Rotoiti are two lakes that form part of a complex of volcanic lakes known collectively as the Rotorua Lakes, in the North Island of New Zealand. A Trophic Level Index is assigned from routine monitoring conducted in the Rotorua Lakes [19] and is used as a basis for the implementation and assessment of management actions for the lakes [4]. An attempt to assess WQ has also been made using the Delphi method for Lakes Rotorua and Rotoiti (Hamilton and Parparov, unpublished data).

Lake Stechlin (Germany) is one of the few remaining oligotrophic clear-water lakes in Germany's South-Baltic Lake District. Long-term monitoring indicates significant changes in the structure and function of the lake ecosystem especially throughout the past two decades. The changes relate to a decrease in oxygen concentration in the deeper hypolimnion accompanied with substantial increase of hypolimnetic phosphorus concentration, and a trend of increasing of chlorophyll a concentration [23,24]. An attempt to assess WQ based on the Delphi method for Lake Stechlin has recently been completed (Kasprzak, unpublished data).

Relevant limnological properties of the study lakes are summarized in **Table 1**. In this study we use the following data sets: for Lake Kinneret - 1991 to 2008, Naroch

Table 1. Limnological variables for the study lakes. For the Naroch Lakes, the first value relates to the period from 1979 to 1982 and the second to the period from 1988 to 2000.

Parameters	Lakes						
	Kinneret[₪]	*Naroch Lakes*†			*Rotorua Lakes*[£]		*Stechlin*[€]
		Naroch	*Miastro*	*Batorino*	*Rotoiti*	*Rotorua*	
Surface area, km^2	160	79.6	13.1	6.3	33.5	80	4.3
Average depth, m	22-24*	8.9	5.3	2.8	31.5	10.5	22.8
Secchi depth, m	2.3	5.1/6.1	1.7/3.8	0.8/1.1	5.3	2.4	8.5
Total Nitrogen mg l^{-1}	0.6	0.9/0.4	1.1/0.7	1.6/0.9	0.3	0.3	0.45
Total Phosphorus μg l^{-1}	20	28/16	52/34	84/44	26	32	10
Chl *a*, μg l^{-1}	18.0	4.9/2.0	20.0/4.9	50/31	6.7	11.7	1.0-4.0
Primary production, g C m^{-2} yr^{-1}	650	78/81	142/218	140/200	431	800	110-205

*- depending on climate conditions; ₪ - [20,30]; † - [22]; £ - [31,32]; € - [23,24]

Lakes - 1979 to 2008, Rotorua Lakes - 1990 to 2005, and Lake Stechlin - 1992 to 2006.

2.3. Methodology of Trophic Level Index and Delphi Methods of Water Quality

2.3.1 Assessment

In the DM, WQ has five grades: from "very bad" (Rating 0 to 20) to "excellent" (Rating 80 to 100) [10,14]. The correspondence between the ecological values of WQI (e.g., nutrient concentration or Secchi depth) and some numeric Rating value ($0 < R < 100$) was established in a form of a rating curve. Long-term monitoring data, including frequency distributions, means and standard deviations of limnological parameters in the studied lakes were provided to separate expert panels for each lake. The experts assessed the "reference state" for each lake as a set of variables corresponding to time periods when lake water resource was suitable for all purposes (e.g., domestic water supply, recreation, fisheries). The panels were asked to choose about 10 parameters most indicative of WQ in the studied lakes, and to construct rating curves for each parameter, spanning the entire range of values observed in the lake. "Acceptable ranges" for the separate water quality indices were restricted to the range 60-100. Rating curves from individual panel members were averaged for each index. Piece-wise approximations of ratings and parameter values were used to construct graphic presentations of the temporal dynamics of the WQIs [16].

In the TLI, lake trophic state is classified by seven types: from "ultramicrotrophic" (rank 1) to "hypertrophic" (rank 7) [2,4]. Traditionally, the users of the TLI establish a correspondence between trophic status and water quality, i.e., by default oligotrophic lakes are considered to be of "good" water quality, and eutrophic lakes of "bad" water quality. Initially, the TLI system included three indices: Secchi depth (SD), concentrations of total phosphorus (TP) and chlorophyll a (Chl) [2]. This system has subsequently been supplemented with total nitrogen (TN) concentration [4]. The correspondence between ecological and Trophic Level values for individual indices (TL_i) was established in the form of a logarithmic function connecting the TL to four "trophic" parameters [2,4]:

$$TL_i = a_i + b_i LOG(Par_i) \tag{1}$$

where 'i' varies from 1 to 4, Par_i means SD, TN, TP and Chl, and 'a' and 'b' are coefficients.

For the objectives of this study, the TL varies from a value of 2 (oligotrophic = "excellent") to 7 (hypertrophic = "very bad") (we omitted the grade of 1, 'ultramicrotrophic', as irrelevant for this study). A similar correspondence in the DM is described by the rating curve R = f(WQI), in which the values of the WQ rating vary from 100 ("excellent") to 0 ("very bad"). The comparison between two systems was quantified as a relationship between the DM ranks and respective TL values:

$$R[TL] = 140 - 20*TL \tag{2}$$

The permissible range of the WQI can then be defined from the following expression: $60 \leq R[WQI] \leq 100$

We assumed that the conservation of aquatic ecosystem in some predefined, "reference", state is a major objective of sustainable water resources management in the study lakes. For the objectives of this study we used an operational definition of "sustainable management" suggested earlier [15], where water resources management should allow use of water resources while maintaining the quality of the resource in question within defined permissible ranges, i.e.,

$$\{WQI_L < WQI < WQI_H\}\hat{U}\{MM_L < MM < MM_H\} \tag{3}$$

where 'L' and 'H' correspond to the lower and upper permissible values of the WQ index and Management Measure, and Û means correspondence. The set of management measures corresponding to (3) forms the basis for a "sustainable management policy".

In the TLI, an aggregated WQ (TLI) is calculated as an arithmetic mean of four "trophic" indices [4]:

$$TLI = (TL_{Chl} + TL_{SD} + TL_{TP} + TL_{TN})/4 \quad (4)$$

Accounting for (2) and (4), the TLI was transformed into rating values as follows:

$$R[TLI] = 140 - 20*TLI \quad (5)$$

We also compared R[TLI] with a similar arithmetic mean value obtained for the same trophic indices assessed with the DM:

$$R[DM] = (R[Chl] + R[S] + R[TP] + R[TN])/4 \quad (6)$$

where R[Chl], R[S], R[TP] and R[TN] are the rating values of Chl, SD, TP and TN, respectively, assessed with the DM. For some of the studied lakes with suitable databases (Kinneret, Stechlin and Naroch Lakes), R[TLI] was also compared with the values of the composite water quality index (CWQI) calculated for entire set of WQIs established with the DM according to the formula [11]:

$$CWQI = \Sigma(R_i*(R_0-R_i)/\Sigma(R_0-R_k)) \quad (7)$$

where R_0 = 100, and $R_{i,k}$ are the rating values for WQI i or k (i, k = 1, 2,...n) from the WQ system established with the DM.

3. RESULTS

3.1. Effects of Water Resource Uses and Environmental Threats on Water Quality Assessment

The multiple-use nature of water resources and the wide range of environmental threats require significant expansion of the variables included in the conventional TLI system, even for the objective of diagnosing a change in the lake trophic state. Therefore, at the initial stage of WQ assessment, the expert panel, using the DM, supplemented the WQI systems in all of the study lakes with additional WQIs beyond the four "trophic" indices in the TLI system. These expanded systems allowed for indications of WQ changes due to specific major water resource uses and associated environmental threats. The additional variables could be categorized generally as

- eutrophication: given by the percentage of cyanobacteria in algal biomass (Lakes Kinneret and Rotorua and Naroch Lakes); percent of oxygen saturation in the near bottom water layer (Naroch Lakes and Lake Stechlin); total phosphorus at the end of stratification period in the deepest water layer (Lake Stechlin); percent of oxygen saturation in surface water (Naroch Lakes); volumetric hypolimnetic oxygen demand (L. Rotorua);
- pollutants indicative of organic pollution or relevant to human health: number of coliform bacteria (Lake Kinneret and Rotorua Lakes) and BOD_5 test (Naroch Lakes);
- food supply for fishes: zooplankton biomass (Naroch Lakes, Lake Kinneret, Lake Stechlin and Lake Rotorua);
- increase in salinity above accepted drinking water supply standards (240 mg l^{-1}): chloride concentration (Lake Kinneret);
- recreation: pH and area occupied by emergent vegetation (Naroch Lakes).

3.2. Rating Curves and Permissible Ranges of WQI in DM and TLI Methods

We illustrate the differences between DM and TLI by comparing the rating curves for Chl established for oligotrophic Lake Stechlin and meso-eutrophic Lake Kinneret (**Figure 1**). In the TLI, lowering of a lake trophic status (i.e., decrease in Chl) is equivalent to gradual improvement of WQ. In contrast to this, the DM shows that both an increase and a decrease of chlorophyll outside of some 'permissible' range (which varies between lakes) should be interpreted as WQ deterioration. Similar distinctions could be revealed for other WQIs under consideration (TN, TP and Secchi depth). This basic difference may lead to contrasting values for the permissible ranges for individual WQIs and management measures.

In the DM, permissible ranges for ecological values of individual WQIs can be described by the following equation:

$$\{WQI_L \leq WQIc \leq WQI_H\}\hat{U}60 \leq R[WQI] \leq 100 \tag{8}$$

where WQI_C is the current value of the index; WQI_L and WQI_H are the lower and upper permissible value of the index, and Û means "correspondence".

The TLI provides a threshold value, which can be described for TN, TP and Chl as follows:

$$\{WQIc \leq WQI_H\}Û60 \leq R[WQI] \leq 100 \quad (9)$$

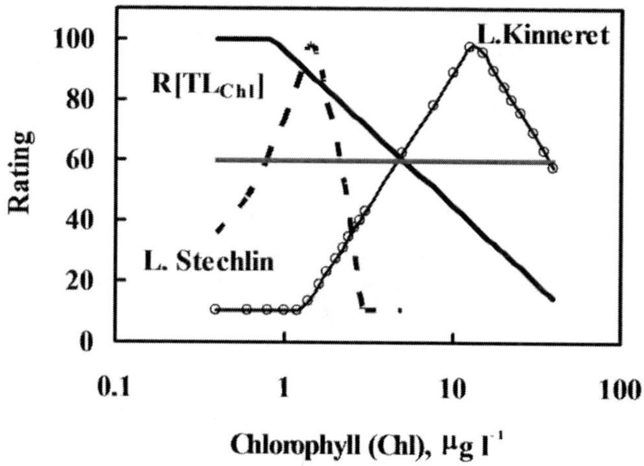

Figure 1. Rating curves of chlorophyll a assessed with the DM for Lakes Kinneret (round symbols) and Stechlin (dashed line), and calculated for the TLI (R[TL_{chl}], solid line) from ((4) and (8)). The horizontal line (R = 60) represents a boundary between acceptable and unacceptable WQ. Note logarithmic scale on X-axis.

For Secchi depth, this 'threshold' value is:

$$\{WQIc > WQI_L\}Û60 \leq R[WQI] \leq 100 \quad (10)$$

The main features of the two approaches can be illustrated using permissible ranges for chlorophyll (Chl) established for the study lakes (**Figure 2**). In the TLI, minimization of concentrations of TN, TP and Chl is indirectly considered as the desirable objective. An increase in Chl would therefore be taken as a sign of eutrophication, and therefore, of WQ 'deterioration'. According to this approach, the concentration limit of Chl for acceptable WQ (R[Chl] ≥ 60) is one-sided (dashed line in **Figure 2**), represented by the oligotrophic category of Chl ≤ 5.2 μg l^{-1} and is identical for all lakes. In the DM, one-sided rating curves are established only for pollution variables (e.g., densities of fecal coliform bacteria), or other components where increases will always be undesirable (e.g., percentage of cyanobacteria in Lake Kinneret). For most WQIs, the DM gives a two-sided, lake-specific permissible range. For example, the desirable Chl ranges are 0.8 to 2.1 μg Chl l^{-1} for oligotrophic Lake Stechlin, 13.0 to 34.0 μg Chl l^{-1} for Lake Miastro, and 7.0 to 27.0 μg Chl l^{-1} for eutrophic Lake Batorino.

Similar distinctions between the TLI and DM exist in the case of use of total phosphorus (TP) as a WQI (not shown). The TLI gives a single-sided permissible limit of TP ≤ 20 mg l^{-1} (identical for any lake), with concentrations higher than 20 mg l^{-1} considered to represent deterioration in WQ. The DM defines a two-sided permissible range for TP: 7 < TP < 19 mg l^{-1} for oligotrophic Lake Stechlin and 30 < TP < 100 mg l^{-1} for eutrophic Lake Batorino.

3.3. Compatibility of Aggregated Estimates of WQ

Conceptual differences in approaches to WQ quantification cause distinctions in the aggregated WQ estimates obtained from the same sets of ecological values. The relationship between R[TLI] and R[DM] in all three Naroch lakes is non-linear (bell-shaped curve). The TLI qualifies as 'acceptable' WQ only for oligo-mesotrophic L.Naroch and a few values from L. Miastro, while WQ in eutrophic L. Batorino is ranked as 'unacceptable' (**Figure 3**). The DM qualifies most of the aggregated WQ values in all three Naroch lakes as 'acceptable'.

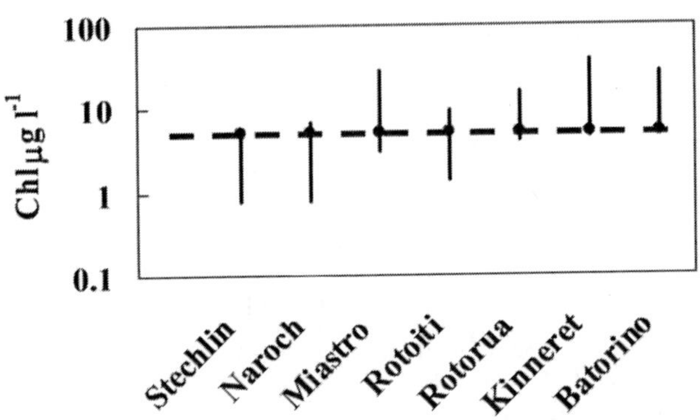

Figure 2. Comparison of permissible ranges for Chl a assessed with the DM (vertical lines) and TLI. The horizontal line represents the threshold values assessed with the TLI: 5.2 μg Chl l^{-1} The lakes on the X-axis are sorted according to their trophic state: from the lowest chlorophyll values in oligotrophic Lake Stechlin to the highest in eutrophic Lake Batorino.

Monitoring of temporal dynamics of WQ is one of the tasks of WQ quantification [15]. Comparison of the dynamics of the calculated aggregated estimates (R[TL], R[DM] and CWQI; ((4) to (7)) indicates different sensitivities in their representation of change over time. For oligotrophic L.Stechlin, the dynamics of both R[TLI] and R[DM], calculated for the four 'trophic' indices ((5) and (6), respectively), do not indicate significant changes in WQ of these lakes over the last 15 years (**Figure 4**). In contrast, the aggregated WQ calculated as CWQI, from an expanded set of WQ variables (7), indicates clear-cut trend in WQ deterioration. For L. Kinneret (not shown), the TLI indicates that WQ remains close to the boundary value that defines the difference between 'good' and 'intermediate' WQ. Conversely, the CWQI dynamics for L. Kinneret indicates gradual WQ deterioration from 'good' to 'intermediate'. Note that WQ deterioration in these lakes (as indicated by the CWQI dynamics) is associated primarily with the in-lake processes usually considered as signs of eutrophication: disruptions of phosphorus and oxygen regimes in hypolimnetic water (Lake Stechlin) and an increase in cyanobacteria abundance (L. Kinneret).

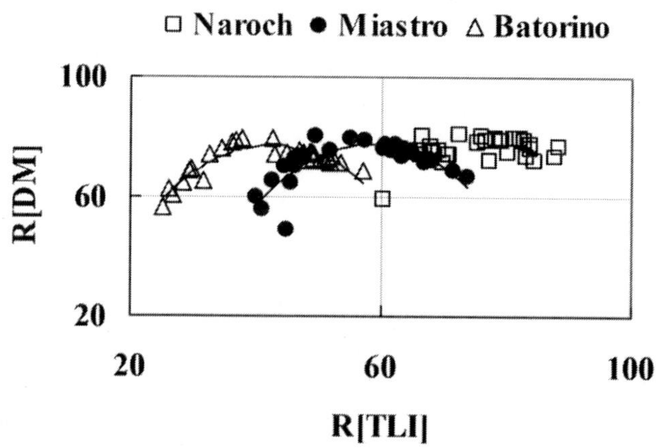

Figure 3. Scatter plots of the relationships between aggregated WQ estimates in the DMi and TLI (R[DM] and R[TLI]), respectively, for the Naroch Lakes (annual average values for 1978-2008). The lines represent best-fit regression lines. For the TLI, the acceptable WQ corresponds to areas to the right of the vertical line given by R[TLI] = 60. For the DM, acceptable WQ corresponds to areas above the horizontal line given by R[DM] = 60.

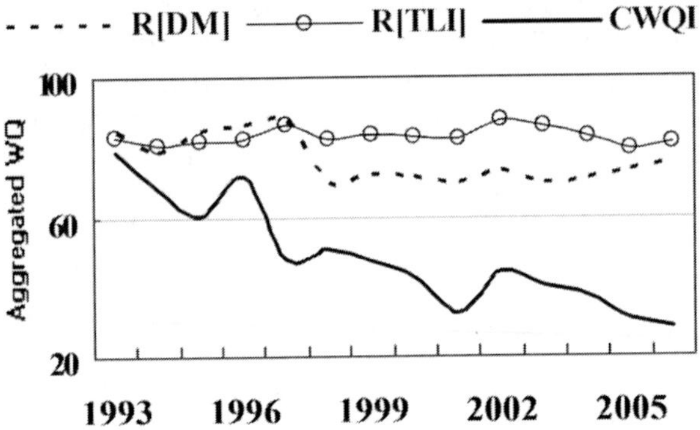

Figure 4. Time series of annual average estimates of aggregated WQ for L. Stechlin: R[TLI] (4&5), R[DM] (6) and CWQI (7).

3.4. Relationships between WQ and Management Policy

Existing models (e.g., [25,26]) relate TP loading to TP concentrations in a lake. Here we applied the model of Rekhow [27] to L. Kinneret as this model specifically includes the potential addition of an internal P load associated with an anoxic hypolimnion. Assuming stability of the average lake depth and hydraulic turnover rate, for L. Kinneret, the following relationship can be established:

$$\text{Pload}\ [\text{g m}^{-2}\ \text{yr}^{-1}] = 15.7*\text{TP}\ [\text{mg l}^{-1}] \qquad (11)$$

Permissible ranges for total phosphorus for L. Kinneret are as follows: TP ≤ 20 g l^{-1} (as obtained with the TLI from (8) and (10) and 7.5 ≤ TP ≤ 36 g l^{-1} (from the DM, [16]).

Substituting these values into (11) allows for an estimate of the permissible ranges of the external phosphorus load (Pload). The estimated maximum allowable Pload (based on TLI allowable TP values) is Pload ≤ 0.31 g m^{-2} yr^{-1} (**Figure 5**), which, given the TLI underlying assumptions, is universal for any lake, independent of its use as a water resource. In contrast, the permissible range of Pload into Lake Kinneret, based on the DM and its acceptable range of TP, is 0.12 to 0.58 g $m^{-2}yr^{-1}$ (**Figure 5**). Management of Lake Kinneret due to Pload regulations within this established permissible range should allow sustaining and conserving the Lake Kinneret ecosystem.

For L. Kinneret, the relationship between aggregated WQ (as R[TLI] and CWQI) was established for the lake water level under the assumption that the water level is predominantly determined by the volume of water pumped from the lake and thus is a management decision. In a water level range of −210 to −214 m of, there was no significant relationship between R[TLI] and lake water level (**Figure 6**).

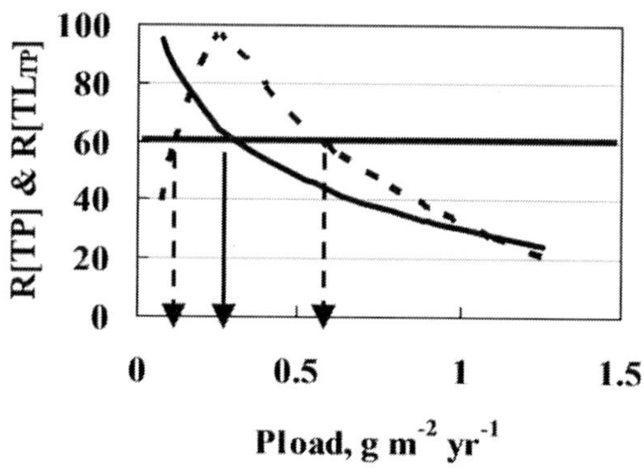

Figure 5. Relationship between management measure (phosphorus loading, Pload) and rating values of WQI (total phosphorus, TP): R[TL_{TP}] – obtained from (11) for L. Kinneret with the TLI (solid curve); R[TP] – obtained with the DM (dotted curve). The arrows indicate permissible ranges for Pload as regards to TP concentration (corresponding to R ≥ 60). Horizontal line corresponds to R = 60.

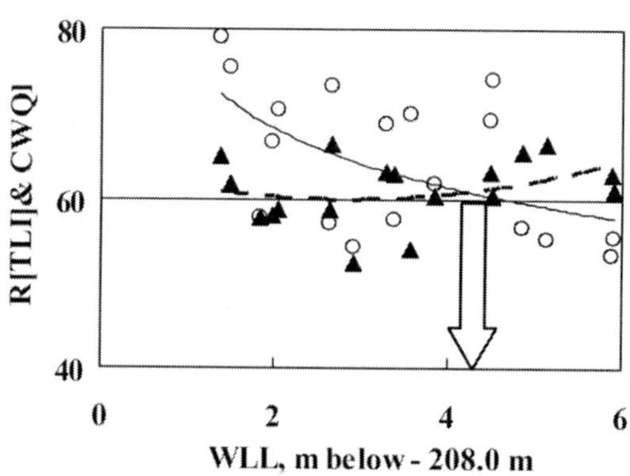

Figure 6. Scatter plots of the relationships between L. Kinneret water level lowering (WLL) as a management measure, and aggregated WQ: R[TLI] and CWQI (solid triangles and open circles, respectively). The best fit regression of R[TLI] vs. WLL (dashed line) is not statistically significant ($R^2 = 0.12$, $P > 0.05$), while the regression of CWQI vs. WLL (solid line) is statistically significant ($R^2 = 0.28$, $P < 0.05$). The block arrow indicates the permissible range of the lake water level lowering estimated with the DM.

Consideration of the aggregated WQ based on the expanded WQ system (CWQI) indicated a statistically significant (at $P < 0.05$) tendency for deterioration of the lake WQ with water level lowering (**Figure 6**). This dependence allowed estimation of a threshold water level value of approximately −213 m.

4. DISCUSSION

The problem of WQ quantification is equivalent to the quantitative solution of the task of "qualification", i.e., the terms like "good/bad", "improvement/deterioration" should be expressed in measurable units. This task is principally different from the task of "classification" (e. g., lake classification according to their mixing or trophic status). The developer of the quantitative trophic classification [28] warned: "An unfortunate misconception concerning trophic state is that the term is synonymous with the concept of water quality. Although the concepts are related, they should not be used interchangeably." Despite this warning, the trophic system of classification, under various forms, is widely used as a system of water quality [4-6,17,18].

Table 2. WQ ranks in Delphi and TLI-system (modified from (4) and (14)).

Water Quality	Delphi		TLI	
	Rank	*Descriptor*	*Rank*	*Descriptor*
Excellent	80-100	Eminently usable for all purposes	2-3	Oligotrophic
Good	60-80	Suitable for all uses	3-4	Mesotrophic
Intermediate	40-60	Main use and/or some uses may be jeopardized	4-5	Meso-eutrophic
Bad	20-40	Unsuitable for main and/or several uses	5-6	Eutrophic
Very bad	0-20	Totally unsuitable for main and/or many uses	6-7	Hyper-eutrophic

Freshwater ecosystems are the objects of intense multipurpose use [29]. The use of the water resource in question is key to the DM (**Table 2**). In this study, we assumed that the objective of sustainable water resources management is conservation of the lake ecosystem in some predefined, "reference", state (3). The TLI approach does not consider the intended use (**Table 2**), and, therefore, the system of WQ indices assessed with this approach tends to be absolute, independent of the uses of the water resource in question, and associated hydroecological problems, Presumably, implementations of the TLI are based on an indirect assumption that eutrophication (caused by excessive entry of nitrogen and phosphorus) is the main hydroecological threat to waterbodies under WQ assessment. This is the underlying distinction between the DM and TLI as regards to WQ quantification, clearly illustrated in Figures 1 and 2. The respective rating curves of almost all WQIs in the DM are bell-shaped curves (**Figure 1**). As a consequence, the ecological values of parameters such as S, Chl, TN and TP should be sustained within limits, (permissible ranges, (8)); excessive increase and decrease of the respective variables should be ranked as WQ deterioration. Therefore, the WQ sets assessed with the Delphi method are more suitable for the needs of sustainable water resources management. Note that differences between permissible ranges assessed with the TLI and Delphi tend to increase with the lake trophic status, especially for Chl (**Figure 2**). In the TLI, the rating value is a monotone function of the concentration of Chl, TN and TP and S (1). It should be noted however, that monotone rating curves are possible also in the DM, usually assessed for pollutants (e.g., E. coli in the Rototua Lakes) or cyanobacteria (%Cyano) in L. Kinneret. However, in some cases, assessment of the monotone rating curves for basic ecosystem parameters may lead to paradoxical results resembling those based on the TLI. According to the monotone rating curves for TN and TP assessed with the TLI for transitional Mediterranean waters [6], the best WQ should correspond to TN, TP and Chl concentrations close to zero (**Figure 1** in [6]). This means that complete distrophication of transient water ecosystems is considered, at least formally, as WQ improvement. This, basic, distinction, is a direct reason for the discrepancies among the relationships between aggregated WQ and management measures. The differences in estimates of the permissible Pload values for L. Kinneret (**Figure 5**) have obvious consequences when translating them to management objectives and measures: the DM allows more intense economic activity in the lake watershed (expressed in terms of Pload), than possible according to the TLI, without unacceptable damage to the lake WQ. The discrepancies between the water level lowering effects on aggregated WQ have a significant impact on the entire process of decision-making in practical water resources management: acceptance of the TLI approach would mean that water level lowering of Lake Kinneret below −214.0 m does not affect lake WQ. In contrast to this, the DM indicates that the lake water level lowering below −213.0 m leads to unacceptable WQ deterioration (**Figure 6**). Similarly to WQ sets, we may conclude that the permissible ranges for management measures assessed with the DM better correspond to contributions of the sustainable water resources management. Respective "threshold" values of the management measures assessed with the TLI (or similar) approach are more suitable for the needs of a management which main objective is preventing of undesirable hydroecological consequences of the anthropogenic effects (e.g., eutrophication) without accounting for the needs of water resource management.

Comparison of the aggregated WQ dynamics obtained with the TLI and DM shows that the simplest average estimates (as R[DM] and R[TLI]) may lead to contrasting values of lake WQ (**Figure 3**). In some cases, a WQ system based on the TLI is not suitable for diagnostics of the eutrophication processes. The TLI approach does not indicate WQ deterioration (as shown by the CWQI dynamics, **Figure 4**) associated with the in-lake processes usually considered as signs of eutrophication: disruptions of the phosphorus and oxygen regimes in hypolimnetic water (L. Stechlin) and an increase in cyanobacteria abundance (L. Kinneret).

Decades of struggle against eutrophication have been expressed in the form of a paradigm linking between lake trophic status and water quality. This misconception can result in a direct threat to the ecosystems of naturally eutrophic waterbodies, while producing a misleading understanding of the possibility of managing water quality based only on regulating activities according to the lake's trophic state. Accounting for the eutrophication problem only, and neglecting of requirements for water resources management prevents the trophic classification system from serving as a "common language" for communication between partners in management

5. CONCLUSIONS

In this study, we evaluated the suitability of the different water quality systems for the needs of sustainable water resources management. We compared two approaches used for quantitative assessment of water quality in natural waterbodies: the trophic level index and the Delphi method.

Comparative analysis of the assessed water quality systems allows us to draw the following conclusions:

1) The basic distinction between the Delphi and Trophic Level based approaches to quantification of water quality lies in the form of correspondence between the ecological and rating/trophic index values of the water quality index. In the trophic based system, the rating value is a strictly decreasing function of the concentration of Chl, TN and TP (increasing function of S). The respective rating curves of almost all water quality indices in the Delphi method account for the needs of sustainable management. Therefore they are bell-shaped curves, i.e., the ecological values of such parameters as S, Chl, TN and TP should be sustained within limits (permissible ranges) that otherwise could result in WQ deterioration. In the trophic state based system, the acceptable limits for all water quality indices are one-sided and are taken by default to be applicable to all lakes. The Delphi method gives a two-sided, lake-specific limit within which sustainable water resources management will be possible.

2) The aggregated estimates of WQ with the Delphi and TLI approaches give comparable estimates for oligoto mesotrophic lakes only. The differences between the two approaches increase with lake trophic state: the trophic status based system always valuate WQ in eutrophic lakes as "bad", independently on water resource uses.

3) The trophic state based system is a classification system, and therefore its implementation does not account for specific characteristics of aquatic ecosystems and the array of different potential uses as a water resource. The trophic state based system assumes reduction of economic activities in catchment areas as the most expedient mean to improve water quality. Therefore, water quality systems assessed with the trophic state based system are more suitable for needs of natural water resources management if eutrophication is a major threat. The Delphi method allows accounting for several water resource uses and therefore it may serve as a more robust and comprehensive tool for WQ quantification.

ACKNOWLEDGEMENTS

Arkadi Parparov, Gideon Gal, David Hamilton, Peter Kasprzak, Alexandr Ostapenia (BMBF), and the New Zealand Foundation of Research, Science and Technology (Contract UOWX0505) and Environment Bay of Plenty.

REFERENCES

1. R. G. Wetzel, "Limnology Lake and River Ecosystems," 3rd Edition, Academic Press, San Diego, 2001.
2. R. E. Carlson, "A Trophic State Index for Lakes," Limnology and Oceanography, Vol. 22, No. 2, 1977, pp. 361- 369.
3. OECD, "Eutrophication of Waters," Monitoring, Assessment and Control, Paris, 1982.
4. N. Burns, J. McIntosh and P. Scholes, "Strategies for Managing the Lakes of the Rotorua District, New Zealand," Lake and Reservoir Management, Vol. 21, No. 11, 2005, pp. 61-72.
5. C. Kaiblinger, O. Anneville, R. Tadonleke, et al., "Central European Water Quality Indices Applied to LongTerm Data from Pre-Alpine Lakes: Test and Possible Improvements," Hydrobiologia, Vol. 633, No. 1, 2009, pp. 67-74.
6. G. Giordani, J. M. Zaldivar and P. Viaroli, "Simple Tools for Assessing Water Quality and Trophic Status in Transitional Water Ecosystems," Ecological Indicators, Vol. 9, No. 5, 2009, pp. 982-991.
7. S. E. Jorgensen and R. A. Vollenweider, "General introduction," In: S. E. Jorgensen and R. A. Vollenweider, Eds., Guidelines of Lake Management, International Lake Environment Committee, Shiga, Vol. 1, 1989, pp. 13-18.
8. R. K. Horton, "An Index-Number System for Rating Water Quality," Journal of Water Pollution Control Federation, Vol. 37, No. 3, 1965, pp. 300-306.

9. W. Ott, "Water Quality Indices: A Survey of Indices Used in the United States," Environmental Monitoring Series, EPA-600/4-78-005, 1978, p. 128.
10. P. M. Brown, N. I. McClelland, R. A. Deninger and R. G. Tozer, "A Water Quality Index - Do We Dare?" Water Sewage World, 1970, pp. 339-343.
11. Parparov and K. D. Hambright, "Composite Water Quality: Evaluation and Management Feedbacks," Water Quality Research Journal of Canada, Vol. 42, No. 1, 2007, pp. 20-25.
12. G. Cude, "Oregon Water Quality Index: A Tool for Evaluating Water Quality Management Effectiveness," Journal of American Water Research Association, Vol. 37, No. 1, 2001, pp. 125-137.
13. M. A. House, "Water Quality Indices as Indicators of Ecosystem Change," Environmental Monitoring and Assessment, Vol. 15, 1990, pp. 255-263.
14. D. G. Smith, "A Better Water Quality Indexing System for Rivers and Streams," Water Research, Vol. 24, No. 10, 1990, pp. 1237-1244.
15. Parparov, K. D. Hambright, L. Hakanson and A. P. Ostapenia, "Water Quality Quantification: Basics and Implementation," Hydrobiologia, Vol. 560, No. 1, 2006, pp. 227-237.
16. K. D. Hambright, A. Parparov and T. Berman, "Indices of Water Quality for Sustainable Management and Conservation of an Arid Region Lake, L. Kinneret (Sea of Galilee), Kinneret," Aquatic Conservation: Marine and Freshwater Ecosystems, Vol. 10, No. 6, 2000, pp. 393- 406.
17. USGS, "Water-Quality and Lake-Stage Data for Wisconsin Lakes, Water Year 2000," U.S. Geological Survey, Open-File Report 01-86, Middleton, Wisconsin, 2001, p. 128.
18. U. S. Environmental Protection Agency, "An Approach to a Relative Trophic Index System for Classifying Lakes and Reservoirs," National Eutrophication Survey, Working Paper No. 24, 1974.
19. P. Scholes and M. Bloxham, "Rotorua Lakes Water Quality 2006 Report. Environmental Bay of Plenty," Environmental Publication 2007/12, Whakatane, New Zealand, 2006, p. 86.
20. C. Serruya, Ed., "Lake Kinneret," Dr. Junk Publishers, The Hague, 1978.
21. T. Zohary, "Changes to the Phytoplankton Assemblage of L. Kinneret after Decades of a Predictable, Repetitive Pattern," Freshwater Biology, Vol. 49, No. 10, 2004, pp. 1355-1371.
22. G. G. Winberg, Ed., "The Ecological System of the Naroch Lakes," Universitetskoye Press, Minsk, 1985.
23. R. Koschel and D. D. Adams, "An Approach to Understanding a Temperate Oligotrophic Lowland Lake (Lake Stechlin, Germany)," Archiv fur Hydrobiologie, Special Issues Advanced Limnology, Vol. 58, 2003, pp. 1-9.
24. T. Gonsiorczyk, P. Casper and R. Koschel, "Long-Term Development of the Phosphorus Accumulation and Oxygen-Consumption in the Hypolimnion of Oligotrophic Lake Stechlin and Seasonal Variations in the Pore Water Chemistry of the Profundal Sediments," Archiv fur Hydrobiologie, Special Issues Advanced Limnology, Vol. 58, 2003, pp. 73-86.
25. R. A. Vollenweider, "Advances in Defining Critical Loading Levels for Phosphorus in Lake Eutrophication," Memorie dell' Istituto Italiano di Idrobiologia, Vol. 33, 1976, pp. 53-83.
26. D. W. Schindler, "Evolution of Phosphorus Limitation in Lakes," Science, Vol. 195, No. 4275, 1977, pp. 260-262.
27. K. H. Rekhow, "Empirical Lake Models for Phosphorus: Development, Applications, Limitations and Uncertainty," In: D. Scavia and A. Robertson, Eds., Perspectives on Lake Ecosystem Modeling, Ann Arbor Sciences, Ann Arbor, 1979, pp. 193-121.
28. R. E. Carlson and J. Simpson, "A Coordinator's Guide to Volunteer Lake Monitoring Methods," North American Lake Management Society, Oklahoma, 1996.
29. "Directive 2000/60/EC of the European Parliament and of the Council of 23 October 2000 Establishing a Framework for Community Action in the Field of Water Policy," Official Journal of the European Communities.
30. Y. Z. Yacobi, "Temporal and Vertical Variation of Chlorophyll a Concentration, Phytoplankton Photosynthetic Activity and Light Attenuation in Lake Kinneret:

Possibilities and Limitations for Simulation by Remote Sensing," Journal of Plankton Research, Vol. 28, No. 8, 2006, pp. 725-736.

31. W. F. Vincent, M. M. Gibbs and S. J. Dryden, "Accelerated Eutrophication in a New Zealand Lake: Lake Rotoiti, New Zealand," New Zealand Journal of Marine and Freshwater Research, Vol. 18, 1984, pp. 431-440.
32. N. M. Burns, J. Deely, J. Hall and K. Safi, "Comparing Past and Present Trophic States of Seven Central Volcanic Plateau Lakes, New Zealand," New Zealand Journal of Marine and Freshwater Research, Vol. 31, No. 1, 1997, pp. 71-87.

CHAPTER 2

Water Allocation Systems

Ralph A. Wurbs[1]

Texas A&M University, USA

1. INTRODUCTION

Water resources are shared by many people who use the water for a variety of purposes. Water allocation systems serve to equitably apportion water resources among users; protect existing water users from having their supplies diminished by new users; govern the sharing of limited water during droughts when supplies are inadequate to meet all needs; and facilitate efficient water use. Effective water allocation becomes particularly important as demands exceed reliable supplies. As water demands increase with population and economic growth, water allocation systems must be expanded and refined.

The institutional framework for water resources development and management involves a hierarchy of water allocation systems. Water resources are allocated between nations by treaties and other agreements. In the United States (U.S.), water is allocated between states through interstate compacts. Water is shared by regional water authorities, municipal utility districts, cities, private companies, irrigation districts, farmers, and individual domestic water users through water rights systems. Water supply entities service their customers in accordance with contracts and other commitments. Federal water development agencies provide reservoir storage capacity for nonfederal sponsors.

This chapter begins with a general overview of institutional systems for allocating the water flowing and/or stored in rivers, lakes, and aquifers to diverse types of water use and numerous water users. The chapter then focuses on the Texas experience in implementing water allocation systems, including both accomplishments and issues still remaining to be resolved.

The state of Texas in the United States serves as a case study that illustrates concepts and strategies that are relevant throughout the world.

With a population of 26 million people and land area of 696,000 km^2, Texas is a large state with diverse geography, economy, climate, hydrology, and water management practices. Texas has a rich heritage of implementing water allocation strategies as a central thrust of its water resources planning, development, and management. The Rio Grande is shared with Mexico, and several major river basins are shared with neighboring states in the U.S. Thousands of government agencies, cities, private companies, and citizens hold rights to use the waters of the state. Water allocation involves multiple overlapping institutional mechanisms. The state has progressed significantly in recent years in improving its water allocation systems. Further refinements continue to be a priority policy emphasis.

The Water Rights Analysis Package (WRAP) modeling system simulates the water allocation systems described in this chapter. WRAP is routinely applied in Texas in regional and statewide planning studies, administration of the water rights permit system, and other water management activities. WRAP is generalized for application anyplace in the world.

2. WATER ALLOCATION MECHANISMS

People in various nations, regions, and local communities have developed their own sets of institutions and practices governing the sharing of water. These water allocation systems have evolved historically and continue to change. Hierarchies of water allocation systems in the U.S. and many other countries generally have the following components or features.

- The waters of international rivers and aquifers are allocated between nations based on international law, customs, treaties, and agreements.
- In the U.S., waters of interstate river basins are allocated between states based on compacts negotiated by the states and approved by the federal government.
- Certain rights are reserved for military installations and other government owned lands and facilities.
- A legally established priority system based generally on variations of the riparian or prior appropriation doctrines guides the allocation of the surface water flowing in streams and stored in reservoirs.
- A separate system of laws and customs guides the allocation of water resources in ground water aquifers.

- An administrative system that grants, limits, and modifies water rights and enforces the allocation of water resources may or may not include formal issuance of written permits to water right holders.
- Water users and water management entities implement various contracts and other formal agreements.
- Sharing of water resources may be governed by cultural traditions and informal agreements that evolve historically over many years.

Water allocation mechanisms typically vary greatly between ground-water and surface-water. From a water law perspective, ground and surface water are usually treated as separate resources. The extent to which the important hydrologic and water management interconnections are recognized varies between geographical regions.

The institutional mechanisms of water allocation are typically viewed from policy, legal, economic, and social perspectives. However, hydrologic science and engineering are also important aspects of developing and maintaining water allocation systems [1, 2, 3].

2.1. Allocation of the waters of international rivers and aquifers

Principles and rules of international water law are found in treaties, international custom, general principals of law, and writings of international institutions [4, 5, 6]. Two hundred and sixty-one international river basins, each encompassing portions of two or more nations, cover about 45 percent of the world's land area excluding Antarctica [7]. Little progress has been made in developing effective water allocation systems in many of these international river basins. Water allocation is even more difficult for groundwater aquifers shared by two or more countries.

Effective joint multiple-nation water management will be a major determinant in achieving stability, peace, and prosperity in many regions of the world in the 21st century [8, 9, 10, 11, 12]. Examples of the many regions with dramatic potential for either cooperation or conflict include the following. The Jordan River shared by Israel, Jordan, Syria, and the Palestinians is a small stream with remarkably great historical and political importance. Shared groundwater is also an important issue this region. The Euphrates and Tigris Rivers flow from Turkey through Syria, Iraq, and Iran. Most of the flow of the Euphrates and Tigris Rivers originate in their upper watersheds in Turkey and is controlled by a system of major dams in Turkey. The Ganges and Brahmaputra River Basins in Nepal, China, India, Bhutan, and Bangladesh, with a history of centuries of water conflicts, contained an estimated 400 million people in 2000 living at an impoverished standard of living. In the Southern African Region encompassing Angola, Botswana, Lesotho, Malawi, Mozambique, Namibia, South Africa, Swaziland, Tanzania, Zambia, and Zimbabwe, every major river is shared by two or more nations. Population growth and economic development are resulting in intensified demands on limited water resources with a long history of controversy.

2.2. Water rights in the united states

A water right is the legal privilege to store, regulate, and/or divert water for beneficial use. Water law is the creation, allocation, and administration of water rights. Institutions and customs for executing water rights play key roles in water management in various regions throughout the world. The following discussion is from a U.S. perspective but has broader applicability in other countries as well.

Books on water rights/law range from a concise book entitled *Water Law in a Nutshell* [13] to a massive eight volume collection [14] first published in 1967 and again as a revised updated edition in 1991 with the individual volumes continuing to be periodically updated and expanded at different times. The American Society of Civil Engineers has published guidelines for developing water right rules and regulations [15, 16, 17]. Other books focus on water rights in particular regions of the world other than the U.S. [18, 19, 20, 21].

Each state in the U.S. has developed its own set of laws, institutions, and practices governing water rights. These systems have evolved historically and continue to change. States in the western and eastern halves of the U.S. have generally adopted different approaches to water rights due largely to the western states having much drier climates. Water allocation and accounting systems tend to be more rigorous in regions where demands approach or exceed supplies. The experience of the state of Texas discussed in this chapter illustrates key aspects of developing and administering water right systems. Regions of Texas are representative of both western and eastern states.

Surface water in streams and lakes is viewed in the U.S. as a renewable resource owned by the state and used by the public. Groundwater is typically viewed as the property of the owners of the land overlying the aquifers. Water rights for groundwater aquifers are very different than for surface water streams and lakes. Although water rights are established primarily at the state level, federal laws govern water rights for military installations, other federal lands such as national parks, and Indian reservations.

For river basins encompassing portions of multiple states, water is allocated between states based on interstate compacts developed by the states, approved by the U.S. Congress, and administered by interstate compact commissions. Disputes may arise in implementing interstate compacts, particularly during droughts. Disputes are settled by the U.S. Supreme Court if compact commissions cannot work out disagreements between the states through negotiation. A major objective of compact commissions is to avoid lawsuits, but disputes in the western states in particular often reach the litigation stage. Various states have experienced lengthy and costly water allocation disputes settled through the U.S. Supreme Court [22]. Shared management of interstate groundwater aquifers and associated development of water allocation mechanisms have not progressed to nearly the extent as surface water.

2.2.1. Surface water rights

Legal rights to the use of stream flow are generally based on two alternative doctrines, riparian and prior appropriation. The basic concept of the riparian doctrine is that water rights are incidental to the ownership of land adjacent to a stream. The prior appropriation doctrine is based on the concept of protecting senior water users from having their supplies diminished by newcomers developing water supplies later in time. In a prior appropriation system, water rights are not inherent in land ownership, and priorities are established based on dates that water is appropriated.

The doctrine of riparian rights common in the eastern U.S. is based on English common law. Under the strictest interpretation of the riparian doctrine, the owner of riparian land adjacent to a stream is entitled to receive the full natural flow of the stream without change in quantity or quality. Since a strict interpretation imposes impractical constraints on water use, the riparian doctrine is normally interpreted to allow riparian land owners to divert reasonable amounts of stream flow for beneficial purposes. Varying definitions of *reasonable amounts* complicate water allocation.

The doctrine of prior appropriation is associated with settling the American West. As settlers moved from the eastern states to the West in the 1800's, farmers and ranchers claimed land, and miners claimed gold and other minerals. Likewise, water was appropriated by the first to arrive and claim the resources for beneficial use. In developing their farms and communities, people needed protection from having their water supplies diminished with later population growth and economic development.

Most of the western states have established permit systems in which a state agency issues permits to water right holders specifying amounts and conditions of water use. Riparian and/or appropriative rights may be incorporated into the original development of the permit system, with additional new permits being issued based on prior appropriation. With growing demands on limited water resources, permit systems will likely continue to be developed in the eastern states, which have more abundant stream flow, similar to those already in place in the drier western states.

Surface water rights in the eight driest western states (Nevada, Arizona, Utah, Idaho, Montana, Wyoming, Colorado, and New Mexico) are based purely on the prior appropriation doctrine. Alaska is also a prior appropriation state though somewhat different. Ten western states with hybrid systems merging riparian and appropriative rights into permit systems include California, Oregon, Washington, Texas, Oklahoma, Kansas, Nebraska, South Dakota, North Dakota, and Mississippi. Hawaii has a unique hybrid system. Water rights in 30 eastern states are based primarily on the riparian doctrine [23].

2.2.2. Groundwater rights

The rights and obligations for groundwater use are generally tied to two legal principles: property ownership and shared ownership of a common public resource. A variety of state approaches to groundwater rights has evolved from these concepts. State groundwater law is based on mixtures of the following doctrines.

Absolute Ownership Doctrine: Landowners own the groundwater under their land and may drill wells and pump as much water as they wish. Texas and several other states have historically adhered to this doctrine but are slowly changing.

Reasonable Use Doctrine: Landowners own groundwater, but their pumping is limited to reasonable use which has been defined in a variety of ways. This doctrine is common in the eastern states.

Correlative Rights Doctrine: In times of shortage, groundwater is shared by overlying landowners in proportion to the amount of land they own. This extension of the reasonable use rule is primarily associated with California.

Prior Appropriation Doctrine: Groundwater is allocated similarly to surface water with priorities assigned based on the dates that users first appropriate the water for beneficial use. This doctrine is common in the western states.

Permit Systems: Systems in which state agencies issue permits specifying the amounts and conditions of water use have been adopted in 33 of the 50 states. The other doctrines may be reflected in the water rights documented by the permits. Texas is among the 17 states that have no state-wide groundwater permit program.

Some states divide groundwater into categories with different water right rules applied to each classification. Percolating groundwater may be legally differentiated from underground streams with definable flow paths. Underground streams are sometimes treated as being similar to surface streams.

The issue of impacts of groundwater pumping on surface stream flow has been addressed to varying extents in different states. In some states, groundwater is classified as either tributary or nontributary. Tributary groundwater hydrologically contributes to surface stream flow. Nontributary groundwater does not. Water right rules and management strategies for tributary groundwater are based on protecting surface water rights.

3. WATER ALLOCATION SYSTEMS IN TEXAS

Texas is a large state located in the south-central U.S. that is representative of both the drier western and wetter eastern states from various perspectives including climate. Mean annual precipitation varies from 20 cm at the city of El Paso in arid west Texas to over 140 cm in the humid eastern extreme of the state.

Texas actually provides two case studies since water allocation in the Lower Rio Grande Valley has distinct differences from the remainder of the state. Water rights are a major consideration in river basin management statewide. Allocation of ground water is very different than allocation of surface water.

Water resources development and management in Texas is governed largely by the need to be prepared for extended droughts. The hydrologically most severe drought of record began gradually in 1951 and ended in 1957 with one of the largest floods on record. Droughts in the 1910's and 1930's were also extended multiple-year dry periods over large areas of Texas and neighboring states.

The drought that occurred during 1995-1996 was much more economically costly than the earlier droughts due to the population and economic growth that had occurred. In terms of annual precipitation, for more than half of the land area of Texas, 2011 was the driest calendar year since the beginning of official observed precipitation records in 1895. The remainder of the state was also very dry during 2011. Severe drought conditions are continuing during 2012 throughout Texas and other states in the U.S.

3.1. Water resources planning, development, and management

The Texas Water Development Board (TWDB) in partnership with 16 regional planning groups outline future water resources needs and challenges in the 2012 State Water Plan, which is presented in a multiple volume report entitled *Water for Texas 2012* [24]. Water demands are projected to increase about 22 percent between 2010 and 2060. Available water supplies with existing infrastructure and current institutional arrangements will decrease about 10 percent during this period due to reservoir sedimentation and depletion of groundwater aquifers. Environmental flow requirements and ecosystem preservation are major concerns as well as meeting municipal, industrial, and agricultural water needs. The regional planning groups have identified several hundred water supply projects with an estimated cost of over $50 billion to address intensifying water needs. The TWDB predicts that annual losses from not meeting water supply needs could result in a reduction in income of approximately $12 billion annually if current drought conditions approach the drought of record, and as much as $116 billion annually by 2060 [24].

The map of Texas in Figure 1 shows the larger rivers and cities of the state. The state encompasses a land area of 696,000 km^2 divided into 15 major river basins and eight coastal basins located along the Gulf of Mexico between the lower reaches of the major river basins. The 1990 population of 17.0 million people increased to 25.7 million in 2011 and is projected by the Texas Water Development Board [24] to increase to 46.3 million by 2060. Fifty-eight percent of Texans live in the state's three largest metropolitan areas: the Dallas/Fort Worth Metroplex in the upper Trinity River Basin, greater Houston area in the San Jacinto River Basin, and city of San Antonio in the San Antonio River Basin. With a 2011 population of 6.5 million people, the Dallas/Fort Worth (DFW) Metroplex is the largest metropolitan area in the southern U.S. and the fastest growing metropolitan area in the U.S. With a 2011 population of 6.1 million, the city Houston and adjacent smaller cities is also one of the largest and fastest growing metropolitan areas in the U.S. San Antonio is the third largest city in Texas with a population of 2.2 million people. Conversely, several of the major river basins of the state have extremely low population densities. Water management practices are very diverse throughout the state.

Total diversions of 22 billion m^3/year from streams, reservoirs, and aquifers statewide in 2010 was for agricultural irrigation (56%), municipal use (27%), industrial use (15%), and livestock (2%). Currently, about half of the water supplied is from groundwater aquifers, and the other half is from surface streams and reservoirs. However, problems caused by decades of groundwater pumping rates significantly exceeding recharge rates in various regions of the state are resulting in major shifts toward greater reliance on surface water.

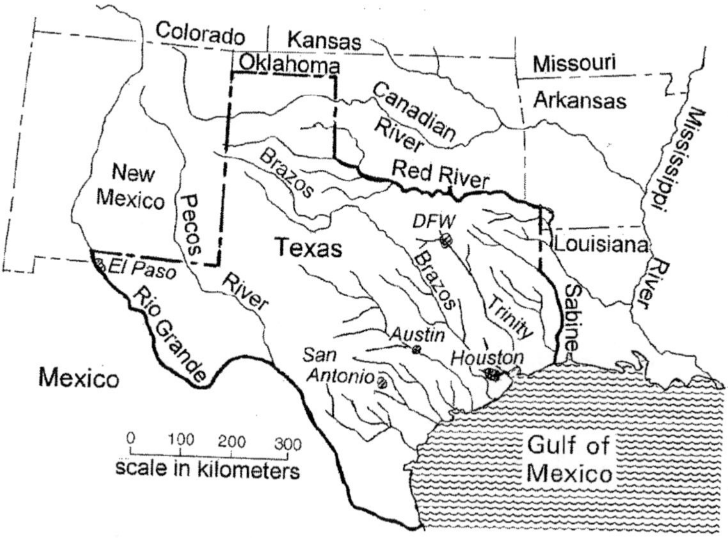

Figure 1. Map of Texas showing major rivers, largest cities, and neighboring states

The naturalized flow volumes during each of the 864 months of 1940-2011 at a gauging station on the Brazos River near Houston are plotted in Fig. 2 to illustrate the tremendous variability that characterizes stream flows throughout Texas. A highly variable resource is allocated to numerous water users. Long-term mean flows may be relatively large, but most of the flow occurs during infrequent flood events. The water management community must deal with droughts with durations ranging from several months to several years. Dams with large storage volumes are essential in order to develop dependable water supplies. The river flows plotted in Fig. 2 are generated by the modeling system described later in this chapter by adjusting gauged flows to represent natural conditions without human water resources development and use.

Conservation and flood control storage capacities totaling 50 and 23 billion m^3 are provided by 211 major reservoirs with individual capacities of 6.2 million m^3 or greater. Although there are several thousand smaller reservoirs, the 211 major reservoirs account for over 95% of the storage capacity in the state. Texoma on the Red River and International Amistad on the Rio Grande are the largest reservoirs with capacities of 6.6 and 6.3

billion m³. The 22 hydroelectric plants in the state are components of electric utility systems dominated by thermal plants. Reservoir releases through hydropower turbines are almost always incidental to downstream water supply needs.

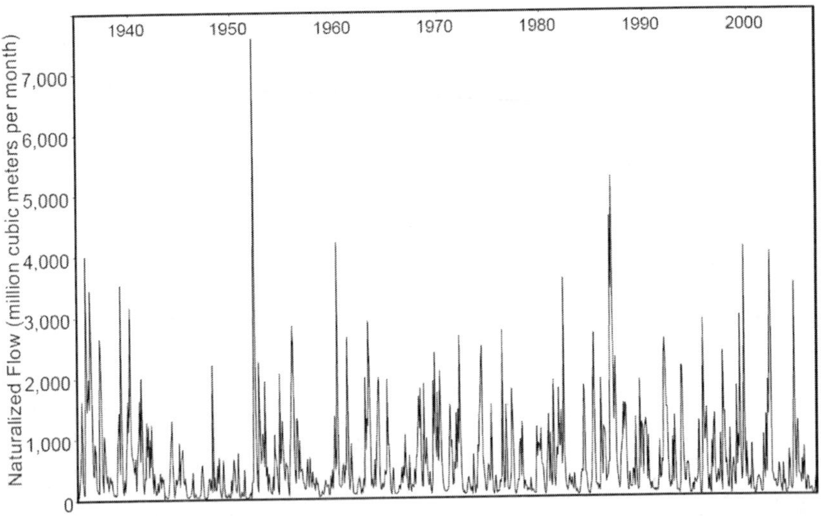

Figure 2. Monthly naturalized flows of the Brazos River illustrating variability

Federal involvement in developing the state's water resources includes 32 U.S. Army Corps of Engineers (USACE) reservoirs in eight river basins and two International Boundary and Water Commission (IBWC) reservoirs on the Rio Grande that account for about 40% and 90% of the conservation and flood control, respectively, storage capacity of the 211 major reservoirs. The U.S. Bureau of Reclamation constructed three reservoirs that are now owned by local entities. River authorities and cities have contracted for the water supply storage capacity of the USACE reservoirs. All costs allocated to water supply are reimbursed by the nonfederal sponsors, who also hold the water right permits. The USACE is responsible for flood control operations.

Nineteen river authorities are responsible for development and management of the water resources of all or portions of major river basins. River authorities construct and operate their own reservoir projects and contract for storage capacity in federal reservoirs. Numerous municipal utility districts and irrigation districts also play key roles in supplying water from rivers and reservoirs. Groundwater conservation districts facilitate groundwater management.

Some type of water rights system has been administered statewide by a centralized agency since 1913, but that agency has changed over time. The Board of Water Engineers was established in 1913; reorganized as the Texas Water Commission (TWC) in 1962; and renamed the Texas Water Rights Commission in 1965 with non-water rights functions being transferred to the Texas Water Development Board (TWDB), which had been previously created in 1957. In 1977, the Texas Department of Water Resources (TDWR) was created by combining the Water Rights Commission, TWDB, and Water Quality Board. In 1985, the TDWR was dissolved, and the TWC and TWDB became separate agencies. The Texas Natural Resource Conservation Commission (TNRCC) was created in 1993 by merging the TWC and Texas Air Quality Board. The TNRCC was renamed the Texas Commission on Environmental Quality (TCEQ) in September 2002. The TCEQ is now responsible for administering the many regulatory programs of the state including the water rights system.

The TCEQ consists of three full-time commissioners appointed by the governor and a professional and administrative staff of over 3,000 employees. Water rights are one of many regulatory responsibilities of the TCEQ. The TCEQ and TWDB interact closely in many of their activities. The TWDB is responsible for developing and updating the State Water Plan and administering an array of financial assistance programs. The TWDB has a governing board of six members appointed by the governor and a staff of about 800 employees.

3.2. Legislatively mandated water management programs

The state-level water allocation systems described in this chapter were created under laws enacted by the Texas Legislature. As discussed later, the statewide water rights permit system for all of Texas except the Rio Grande was implemented pursuant to the Water Rights Adjudication Act of 1967. The water rights adjudication process mandated by the 1967 Act was completed shortly before a severe statewide drought in 1995-1996 which motivated enactment of the 1997 Senate Bill 1.

The Brown-Lewis Water Management Plan enacted by the Texas Legislature as its 1997 Senate Bill 1 is considered to be a milestone in the history of water resources management in Texas. The designation *Senate Bill 1* is traditionally applied each legislative session to highlight legislation of the upmost importance. The 1997 Senate Bill 1 and subsequent amending legislation authorized the Water Availability Modeling (WAM) System described later in this chapter, a regional and statewide water resources planning process noted below, and other related water management activities.

The 1997 Senate Bill 1 established a program for developing regional water plans that are integrated into a statewide planning process administered by the Texas Water Development Board (TWDB). Committees of local water interests have been established to prepare plans for the orderly development, management, and conservation of the water resources of each of 16 regions. The TWDB provides funding, administrative, and technical support to the regional committees. Consulting firms perform much of the technical work. The 1997 Senate Bill 1 mandated that initial regional plans be completed and incorporated into a statewide plan by 2002. Continuing planning is organized based on updated plans being reported at cycles of not to exceed five years. The 2002, 2007, and 2012 state water plans consists of 16 regional reports and a statewide report which are publically available at the TWDB website.

In evaluating water right permit applications, the TCEQ requires that proposed actions must be consistent with pertinent regional plans. The WAM System is routinely applied in both regional and statewide planning and administration of the water right permit system.

The Texas Instream Flow Program [25] established by Senate Bill 2 (SB-2) enacted by the Texas Legislature in 2001 and expanded by Senate Bill 3 (SB-3) in 2007 is administered jointly by the TCEQ, TWDB, and Texas Parks and Wildlife Department (TPWD). The purpose of the SB-2 program is to perform scientific studies to determine flow conditions necessary for supporting a sound ecological environment in the river basins of Texas. SB-3 mandated a new regulatory approach for protecting environmental flows through a local stakeholder process culminating in rules to be administered through the TCEQ. The objectives of the 2001 SB-2 and 2007 SB-3 are being accomplished largely through the ongoing work of Bay and Basin Expert Science Teams (BBEST) and Bay and Basin Area Stakeholder Committees (BBASC) for the individual river basins, representing the scientific and water management/use communities, with technical support from consultants. The TCEQ, TWDB, and TPWD provides administration oversight and technical support.

Environmental instream flow requirements are determined by the BBEST teams for their assigned streams within a framework of subsistence flows, base flows, high flow pulses, and overbank flood events needed to support riverine ecosystems, wetlands, and freshwater inflows to bays and estuaries. The Stakeholder Committees consider the BBEST recommendations within the content of municipal, industrial, and agricultural water needs as well as environmental water needs. The TCEQ is responsible for final approval and incorporation of the recommendations of these groups and the public into the water rights permit system. As of 2012, the endeavors of the Texas water management community, under the leadership of the three-agency partnership and the BBEST and BBASC groups, to better incorporate environmental considerations into the water rights permit system is well underway and is expected to continue for many years into the future.

3.3. Groundwater management

Most of Texas is underlain by 9 major and 21 minor aquifers that currently supply about half of the water used in the state. Depleting ground water reserves due to excessive pumping is a major problem throughout the state and is forcing a shift from groundwater use to a greater reliance on surface water supplies. Groundwater rights in Texas have historically been based on the common law rule allowing land owners to pump as much water as they wish from under their land. However, increased state regulation of groundwater is evolving over time primarily through the establishment of groundwater conservation districts.

In 1949 and 1985, the Texas Legislature passed laws authorizing creation of groundwater conservation districts with local voter approval. The 1949 legislation allows local residents to petition the state to have a district created. The 1985 amendment authorizes the TCEQ to designate and initiate formation of districts for areas with critical problems. Local voters can still veto a proposed district, but if they do, state funding for water projects can be withheld. Twelve districts existed prior to 1985. As of 2012, 96 groundwater districts covering over half of the land area of the state are operational. Three additional districts are currently in the process of being confirmed by voters through local elections at the county level. A total of 174 of the 254 counties in Texas are within a groundwater conservation district.

The primary purposes of the districts are to encourage water conservation and to protect water quality. Most districts direct their efforts toward prevention of waste, water conservation education, recharge projects, and data collection. Some are now moving toward stricter regulation of groundwater use. The districts tread a narrow path between private ownership of groundwater and state responsibility to protect the water resource. Texans are reluctant to allow anyone to tell them how much water they can pump from under their own land. Governmental regulation of pumping has been driven by necessity as depleting aquifers resulted in major problems. The Harris-Galveston Coastal Subsidence District and Edwards Underground Water Authority have developed the strongest regulatory programs.

The Harris-Galveston Coastal Subsidence District was created in 1957 in response to severe subsidence in the vicinity of the cities of Houston and

Galveston. Due to decades of over-drafting groundwater, the ground surface has been lowered over three meters in places in this low-lying, heavily urbanized coastal region. Houston and neighboring smaller cities continue to shift from ground water to surface water supplies. Groundwater pumping is regulated by the Subsidence District through a permit program.

The Edwards Aquifer Authority was created in 1993 largely due to a federal court ruling related to protection of endangered species under the Endangered Species Act. The Edwards is a limestone aquifer shared by San Antonio, several smaller cities, and extensive irrigated farming interests. San Antonio is the largest city in the U.S. that relies solely on groundwater for its water supply. Springs fed by the Edwards Aquifer maintain the flow of several rivers and support ecosystems that include several endangered species. The Edwards Aquifer Authority administers a water rights permitting system to facilitate aquifer management. Pumping limits are reduced whenever water levels in selected wells fall below specified criteria.

All of the groundwater conservation districts are required to develop, implement, and periodically update management plans for the effective management of their groundwater resources. These plans are subject to approval by the TWDB and are publically accessible through the TWDB website.

Although interstate and international aquifers are important, allocation agreements do not exist. The U.S. and Mexico have different legal reasons for their lack of management of ground water [26]. The Mexican federal government has the authority to enforce comprehensive management and regulation of ground water but has not chosen to do so. In the U.S., the legal framework is inadequate and chaotic. Another constraint to joint U.S. and Mexico management of ground water is the lack of clarity of international ground water law [26].

3.4. Allocation of the waters of the rio grande

The Rio Grande (Spanish for Large River) is unique relative to the other river basins of Texas from several perspectives. It is shared by two nations. The Lower Rio Grande Valley accounts for the majority of the surface water irrigation in Texas. The intensive agricultural production of the region depends almost exclusively on the Rio Grande with little use of groundwater. Other major irrigation regions of the state rely primarily on groundwater. The water rights system for the Lower Rio Grande was developed separately and has distinct differences from the remainder of the state, particularly in regard to the priority system and water master operations. Fort Quitman shown in Fig. 3, located 140 km downstream of the city of El Paso, is used as the divide between the lower and upper portions of the basin in both the international and state water allocation systems.

The Rio Grande Basin is shared by Mexico and three states in the U.S. Water allocation is governed by two international treaties and two interstate compacts. Allocation of the Texas share of the waters of the Rio Grande to irrigators, cities, and other users is based on a water rights permit system governed by state law. The Rio Grande Compact approved by the legislatures of Colorado, New Mexico, and Texas in 1939 allocates the uncommitted waters of the Rio Grande above Fort Quitman. The Pecos River Compact adopted in 1948 allocates the waters of that tributary between Texas and New Mexico.

3.4.1. Mexico-united states treaties

A 1906 treaty between the U.S. and Mexico provides for delivery of 74 million m^3/yr of Rio Grande water to Mexico in the El Paso-Juarez Valley above Fort Quitman. Elephant Butte Reservoir in New Mexico, operated by the Bureau of Reclamation, and the American and International diversion dams near El Paso, operated by the IBWC are used to implement the water allocation provisions of the treaty. The treaty further provides that if water is unavailable, the amount allocated to Mexico shall be diminished in the same proportion as water delivered to irrigate lands in the United States. This provision has been invoked in about a third of the years since 1951.

The Water Treaty of 1944 expanded the International Boundary Commission to the International Boundary and Water Commission (IBWC), provided for the distribution of waters of the Rio Grande from Fort Quitman to the Gulf of Mexico between the two nations, and authorized construction of International Amistad and Falcon Reservoirs.

The 1944 treaty administered by the IBWC also includes provisions for allocation of the waters of the Colorado River. The Colorado River drains 629,000 km^2 in seven western states and flows into the Gulf of California in Mexico. Although both are included in the same 1944 treaty, water allocation for the Colorado River and Rio Grande are very different. The Rio Grande serves as a boundary between the two countries. The Colorado River flows from the U.S. to Mexico with most of its watershed area being located in the U.S.

The IBWC is composed of a Mexico Section with commissioner and technical and administrative staff located in Ciudad Juarez and a U.S. Section with offices across the river in El Paso. Amistad and Falcon Reservoirs are operated by the IBWC primarily for flood control and water supply for the Lower Rio Grande Valley. Hydroelectric power generation uses releases for downstream water supply. The U.S. share of the conservation storage is used to meet demands in the lower basin in accordance with the state water rights system.

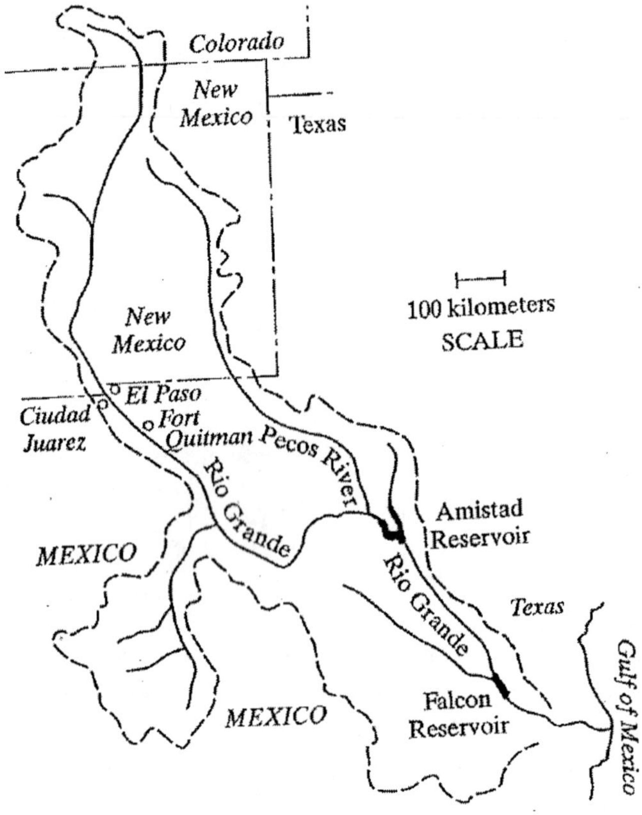

Figure 3. Rio Grande Basin

Conservation and flood control capacities are 4.17 billion m^3 and 2.15 billion m^3 in Amistad Reservoir and 3.29 and 0.63 billion m^3 Falcon Reservoir. In accordance with the 1944 treaty, the U.S. has 56.2% and 58.6% of the conservation storage capacity of Amistad and Falcon, respectively, with Mexico owning the remainder. The IBWC operates Anzaldus and Retamal Dams on the lower Rio Grande to facilitate diversions. The travel time for releases from Falcon Reservoir to reach the most downstream diversion locations is about a week.

The IBWC maintains a continuous accounting of the volume of stored water owned by each of the two countries. Stream flows into Falcon and Amistad

Reservoirs are allocated between the two countries. Flows on a number of major tributaries named in the treaty are gauged and allocated as specified by the treaty. All other flows not otherwise allocated are divided equally between the two countries. Mexico receives all the flows from several specified Mexican tributaries and two-thirds of the flows from other specified tributaries. The U.S. receives all the flows from certain U.S. tributaries and one-third of the flows from the other specified tributaries. Computations are performed routinely to allocate the reservoir inflow and evaporation volumes, which are combined with recorded releases to determine the amount of water that each country has in storage.

The 1944 treaty gives the U.S. a right to one-third of the flow reaching the Rio Grande from six Mexican tributaries, provided that this third shall not be less, as an average amount in cycles of five consecutive years, than 431,721,000 m^3 annually. Other provisions relate to special conditions. A significant deficit accumulated over several years during the 1900's-2000's in meeting these requirements. Discussions addressing the stream flow deficit owed by Mexico to the U.S. included requests by the U.S. that operating policies for reservoirs in Mexico be modified to mitigate the deficit.

3.4.2. Allocation of the texas share of the lower rio grande

The Texas share of the waters of the Rio Grande below Fort Quitman was allocated among numerous water rights holders in conjunction with a massive lawsuit, commonly called the Lower Rio Grande Valley Water Case. The lawsuit was filed in 1956, the trial was held in 1964-1966, and the final judgment of the appellate court was filed in 1969. In 1971, the Texas Water Rights Commission adopted rules and regulations implementing the court decision. Assorted versions of riparian and appropriative rights were combined into a permit system. The litigants in the Rio Grande law suit included 42 water districts and 2,500 individuals. More than 90 lawyers appeared before the court. The expense and effort demonstrated the impracticality of a purely judicial determination of water rights for the entire state and led to enactment of the Water Rights Adjudication Act of 1967.

The lawsuit resulted in water rights being divided into three categories. Municipal rights have the highest priority. Irrigation rights are divided into Class A and Class B rights, with Class A rights receiving more storage in Falcon and Amistad Reservoirs storage accounts in the allocation procedure. Although this weighted priority system for irrigation rights has little significance during years of plentiful water, its effect in water-short years is to distribute the shortage among all users, with the greater shortages occurring on lands with Class B water rights.

Most of the U.S. share of the water regulated by Amistad and Falcon Reservoirs is used in the very productive agricultural region of Texas below Falcon Reservoir. Irrigation districts, individual farmers, and cities communicate their water needs to the TCEQ Rio Grande Water Master Office, which in turn schedules releases from Falcon and Amistad Reservoirs with the IBWC.

The Water Master Office maintains an accounting of the amount of water used and the amount of water in reservoir storage allocated to each of about 1,600 water rights accounts.

The allocation rules administered by the TCEQ Water Master first provide a reserve of 278 million m^3in Amistad and Falcon Reservoirs for domestic, municipal, and industrial uses. Next, available water is allocated to an operating reserve that provides for seepage and evaporation losses, adjustments required as the IBWC computations of Mexico-U.S allocations are updated periodically, and emergency requirements.

The remaining water in storage is allocated among all the irrigation permit holders. The storage is basically allocated in proportion to annual diversion rights, except the Class A rights are multiplied by a factor of 1.7 to allow them a greater storage allocation than Class B rights. Other provisions include limiting each storage allotment to not exceed more than 1.41 times its authorized diversion right. If an irrigation right does not use water for two consecutive years, its storage account is reduced to zero.

3.5. Interstate river compacts

Texas participates in five interstate river compacts executed by the member states and approved by the U.S. Congress. The rivers and the dates the compacts became effective are Rio Grande, 1939, Pecos, 1948, Canadian, 1952, Sabine, 1954, and Red, 1980.

The purposes of the compacts are to provide for equitable allocation of water between the states and to facilitate cooperative planning and management. Commissions with a representative from each member state administer the compacts. The commissioners have minimal staffs and rely on the state water agencies for technical support.

Texas filed a lawsuit against New Mexico in 1974, claiming that Texas did not receive its full share of water allocated by the 1948 Pecos River Compact due to over-use in New Mexico [27]. In 1988, based on the recommendations of a court appointed Special Master, the Supreme Court awarded Texas a cash settlement for past damages and established a Pecos River Master to administer future water allocations under the compact.

3.6. Statewide surface water rights permit system

All of Texas except the Lower Rio Grande below Fort Quitman has a consistent surface water rights permit system which is different than the system just described for the Lower Rio Grande. The TCEQ administers both systems. River authorities, municipal utility districts, cities, irrigation districts, farmers, companies, and citizens hold about 8,000 permits statewide. The water is owned by the state, with rights granted to organizations and people to use prescribed amounts for prescribed purposes under prescribed conditions.

3.6.1. Historical evolution of water rights

The riparian doctrine was introduced in Texas during the rule of Spain and later Mexico and after independence in 1836, in a somewhat different form by the Republic of Texas. In 1840, the state of Texas adopted the common law of England with another variation of riparian water rights. The extent to which riparian rights allow large irrigation developments or other large amounts of use depends upon the laws in effect when the land was originally transferred from public to private ownership. Riparian rights are different depending on whether the land can be traced to land grants from Spain, Mexico, Republic of Texas, or the state of Texas.

The prior appropriation doctrine was adopted by legislative acts in 1889 and 1895. After 1895, public lands which transferred into private ownership no longer carried riparian water rights. Land already privately owned kept its riparian rights. At first, appropriation simply involved water users filing sworn statements with county clerks describing their use. Since 1913, more strictly administered procedures have been followed based on a statewide appropriation system administered by a centralized state agency.

All appropriation statutes recognized existing riparian rights. Riparian landowners could also acquire appropriative water rights. In 1926, the courts divided stream flow into *ordinary normal flow* and *flood flow.* Riparian rights were limited to normal flow and therefore are not applicable to flood waters impounded by reservoirs. Although difficult to apply in actual practice, this distinction was the basis for correlating riparian and appropriative rights from 1926 until the riparian rights were merged into the appropriative system pursuant to the Water Rights Adjudication Act of 1967.

Thus, an unmanageable system evolved, with various conflicting rights and many rights being unrecorded. The 1951-1957 drought motivated the previously noted Lower Rio Grande Valley Water Case which clearly demonstrated the impracticality of a purely judicial adjudication of water rights statewide. Thus, the Water Rights Adjudication Act was enacted by the Texas legislature in 1967, with the stated purpose being to require a recording of all claims for water rights, to limit the exercise of those claims to actual use, and to provide for the adjudication and administration of water rights. All riparian water rights were merged into the prior appropriation system, creating the present permit system applicable to all of Texas except the Lower Rio Grande. The water rights adjudication process required for transition to the permit system was initiated in 1968 and completed by the late 1980's.

3.6.2. Prior appropriation permit system

Water rights are granted by a state license, or permit, which allows the holder to divert a specified amount of water annually at a specific location, for a specific purpose, and to store water in reservoirs of specified capacity. Any organization or person may submit an application to the TCEQ for a new water right or to change an existing water right at any time. The TCEQ will approve the permit application if unappropriated water is available, a beneficial use of the water is contemplated, water conservation will be practiced, existing water rights are not impaired, and the water use is not detrimental to the public welfare. Proposed actions reflected in water right permit applications must be consistent with regional water plans.

The water authorized to be appropriated under the terms of the particular permit is not subject to further appropriation unless the permit is canceled. A permit may be canceled if water is not used during a 10-year period. Special term permits may also be issued allowing water use for specified periods of time. The Rio Grande and segments of other rivers are over-appropriated with no new rights for additional water use being granted. However, unappropriated water is still available for appropriation in other regions of the state.

A permit holder has no actual title of ownership of the water but only a right to use the water. However, a water right can be sold, leased, or transferred to another person. The Lower Rio Grande Valley has been the only region of Texas with an active water market historically. In 1993, the legislature established a statewide water bank to be administered by the TWDB. Although transfers can be accomplished independently of the water bank, the program was created to encourage and facilitate water marketing, transfer, and reallocation.

The Texas Water Code lists beneficial uses in order of priority as follows: (1) domestic and municipal, (2) industrial, (3) irrigation, (4) mining, (5) hydroelectric, (6) navigation, (7) recreation and pleasure, and (8) other beneficial uses. These priorities are invoked only when a conflict exists between water use applications. Under the prior appropriation system, after a permit is in effect, priorities are based on the date specified in the permits. During the 1968-1980's adjudication process, priority dates were established based on historical water use. Since then, priorities are based on the dates that the permits are issued. In times of emergency, cities may take water even if non-municipal users are adversely affected, regardless of priority dates.

4. WATER RIGHTS ANALYSIS PACKAGE (WRAP) MODELING SYSTEM

The Water Rights Analysis Package (WRAP) computer modeling system is generalized for application to river/reservoir/use systems located anywhere in the world, with model-users developing input datasets for the particular river basin of concern. For applications in Texas, publicly available WRAP input datasets from the TCEQ Water Availability Modeling (WAM) System [28] are altered as appropriate to reflect proposed water management plans of interest, which could involve changes in water use or reservoir/river system operating practices, construction of new facilities, or other water management strategies.

WRAP simulates water resources development, management, regulation, and use in a river basin or multiple-basin region under a priority-based water allocation system. In WRAP terminology, a water right is a set of water use requirements, reservoir storage and conveyance facilities, hydropower projects, operating rules, and institutional arrangements for managing water resources. Stream flow and reservoir storage is allocated among users based on specified priorities, which can be defined in various ways. Simulation results are organized in optional formats including time sequences of many different variables, summaries, water budgets, frequency relationships, and various types of reliability indices.

The public domain WRAP software and documentation [29, 30, 31, 32, 33] may be downloaded free-of-charge from http://ceprofs.tamu.edu/ rwurbs/ wrap.htm. The WRAP website links to TCEQ WAM and TWRI websites that provide reports, datasets, and other related information. Wurbs provides a concise summary of WRAP [34] and comparison with other generalized modeling systems [35].

WRAP modeling capabilities that have been routinely applied in the Texas WAM System consist of using a hydrologic period-of-analysis of about 60 years and monthly time step to perform water availability and reliability analyses for municipal, industrial, and agricultural water supply, environmental instream flow, hydroelectric power generation, and reservoir storage requirements. Recently developed additional WRAP modeling capabilities include: short-term conditional reliability modeling [36]; daily time step modeling capabilities that include flow forecasting and routing and disaggregation of monthly flows to daily; simulation of flood control reservoir system operations [31]; and salinity simulation [37]. Further improvements to WRAP currently underway, as of 2012, are focused on better integrating environmental flow requirements into comprehensive water management.

The generalized WRAP modeling system was developed and continues to be expanded at Texas A&M University (TAMU). Early versions dating back to the 1980's were developed under the sponsorship of the Texas Water Resources Institute (TWRI) and U.S. Department of Interior. Efforts at TAMU to expand and improve WRAP from 1996 through the present have been sponsored primarily by the TCEQ. However, the TWDB, U.S. Army Corps of Engineers, Brazos River Authority, and other agencies have funded specific improvements to WRAP. Model development has been an evolutionary process with extensive interactions between professionals from the agencies and consulting firms applying the model to specific river basins and university researchers responsible for improving the modeling methodology and computer software.

4.1. Texas water availability modeling system

As previously noted, water management legislation known as Senate Bill 1 enacted by the Texas legislature in 1997 directed the water agencies to establish a regional planning process and a water availability modeling system. The Texas WAM System was implemented by the TCEQ, its partner agencies (TWDB and TPWD) and contractors (ten consulting engineering firms and two university research entities) during 1997-2002 pursuant to the 1997 Senate Bill 1. The WAM System has continuously to be improved and expanded since 2002 along with being routinely applied by the water management community [28].

The Water Availability Modeling (WAM) System consists of the generalized WRAP and input datasets for the 23 river basins of Texas. The WAM datasets include naturalized stream flows at a total of about 500 gauged sites, watershed parameters for distributing these flows to over 12,000 ungauged locations, 3,450 reservoirs, various other constructed infrastructure, operating plans that in many cases are quite complex, two international treaties, five interstate compacts, various contractual agreements, and water use requirements associated with about 8,000 water right permits reflecting two different water right systems.

Prior to creation of the WAM System, many water right permit holders incorrectly assumed that the amount of water specified in their permits would always be available to them. Senate Bill 1 required that the TCEQ notify all permit holders regarding the reliabilities associated with their permits. All water needs cannot be met during severe droughts.

The TCEQ requires that permit applicants, or their consultants, apply the WRAP/WAM system to assess supply reliabilities and flow and storage frequencies associated with their proposed actions and the impacts on all of the water users in the river basin. TCEQ staff applies the modeling system to evaluate the permit applications. The TWDB and regional planning groups apply the modeling system in their planning studies. The TCEQ requires that water right permit applications be consistent with regional plans. River authorities and other water suppliers apply the modeling system in operational planning studies.

4.2. Lessons learned in implementing the wrap/wam system

Developing and applying computer models have typically been viewed in terms of the engineering and scientific concepts and methods incorporated in the models. However, modeling has important institutional as well as technical dimensions. Lessons learned from development and application of the Texas WAM System demonstrate the importance of the following two institutional dimensions of river/reservoir system modeling.

1) Modeling water rights, contractual agreements, treaties, interstate compacts, and other complex institutional aspects of water resources development, management, allocation, and use may be a key consideration in developing and applying a modeling system.

2) Effective implementation of a modeling system may require a partnership effort of an entire water management community that includes water users, political officials, legislatures, environmental and other non-governmental special interest groups, government agencies, consulting firms, and university researchers.

The following general observations characterize the Texas experience in implementing a water availability modeling system.

- Droughts motivate political concern, improvements in water management, and development of computer modeling systems.

- Partnerships and consensus building are key aspects of water resources planning and management. Likewise, a water management community may work together to effectively implement a shared modeling system. Model development and application may be an institutional partnership effort.

- Administration of water allocation systems has become a central focus of river basin management. Regulatory and planning functions are integrally related. Shared modeling tools can facilitate integration of planning and regulatory functions.

- Modeling systems include computer programs, databases, organizations, people, and decision-making processes. Compilation and management of voluminous data is a central governing concern. A modeling system is constructed rather than just a model.
- Model development is a dynamic evolutionary process. As long as a computer simulation model such as WRAP continues to be applied, its development is never completed. Model development is a process of continual expanding and improving.
- Water availability modeling is essential for effective water management.

5. WATER ALLOCATION ISSUES

Allocating water resources that are highly variable both temporally and spatially among a myriad of water management entities and numerous water users within an institutional setting that has evolved historically over many years is necessarily complex. The following concerns highlighted by the Texas experience are illustrative of the numerous complexities in creating and administering water allocation systems.

For most of Texas, the water right permit system is administered without water master operations. Upon request, the TCEQ takes enforcement action to stop reported unauthorized water use in violation of water rights permits. However, water users are not closely monitored except during droughts or emergency conditions. This approach is similar to most western states. Several western states have water-master operations, but most states do not. The TCEQ during 2012-2013 is investigating the feasibility of expanding water master operations in Texas.

The TCEQ Lower Rio Grande Water Master Office maintains a precise accounting of water use, working closely with irrigators, cities, and the International Boundary and Water Commission. With completion of the adjudication process during the late 1980's, plans were developed to establish water-master operations in all of the major river basins of Texas. The South Texas Water-Master Program was created in the late 1980's with responsibilities for the Guadalupe, Nueces, and San Antonio River Basins. However, water users are reluctant to have requirements imposed upon them for installing meters and monitoring and regulating diversions. Political pressures have prevented the establishment of water-master offices in other river basins. However, the Texas Legislature in 2012 directed the TCEQ to solicit public input and develop recommendations for establishing water master operations for other river basins.

Since stream flow, evaporation, reservoir sedimentation, water use, and other factors are highly variable, and the future is unknown, water availability must be viewed from a reliability, likelihood, or percent-of-time perspective. Tradeoffs occur between the amount of water to commit for beneficial use and the level of reliability that can be achieved. Beneficial use of water is based on assuring a high level of reliability. However, if water commitments are limited as required to assure an extremely high level of reliability, the amount of stream flow available for beneficial use is constrained, and a greater proportion of the water flows to the ocean or is lost through reservoir evaporation. The optimal level of reliability varies with type of water use. Water allocation decisions necessarily require qualitative judgment in determining acceptable levels of reliability both in terms of the reliability of the proposed new or increased water use and the impacts on the reliabilities of all of the existing water users.

Many of the existing water rights adjudicated pursuant to the 1967 Water Rights Adjudication Act and the Lower Rio Grande Court Case as well as some recently established rights have supply reliabilities that are much lower than the water management community considers desirable. However, the TCEQ has applied more stringent criteria in approving newer water right permits or modifications to permits. In evaluating water right permit applications for agricultural irrigation, the TCEQ now applies the criterion that at least 75 percent of the proposed demand should be supplied at least 75 percent of the time as determined by the WRAP/WAM System. The TCEQ criterion for municipal permit applications is that 100 permit of the demand should be supplied 100 permit of the time based on the premises reflected in the WRAP/WAM model including historical hydrology. However, these criteria may be modified depending upon backup water supply sources. For example, with depleting groundwater reserves, a transition to a surface water source may be worthwhile, even if the reliability of the surface water source is low, if groundwater can still be used as a backup supply.

Although Texas and other western states are viewed as adopting the prior appropriation doctrine, strictly speaking a pure prior appropriation system is not feasible and does not exist. Although the effect may be very small, developing additional supplies for new users always affects downstream supplies. Also, in drought situations, water supply shortages are shared, to some degree, by water users regardless of the relative seniority of their rights. Sharing of water during drought typically depends on political negotiations, alternative demand management and supply augmentation measures available to different entities, and other factors in addition to the water rights permit system.

Assigning water right priorities to maintaining reservoir storage levels relative to diversion rights is another issue. Protecting reservoir inflows is crucial to providing a dependable water supply. Each drawdown could potentially be the beginning of a several-year drawdown that empties the reservoir. However, forcing appropriators, with rights junior to the rights of the reservoir owner, to curtail diversions to maintain inflows to an almost full, or even an almost empty, reservoir is difficult and often is not the optimal use of the water resource. If junior diversions are not curtailed, the reservoir will likely later refill anyway, without any supply shortages occurring. Handling of the storage aspect of water rights is not yet precisely defined in Texas except for the Lower Rio Grande. The Lower Rio Grande is simpler in this regard because essentially all of the water users are supplied by two large reservoirs. Elsewhere, numerous reservoirs are owned and operated by various entities in the same river basin.

Although some recently issued permits specify the amount of the diversion to be returned to the stream, most permits do not. Return flows can significantly impact the availability of water to downstream users. This issue is currently being addressed particularly as related to programs to encourage reuse of wastewater effluents.

From the perspective of hydrology and water resources management, groundwater and surface water are closely interrelated. Use of one resource often has significant impacts on the other. However, there is only limited governmental control over the use of groundwater in Texas. Consequently, conjunctive management of ground and surface water resources is difficult. Depleting groundwater reserves are forcing a shift toward greater groundwater regulation.

Water availability depends upon water quality as well as quantity. The water supply capabilities of several major river systems that include very large reservoirs in Texas and neighboring states are severely constrained by salinity originating from natural salt deposits in geologic formations underlying the upper watersheds of the rivers. Salinity simulation capabilities have been added to WRAP for assessing the impacts of salinity on water availability [37]. Salinity problems are addressed in the regional plans [24]. However, incorporation of salinity into the water allocation systems is a complex issue yet to be resolved.

The Texas Water Code has required consideration of environmental flow needs in the water rights permitting process since 1985. Such needs include maintenance of aquatic habitat and species, water quality, public recreation, wetlands, and freshwater inflows to bays and estuaries. Although such needs have been considered in issuing permits since 1985, most water rights in the state have been granted without specifying instream flow requirements. Under mandates enacted by the Texas Legislature in 2001 and 2007, developing methodologies for establishing environmental flow criteria, establishing requirements for each river reach in the state, and incorporating them into the water rights system is currently a major focus.

6. SUMMARY AND CONCLUSIONS

As water demands increase with population and economic growth, water allocation systems become increasing important worldwide. This chapter illustrates mechanisms for allocating water and highlights issues that must be addressed in their implementation. The Texas experience illustrates fundamental principles, issues, strategies, and complexities involved in developing and administering water allocation systems. Allocation of water resources among nations, states, regions, types of use, and numerous water users is a governing concern in water management in Texas as well as throughout the world.

Water allocation practices in Texas have evolved historically over centuries, with significant improvements occurring in recent years that continue to be refined. Texas shares water with Mexico and several neighboring states in the United States. Thousands of government agencies, cities, private companies, and citizens within Texas hold rights to use the waters of the state. Legal rights to use surface water differ from those for ground water. Surface water allocation for the Lower Rio Grande is different than for the rest of the state. With growing demands on limited water resources, expanding and refining water allocation systems has become a central governing concern in water management. The following observations regarding development and administration of water allocation systems in Texas are generally applicable in various other regions throughout the world.

- Water allocation systems include hierarchical systems of treaties and compacts between nations and states, ground water rights systems, surface water right systems, reservoir project ownership and operating practices, and contracts and agreements among a myriad of water management entities and water users. These water allocation mechanisms are overlapping and interconnected. Water allocation is integrally connected with other water resources planning and management activities.

- Water allocation practices evolve historically. In Texas, with growing demands on limited water resources, by the 1950's a disorganized myriad of water allocation practices had become a governing constraint to effective water management. Major change involving better water allocation mechanisms was necessary. The evolution of water allocation systems is continuing now and will continue in the future.

- Droughts motivate political action to improve water allocation systems. In Texas, major droughts led to milestone water management legislation being enacted in 1967 and 1997. The current drought that began in 2010 is expected to motivate additional legislative action that will contribute to the continued evolution of water allocation systems.

- The historical competition between agricultural, municipal, industrial, energy, and environmental water uses continues. There is a continuing shift from agricultural to municipal water use. Establishing and implementing environmental instream flow requirements is currently a major focus in Texas.
- Capabilities for assessing water availability and supply reliability within the framework of governing water allocation systems are essential for effective water resources planning and management. The Water Rights Analysis Package (WRAP) modeling system significantly contributes to effective water planning and allocation in Texas. WRAP is generalized for application to river/reservoir systems located anyplace in the world.

REFERENCES

1. Santos Roman DM. Systematization of water allocation systems: an engineering approach.PhD dissertation, Texas A&M University, College Station; 2005.
2. Rice L., White MD. Engineering Aspects of Water Law. Malabar: Krieger Publishing;1991.
3. Wallace J., Wouters P., editors. Hydrology and Water Law: Bridging the Gap. London: International Water Association (IWA) Publishing; 2006.
4. Tofan C., Strambu S., editors. International Water Law, 2 volumes. Oisterwijk: ICAPress; 2008.
5. Beach LB., Mammer J., Hewitt JJ., Kaufman E., Kurki A., Oppenheimer JA., Wolf AT.Transboundary Freshwater Dispute Resolution: Theory, Practice, and Annotated References. New York: United Nations University Press; 2000.
6. Pricoli JD., Wolf AT. Managing and Transforming Water Conflicts. Cambridge :Cambridge University Press; 2009.
7. Wolf AT., Natharius JA., Danielson JJ., Ward BS., Pender JK. International River Basinsof the World, International Journal of Water Resources Development 1999; 14(4)387-427.
8. Grover VI., editor. Water: A Source of Conflict or Cooperation? Enfield: Science Publishers;2007.
9. Barlow M. The Global Water Crisis and the Coming Battle for the Right to Water.Toronto: McClelland & Stewart, 2007.
10. Baillat A. International Trade in Water Rights: The Next Step. London: IWA Publishing; 2010.
11. Roth D., Boelens R., Zwarteveen M., editors. Liquid Relations: Contested WaterRights and Legal Complexities. New Brunswick: Rutgers University Press; 2005.
12. Selby J. Water, Power, and Politics in the Middle East. London: I.B. Tauris; 2003.
13. Getches DH. Water Law in a Nutshell. 4th Edition. St. Paul: West Publishing, 2009.
14. Beck RE., Kelley AK., editors. Waters and Water Rights, Cumulative Index, Volume8, 2008 Edition. San Francisco: Matthew Bender (LexisNexis), 2008.
15. Draper, SE., editor. Model Water Sharing Agreements for the Twenty-First Century.Reston: American Society of Civil Engineers, 2002.
16. Eherat JW., editor. Riparian Water Regulations: Guidelines for Withdrawal Limitationsand Permit Trading. Reston: American Society of Civil Engineers; 2002.

17. Dellapenna JW. Appropriative Rights Model Water Code, Reston: American Societyof Civil Engineers; 2007.
18. Fischer D. The Law and Governance of Water Resources. Cheltenham: Edward ElgarPublishing; 2009.
19. Hu D. Water Rights: An International and Comparative Study. London: IWA Publishing2006.
20. Kissling-Nag I., Kuks S., editors. The Evolution of National Water Regimes in Europe:Transitions in Water Rights and Water Policies. London: Kluwer Publishers;2004.
21. van Koppen B., Giordano M., Buttersworth J. Community-based Water Law and WaterResource Management Reform in Developing Countries. Oxfordshire: CAB International;2007.
22. Longo PJ. Water Across Borders: Judicial Realities. In: Longo PJ., Yoskowitz DW.Eds. Water on the Great Plains: Issues and Policies. Lubbock: Texas Tech UniversityPress; 2002.
23. Water Resources Planning: Manual of Water Supply Practices, Manual M50. Denver:American Water Works Association; 2001.
24. Texas Instream Flow Studies: Technical Overview, Report 369. Austin: Texas Commissionon Environmental Quality, Texas Parks and Wildlife Department, Texas Water Development Board; 2011.
25. Eaton DJ., Hurlbut D. Challenges in the Binational Management of Water Resourcesin the Rio Grande/Rio Bravo. U.S.-Mexican Policy Report No. 2. Austin: Lyndon B.Johnson School of Public Affairs, University of Texas; 1992.
26. Hall GE. High and Dry: the Texas-New Mexico Struggle for the Pecos River. Albuquerque:University of New Mexico Press; 2002.
27. Wurbs RA. Texas Water Availability Modeling System. Journal of Water Resources Planning and Management. American Society of Civil Engineers. 2005;131(4) 270-279.
28. Wurbs RA. Water Rights Analysis Package (WRAP) Modeling System ReferenceManual. 9th Edition. TR-255. College Station: Texas Water Resources Institute; August2012.
29. Wurbs RA. Water Rights Analysis Package (WRAP) Modeling System Users Manual.9th Edition. TR-256. College Station: Texas Water Resources Institute; August 2012.
30. Wurbs RA., Hoffpauir RJ. Water Rights Analysis Package Daily Modeling System.TR-430, College Station: Texas Water Resources Institute; August 2012.
31. Wurbs RA. Fundamentals of Water Availability Modeling with WRAP. 6th Edition.TR-283, College Station: Texas Water Resources Institute; September 2011.
32. Wurbs RA. Salinity Simulation with WRAP. TR-317, College Station: Texas Water ResourcesInstitute; July 2009.
33. Wurbs RA. Chapter 24 Water Rights Analysis Package (WRAP) Modeling System. In:Singh VP., Frevert D.K. (eds.) Watershed Models. Baton Raton: CRC Press; 2006.
34. Wurbs RA. Chapter 1 Generalized Models of River System Development and Management.In: Uhlig U. (ed.) Current Issues of Water Management. Rijeka: InTech2011.
35. Wurbs RA., Schnier ST., Olmos HE. Short-Term Reservoir Storage Frequency Relationships.Journal of Water Resources Planning and Management. American Societyof Civil Engineers. 2012;138(6).
36. Wurbs RA., Lee and CH. Salinity in Water Availability Modeling. Journal of Hydrology. Elsevier Science, 2011;407(2) 451-45

CHAPTER 3

Economically Precise Water Resource Management for Domestic Usage in India

P. Babu, C. Rajasekaran

Department of ECE, K. S. Rangasamy College of Technology, Tiruchengode, India

ABSTRACT

In the present years water resource that can be utilized for domestic usage is getting reduced dramatically. In order to have sufficient water for the utilization of future generation, it is our duty to conserve and utilize the available water in a very efficient way. This paper is aimed to help the people to conserve water; it is done with the help of an embedded system that is reconfigurable depending on the number of persons in the family. The system consists of three modules. Each has a different purpose, one module senses the water level present in the tank and intimates the user regarding when the water is needed to be restored. The second module deals with the division of water based on the type of application for which is intended. The third module deals with intimating the municipality regarding the availability of water. All the three modules combine to work in conserving the water; as a result, we can consume nearly 30% per year per head. So this would result in a large amount of conservation of water. Thus the saved water can be reused for several applications.

Keywords: Water Resource Management, Raspberry Pi, Water Level Sensing, GSM Modem

1. INTRODUCTION

Enormous growth of developing world has led to huge need of water. Due to unequal distribution of water, sufficient amount of water is not received by few people. Water is said to be another basic need for all the living beings in the earth. Process automation system based upon utilization of an industrial PLC and PC systems including all the network components represents the best way to improve the water distribution technological process [1] . SCADA is one of the emerging technologies which are used for complete monitoring. The overall system is connected to PLC with the help of RS-232 cable [2] . This proposed automated water distribution system is used to distribute the municipal water equally to all street pipe line [3] so that everyone will get the equal amount of water. Here we have explored an idea as solution for the issues such as improper water supply, water quality and over consumption by people. The proposed system is fully automated.

2. WATER USAGE AND SCARCITY

In India, the design of water supply systems has been done using certain standards. Currently the standard being used is BIS 1172: 1993, reaffirmed in 1998. This specifies a consideration of use of the following:

For communities with a population of between 20,000 to 100,000—100 to 150 litres per head per day.

For communities with a population of over 100,000—150 to 200 litres per head per day.

In its previous avatar there was also an attempt made in IS 1172 to understand the break-up of this demand which was then put as 135 litres per person per day. The break-up was as follows:

- Bathing: 55 litres.
- Toilet flushing: 30 litres.
- Washing of clothes: 20 litres.
- Washing the house: 10 litres.
- Washing utensils: 10 litres.
- Cooking: 5 litres.
- Drinking: 5 litres.

3. METHODOLOGY

According to our work, we are using Raspberry Pi, GSM Modem, Float Sensor and Solenoid valve are the main components of this system. Raspberry Pi is the advance version of ARM11 processor. Using putty configuration it is interfaced on computer and commands are written using Python. We are using latest version of Raspberry Pi that is Model B+ because it supports to the Internet of Things (IOT). GSM Modem is interfaced for communication purpose. Relays are used for the connection of the Raspberry pi and Solenoid valve. Thus, first of all we design water level sensing circuit using Float sensor which measures the level of water in the main tank. Now we send this information of water level to the user through GSM Modem. Solenoid valves are used to split the amount of water into different purpose washing, cooking, bathing, gardening, etc.

The system is implemented when the user get the limited amount of water in home. This water is automated for each application such as washing, cooking, bathing, gardening, etc. The overall block diagram representing our idea is shown in Figure 1.

Raspberry PI is the heart of the project. We have proposed to use "Raspberry Pi B+". The work of the Raspberry in the system is to monitor the level of water stored in the tank and allocate the water accordingly to the intended application and also it will get the input from the user so that the process can be reconfigured.

The water from the corporation reaches each user's home as usual and stored in the main water tank. The water level sensing circuit is interfaced to the raspberry pi processor with the help of inverting amplifier circuit. The level of water stored in the main tank is continuously monitored by the processor and display the water level in the Liquid Crystal Display (LCD).

The solenoid valve distributes the water stored in the main tank for different purpose. Solenoid valve operated in 12 V and Raspberry pi processor operated in 5 V. So the interfacing of solenoid valve and raspberry pi processor is achieved by the relay and driver circuit.

If the numbers of users in the family get increased, the system is reconfigurable. The reconfiguration is achieved by the keypad and GSM modem. The user sends the message to the municipality regarding their water needs.

The relay circuit is designed to interface the Raspberry pi and solenoid valve. The relay ON and OFF is controlled by the pair of switching transistors (BC 547) shown in Figure 2. The relay is connected in the Q2 transistor collector terminal. The operation of a relay circuit is mentioned in Table 1.

Float is one type of transducer which is used to measure the water level in the tank shown in the Figure 3. The float changes the resistance value depending on the water level. This change in resistance is converted into corresponding voltage signal which is given to inverting input terminal of the comparator. The reference voltage is given to non-inverting input terminal.

The comparator is constructed by the operational amplifier LM 741. The comparator compares with reference water level and delivered the error voltage at the output terminal. Then the error voltage is given to next stage of gain amplifier which is constructed by another operational amplifier LM 741. In the gain amplifier the variable resistor is connected in the feedback path, by adjusting the resistor we can get the desired gain. Then the final voltage is given to ADC for converting the analog signal to digital signal. The water level in the tank is identified by microcontroller.

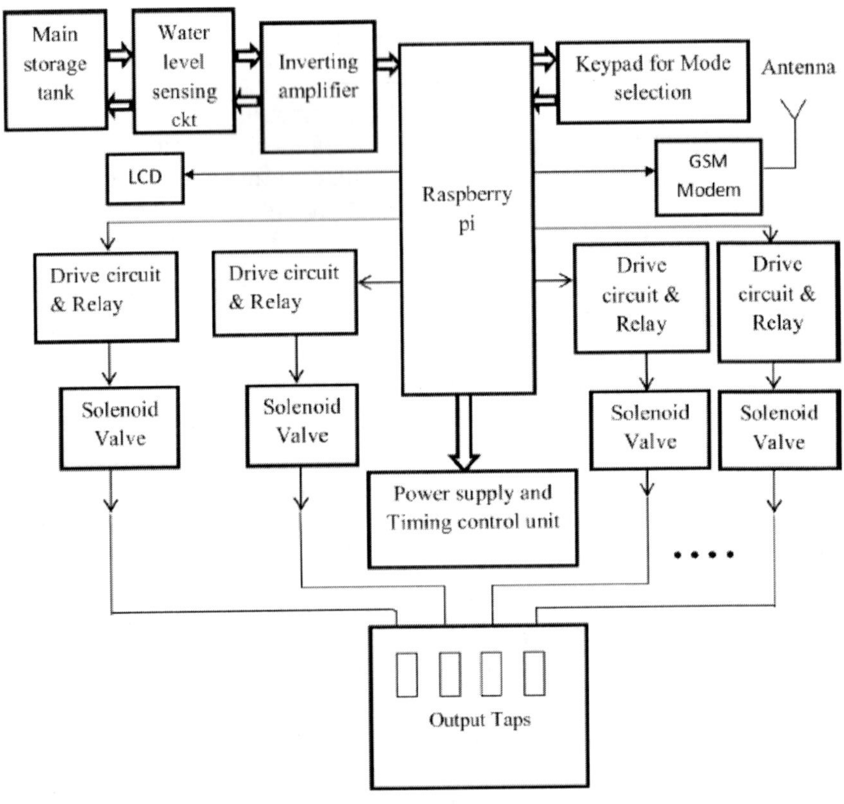

Figure 1. Block diagram of water automatic system.

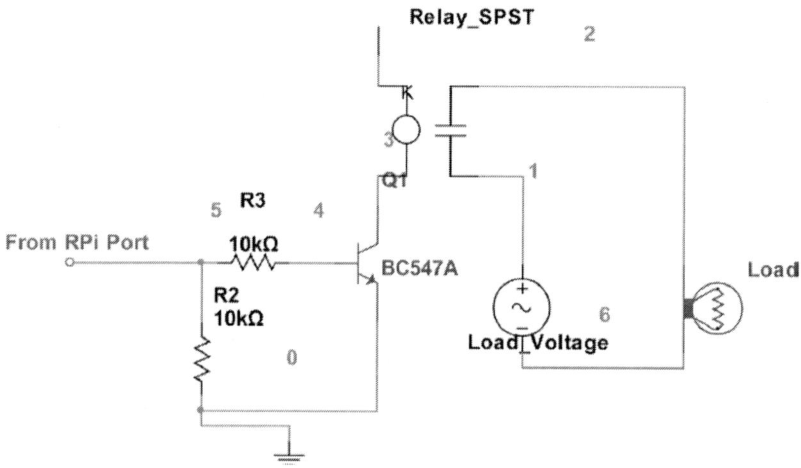

Figure 2. Relay circuit-SPST.

Table 1. Operation of a relay circuit.

Voltage Signal from Raspberry PI B+	Transistor Q1	Transistor Q2	Relay
1	ON	OFF	OFF
0	OFF	ON	ON

The water required for the various uses can be given with the help of Solenoid Valve. The work of solenoid valve is to control the flow of water based on the intended application. The function of keypad utilized in the project is to get the information from the user regarding the number of available people in the home, so that water available can be divided for all the members in the family. Also the keypad helps to get the amount of water which is needed. The various needs are like Cooking, Bathing, Washing, Gardening etc.

4. HARDWARE DESCRIPTION OF PROPOSED SYSTEM

The Hardware of the proposed water automated system consist of three modules shown in Figure 4.The three modules are water level sensing, division of water for each intended application, intimating to the corporation when the number of user in the family get increased.

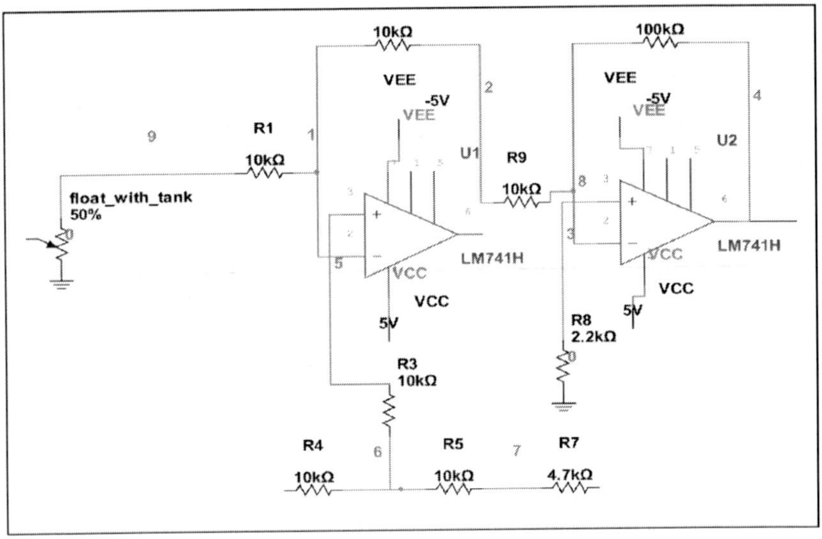

Figure 3. Water level sensing circuit.

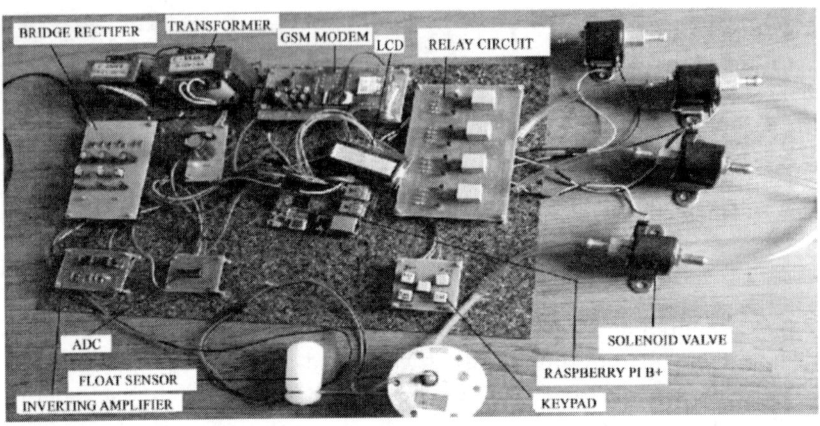

Figure 4. Hardware module of water automatic system.

Float sensor is used to measure the water level in the main storage tank. Float sensor continuously monitor the water level with the help of variable resistance (10 K, 100 K). The process of water level sensing is shown in Figure 5.

The solenoid valve is used to distribute the amount of water to user needs such as washing, cooking, bathing, gardening, etc. If the number of application is increased, redesign the hardware for the extended application. The proposed method of distribution of water is shown in Figure 6. The amount of water stored in the main storage tank distributed in the taps. In the proposed water supply system each man uses only 135 litres. So each tap release only limited amount of water. For example, bathing tap flows at the rate of 55 litres, Cooking tap flows at the rate of 5 litres, Washing taps flows at the rate of 40 litres and Gardening tap flows at the rate of 35 litres. If the number of application (i.e. user needs or taps) is increased, the system is reconfigurable.

The GSM modem communicates to the user and corporation. Depending on the situation, requirement of water is intimated to the corporation. In the GSM modem, SIM 900 is used. The process of indicating to the corporation through GSM modem is shown in Figure 7.

5. HARDWARE DESCRIPTION OF PROPOSED SYSTEM

The existing water supply system does not distribute the water equally. This is shown in the Figure 8. The first user receives the 800 litres. The second user receives 600 litres. The user does not get the equal amount of water.

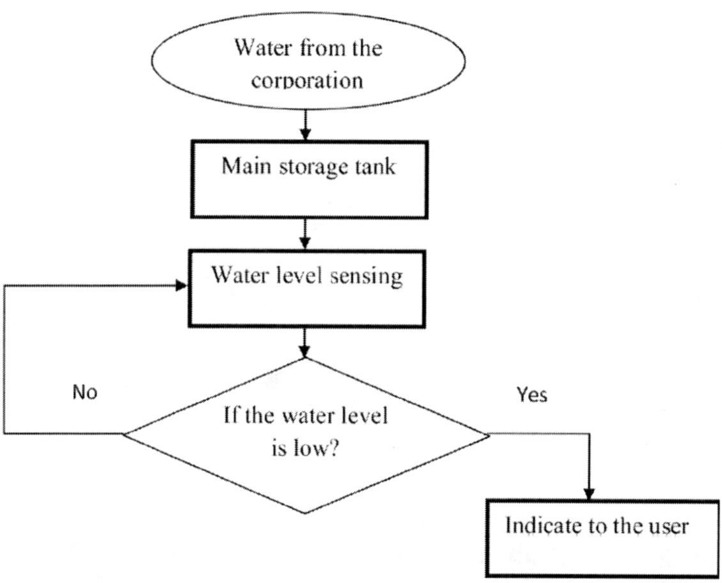

Figure 5. Flow chart of water level sensing.

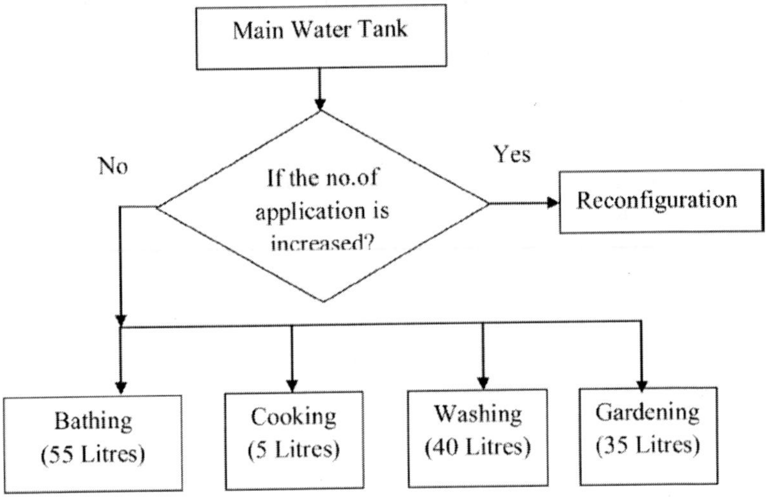

Figure 6. Proposed method of water distribution.

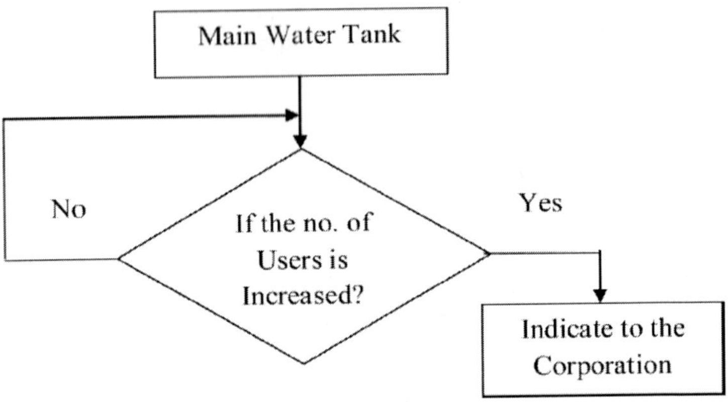

Figure 7. Process of indicating to the corporation.

This varies depending upon the users. But the proposed system distributes the water equally to all the users. In the Figure 8, each family gets up to 600 litres of water in this proposed system. Each family saves up to 200 - 400 litres every day.

The IS 1172 mentioned that 135 litres is sufficient for one person per day. The break-up of 135 litres is shown in Figure 9. But every person uses 150 - 200 litres every day. Approximately every person wastes 15 - 65 litres every day (Table 2).

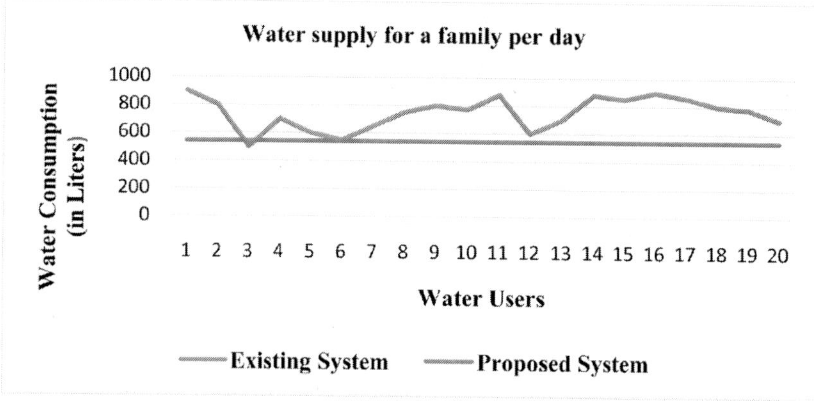

Figure 8. Comparison of existing and proposed water supply system.

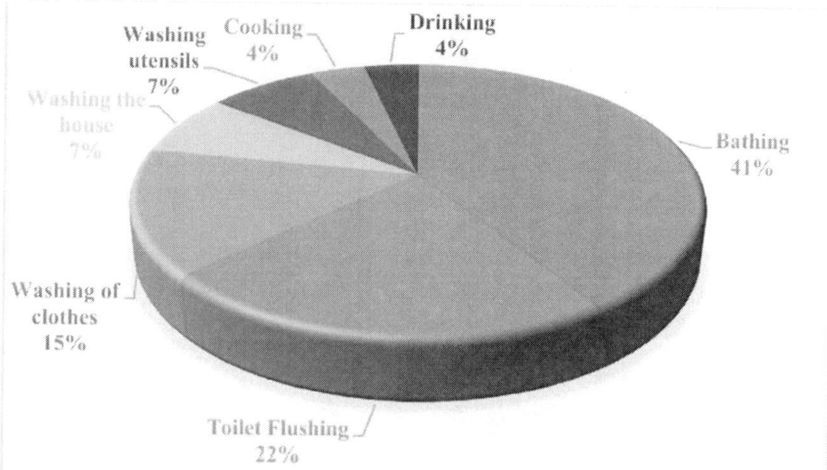

Figure 9. Water usage of a person per day by IS 1172.

After the distribution of corporation water, the water consumption of an individual person and family has shown in Figure 10 and Figure 11. In existing system the family uses 800 litres of water in a day. But the proposed system use only 540 litres. Also the comparison of week, month and year is plotted in the graph.

The existing system uses 292,000 litres in one year. But the proposed system will use 200,000 litres. In one year 92,000 litres of water is saved by each of the family members. Approximately 1/3 (i.e. 33%) of the water is saved in a year by our proposed system.

Table 2. Water consumption of a person and family.

	Existing system water usage (in liters)		Proposed system water usage (in liters)		Saved water (in liters)	
	One person	One family (4 members)	One person	One family (4 members)	One person	One family (4 members)
Day	150 - 200	600 - 800	135	540	15 - 65	60 - 260
Week	1050 - 1400	4200 - 9800	945	3780	105 - 455	420 - 1820
Month	4500 - 6000	18.000 - 24,000	4050	16.200	450 - 1950	1800 - 7800
Year	54.750 - 73.000	219.000 - 292.000	49.275	197.100	5475 - 23.725	21.900 - 94.900

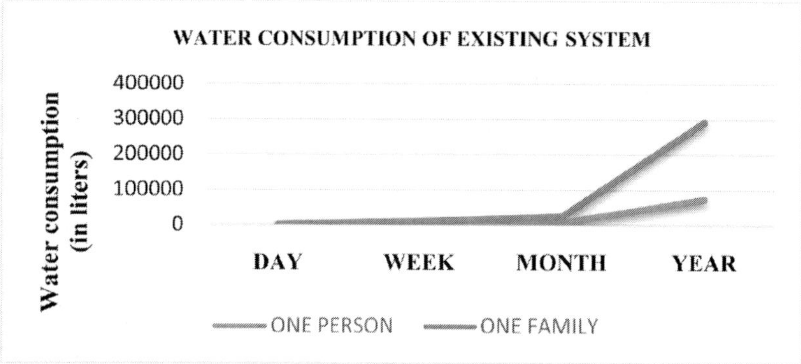

Figure 10. Water consumption of existing system.

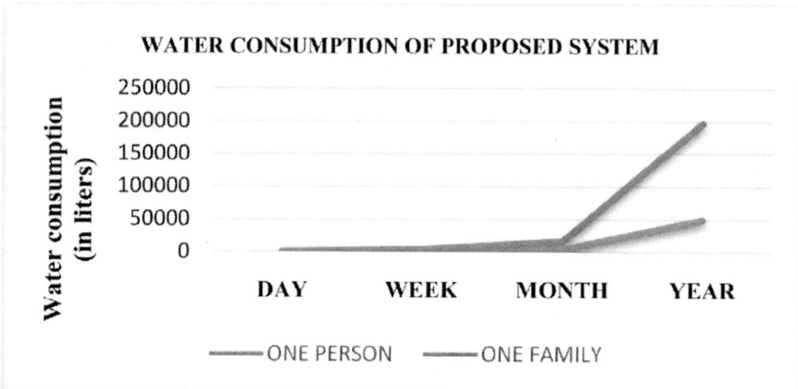

Figure 11. Water consumption of proposed system.

6. CONCLUSION

Automated water distribution and performance monitoring system focuses on various entities such as proper supply, over consumption alert and water quality assurance. Automation system provides continuous water flow according to the set point. This system is automatic so it reduces lots of man power. The automation implemented in water distribution system ensures to avoid wastage of water and reduces times so that people could get equal amount of water. This system is excellent and cost effective to prevent the drinking water from the theft during the distribution period. The way in which the water is being utilized is not up to the mark through which water can be conserved, so in order to achieve self-sufficiency in water, we need to be more cautious in conserving the water. Approximately 1/3 (i.e. 33%) of the water is saved by the proposed automatic water distribution system in one year, thus resulting in large saving of water which can be used to for other applications like home usage, agriculture, etc.

REFERENCES

1. Gouthaman, J., Bharathwajanprabhu, R. and Srikanth, A. (2011) Automated Urban Drinking Water Supply Control and Water Theft Identification System. Proceedings of the International Conference of the IEEE Student's Technology Symposium, Kharagpur, 14-15 January 2011, 87-91. http://dx.doi.org/10.1109/techsym.2011.5783807
2. Stancil, E., Stoian, I., Kovacs, I., Gyurka, B.Z. and Balogh, S.Z. (2008) Urban Water Supply Distributed System. Proceedings of the International Conference of the IEEE Automation, Quality and Testing, Robotics (AQTR), Cluj-Napoca, 22-25 May 2008, 316-320. http://dx.doi.org/10.1109/aqtr.2008.4588936
3. Baranidharan, T., Chinnadurai, A., Gowri, R.M. and Karthikeyan, J. (2015) Automated Water Distribution System Using PLC and SCADA. International Journal of Electrical and Electronics Engineers, 7, 355-360.

CHAPTER 4

Sustainable Water Resources Management in Arid Environment: The Case of Arabian Gulf

Elnazir Ramadan

Department of Geography, Sultan Qaboos University, Muscat, Oman

ABSTRACT

Over the past decades the Arabian Gulf region has witnessed a great economic development and social transformation. The region is facing a water insufficiency problem that is one of the biggest in the world. The level of available renewable water in the region is one fifth of what the rest of the world enjoys on a per capita basis. The population of the region is growing 55% more quickly than the population at the rest of the world. By 2020, water needs is predicted to be around 341 million imperial gallons per day. The threat of water scarcity ensures that investment in developing freshwater supply, along with the recycling and reuse of waste and sea water is an urgent priority across the region moreover strategies for sustainable water use have to be adopted. This means that if strategies for rational water use that entail educational component is not established the water storage is going to run out. This paper reviews the existing situation identifies the gap and proposes an appropriate institutional framework which involves assignment of responsibilities among various levels. Ensures stakeholders participation, accommodates adaptive change and remain self sustainable.

Keywords: Arabian Gulf ; Water scarcity; Sustainable management; Arid environment

1. INTRODUCTION

The availability of adequate freshwater has become a limiting factor of the quality of life, worldwide. More than availability, the problem is often the rational use of water than ensure its continuality. In the semi-arid and arid regions water scarcity was always a dominant problem moreover, the interference with the natural hydrologic cycle as a result of overexploitation of both surface and ground waters and of changes in land usage resulted not only in the reduction of the available water amounts but also in the deterioration of the water quality due to pollution from urban, industrial and agricultural practices and salinity build-up in soil and water. Such a situation is in all cases a pretext for discord and assignment of blame on those supposedly responsible for the deterioration of the water accessibility , especially in view of the possible high cost and technical complexity of the measures that need to be taken for remedying and alleviating the situation. Arabian Gulf countries are located in an arid area with limited water resources [1]. Hydrological investigations point to limited resources of **underground water**. The best choice for providing fresh water in the Arabian Gulf countries is through seawater desalination with ground water as a backup. About 65% of desalination plants that are in operation worldwide are located in the Arabian Gulf countries, most of which are the dual-purpose multistage flash (MSF) plants, producing power and water. Integrated water resource management (IWRM) is the need to take a holistic approach to ensure the socio-cultural, technical, economic and environmental factors are taken into account in the equitable development and management of water resources.

2. SUSTAINABLE AND INTEGRATED WATER RESOURCES MANAGEMENT

The relationship between "Sustainable" and "Integrated" Water Resources Management is essentially that sustainability is the general goal whereas Integrated Water Resources Management (IWRM) is a strategy for pursuing this goal. Water is a resource under considerable pressure. Effective and sustainable management of water resources is vital for ensuring sustainable development [2]. However, efforts of water resource management seem to demonstrate inappropriate practices, especially when compared to water consumption trends in Arabian Gulf states. Being a major and vital ingredient to human kind, water resources influence all sectors. However, there have been increased problems over time that subject water resources to a number of crisis and pressures. Poor water resources management have stimulated and sustained a number of problems related to health, socio-economic and **environment**, which need to be solved. Integrated water resources management (IWRM), is a process, a change, and an approach that mainstream water resource use and management into the national economic in an equitable manner without compromising the sustainability of vital ecosystems. Successful community involvement demonstrates the importance of IWRM. Public/ community involvement is crucial for a successful and sustainable water resource management. It has been emphasized that natural resources management related policies including water requires the use of knowledge, experience and opinions of local communities who are the key stakeholders in resource conservation.

3. WATER RESOURCES OF THE REGION

In the area under discussion the arid climate extending over the Gulf States is characterized by large geographic and temporal disparities of the precipitation distribution, with relatively more annual amounts of precipitation in winter season, a relatively variable regime, with alternating drought and excessive rain periods further acerbates the situation. Scarcity of water resources in the region is evident some studies earlier this year said that the imminent shortage of water resources in the region has been compounded by the real estate boom, with new construction projects taking a larger share of resources. This is alarming since this region is already the driest in the world. The Gulf remains the largest market for water desalination in the world and local municipalities are seriously examining ways to double existing capacity to meet regional demand [3].

The arid climate of the region where rainfall is sparse, limits conventional water resources and recent studies show that temperatures are increasing. In August 2009, a temperature in some areas of the region has exceeded 47.3°C. The available water resources in the Arabian Gulf countries are mainly the ground water which is not enough to satisfy the water demand due to the rapid urban and industrial developments. Due to the improper planning and mismanagement of the water abstraction from the wells, the ground water is very much reduced and in many countries the withdrawal of water is now done from the non renewable layers. Abstracting water from the non-renewable layer causes a serious problem because this layer was formed during very long decades and abstraction from this layer will definitely deteriorate the ground water storage. Desalination of seawater is one of the main alternatives for the substitution of water shortage in the Arabian Gulf countries and other countries. Although desalinating the seawater is costly, it is still an important option for compensating for the water shortage. Most of Gulf countries built power and desalination plants for water and power production. We should be aware of the fact that the effluent discharges from the plant back to the sea may have a negative impact on environment.

The Gulf remains the largest market for water desalination in the world and local municipalities are seriously examining ways of looking to double existing capacity to meet regional demand. However, it is North Africa that will experience the greatest growth in desalination, Algeria and Libya experiencing a 300 per cent growth. Water desalination, specifically membrane processes based on reverse osmosis, is becoming more efficient as more technologically advanced methods are being developed. The price of desalinating one cubic meter of water has come down more than a third of what it used to be 10 years ago.

Table 1: Water production in Gulf States.

	2001	2002	2004	2007	2010
UAE	165.1	182.4	204.9	229.1	229.1
Bahrain	27.9	29.8	30.4	32.1	33.87
KAS	293.14	214.98	233.99	270.46	275.14
Oman	20.79	21.54	22.72	23.98	26.19
Qatar	32.3	33.5	34.8	35.6	35.6
Kuwait	82.134	107.27	90.89	95.17	97.87
Total	621.364	589.49	617.7	686.41	697.77

4. WATER RESOURCES DEFICIENCY

The problem of water shortages in the Gulf States is quite evident despite the fact that these countries adopting advanced technology. **Water scarcity** in the region is rapidly becoming part of a widespread environmental concern for the region. water deficit in the Arab Gulf countries is estimated to be 16 billion cubic meters, this would have an impact on the quality of ground water. Consumption per capita, which is currently at 1,100 cubic meters annually, is projected to fall to 550 cubic meters a year by 2050. While Gulf countries spend around $133bn in water and wastewater management yearly, more effective strategies are needed, such as convincing companies and even residents to set up their own sustainable infrastructure.

The region has the lowest availability of actual renewable water resources per capita in the world. It is thus imperative that gulf governments invest in technologies that can ensure adequate supply for the region's vibrant industries and its more than 300 million inhabitants [4].

Building and industrial projects alone in the region will consume more than 112 billion liters of drinkable water from 2008 to 2009 this part of the world hosts 5% of the global population and yet possesses only 1% of the world's renewable fresh water. Desalination is likely to become one of the world's biggest industries. Growing communities and new industries must have dependable water supplies in order to prosper. Desalting systems have long proven effective in Kuwait, Bahrain, Qatar, the United Arab Emirates, Oman, and Saudi Arabia. Where once there were bleak villages on barren deserts there are now bright modern cities with tree lined streets. There are homes with lush gardens. In the countryside there are productive farms. The big desalting plant at Jubail, Saudi Arabia, is a model for the world. A pipeline carries a river of **freshwater** 200 miles inland to the capital city of Riyadh, and desalted seawater has given a large region an entirely new future filled with opportunities. There are more than 7,000 desalination plants, mostly small ones, in operation worldwide. About two thirds are located in the Middle East, and others are scattered across islands in the Caribbean and elsewhere. Aruba's high-tech water plant has for many years met the needs of a thriving tourist industry.

The largest plant in the United States is the pioneering $158 million project of the Tampa Bay Water agency. The project was left to contract in 1999 and after overcoming some technical problems in its early years is now performing well and causing no significant environmental problems. But no U.S. water agency has yet undertaken a really big project comparable to those found along the Arabian Gulf.

5. TOWARDS A STRATEGY FOR WATER MANAGEMENT IN THE REGION

Certainly, water conservation programs should come first as a strategy for regions facing water problems. Many jurisdictions are already imposing water use limits. Other communities try drilling wells deeper and deeper until their aquifer is maxed out, or they propose to pipe water from distant streams. But such shortsighted strategies can do incalculable damage to the environment.

Table 2: Water consumption in Gulf States.

	2002	2003	2004	2007	2010
UAE	142.8	149.4	196.9	212.5	229.1
Bahrain	27.9	29.8	30.4	32.2	33.82
KAS	------	252.7	279.52	318.24	207.28
Oman	20.69	21.5	22.85	23.95	26.14
Qatar	-------	33.5	34.8	35.6	35.6
Kuwait	82.11	102.06	91.05	94.99	97.88
Total	273.5	588.96	655.52	717.48	629.82

Water conservation programs should be carried out by the agency responsible for water resources management. The execution of such plans for the various sectors (such as municipal, agricultural, or industrial) should be co-ordinated by the relevant government body in each sector. Close co-ordination and partnership should be institutionalized between the agenciesresponsible for water supply, demand management, and education, media, and awareness. Unfortunately, the ministries of education, Awqaf, and Islamic Affairs rarely participate in water conservation programs. in the region, although this is essential for effective awareness activities.

Equally, the public's involvement and its co-operation in designing and implementing conservation measures are essential to the success of water conservation programs. The public includes consumers, service providers, managers, and planners as well as policy-makers. Raising public awareness using Islamic conservation concepts should always be integrated with the use of other communication tools and channels.

To achieve greater co-operation and involvement, the public must understand the water supply situation, including the cost of delivery, the overall water resources situation, and the need to conserve water resources and to maintain them for future generations. This increased understanding is the first step in any successful public awareness activity. However, the credibility of this information is essential. Because honesty is a core principle in Islam, the public expects the truth from imams and other Islamic sources. Most water conservation activities require changes of behavior and attitudes, which is usually a slow process. Therefore, ad hoc and occasional public awareness activities are not effective. Water authorities should plan continuous, long-term activities in close collaboration and coordination with ministries of Education and Islamic Affairs. Some water conservation activities involve costs that must be paid by the public, such as fixing water taps, upgrading irrigation systems, or modifying industrial production lines. These costs of water conservation programs must be offset by some incentives. In addition to the physical incentives, the spiritual incentives offered by Islam can be of value.

Water conservation activities and awareness campaigns typically focus solely on domestic users. This is shortsighted, and the focus should be on all water users. Mosques are ideal places for awareness campaigns, since all kinds of people meet there at least weekly. However, imams should be aware of the need to address all sections of the population. In the GCC countries, a shift in societal values from a development-oriented to a conservation-oriented view of water resources is occurring. It is believed in the GCC countries that conservation of natural resources in general, and water resources in particular, are a principal component of Islamic teachings. It is also believed that the most important and effective way to make the public aware of conservation from an Islamic perspective is through the media and the educational system [5]. Islamic messages are being used in the preparation of posters and video clippings for these campaigns. On the occasion of World Water Day 1998, and upon request of ministries of Islamic Affairs, imams were requested to devote their Friday speeches to the theme of Islam and water conservation [6]. But such occasional public awareness water conservation campaigns need to be integrated through a comprehensive and long-term plan of action that targets behavioral change [7], otherwise their effect will limited water treatment and reuse is now one of the top priorities for governments across the GCC as populations increase and the demand on existing infrastructure and potable water generating capacity is stretched to the limits. More than funds will be invested in the wastewater and water reuse sector over the coming 20 years as Gulf states embrace the concept of full sewerage coverage and the need for conservation of scarce water resources. The investment drive in nearly every GCC state will be carried out by the private sector either in tandem with the government in public-private partnerships (PPPs) or through fully-fledged privatization.

Technology will continue to play an important role. Most states are still reluctant to utilize the more recent membrane bioreactor (MBR) technology, preferring to stick to conventional wastewater treatment applications, the exception being some of the smaller real estate players in Qatar and UAE. However, MBR technology is expected to take hold in the region. Sharjah is building a small-scale pilot STP for Ashghal using MBR and reverse osmosis technology. There are no plans, at the moment at least, for states to connect the treated wastewater network to potable water networks. 'It's a psychological fear,' said Salah al-Mutawa, manager of the treated sewage effluent department at Bahrain's Works & Housing Ministry. 'People aren't generally aware of the treatment process' [8,9].

Education and increasing public awareness are now firmly on the agenda. Both Dubai and Doha are thinking about public visitor centres to teach people about the benefits of water reuse and the processes involved in treatment. Cultural and religious factors will remain an obstacle to the implementation of a full-water reuse cycle, although states are now looking at using treated wastewater to recharge aquifers [10,11].

6. CONCLUSION

Integrated water resources management is the most feasible option to overcome the serious crisis of water shortage. Water conservation should be an integral part of this option, with a clear focus on public awareness and participation without which the chances of success will be lessened.

Access to information related to water conservation and public awareness activities is lacking in Gulf countries both because of the limited number of these activities, and because of poor information management and exchange. Such access can be improved by adopting a strategy to identify sustainable water management practices, documenting the experiences in water conservation and public awareness. This Strategy should be made available to water specialists as well as to the public by networking at the regional and national levels.

Gulf States can meet the challenge of global warming with a threepronged approach: reducing emissions of carbon pollution, minimizing human stresses on ecosystems, and adapting to the challenges to come. Reducing the dependence on fossil fuels by developing clean energy sources would reduce global warming gas emissions and create jobs and new economic opportunities for region. By implementing the best practices in land and water resource use, policy-and decision-makers can minimize ecologically harmful side effects of climate change. And finally, elected officials and government leaders can plan ahead by increasing their flexibility and adaptive capacity in managing the state's precious water resources, agriculture, forests, ecosystems, and coasts.

REFERENCES

1. Al Zawad (2008) Impacts of Climate Change on Water Resources in Saudi Arabia.
2. Water and Energy Linkages in the Middle East (2009) Stockholm International Water Institute, Sweden.
3. Afifi, Madiha Moustafa (1996) Egyptian National Community Water Conservation Programme. Environmental Communication Strategy and Planning for NGOs ,Jordan Environment Society, Jordan.
4. Ayesh, Mohammed (1996) Awareness Project in Water. Environmental Communication Strategy and Planning for NGOs, Ma'ain, Jordan Environment Society, Jordan.
5. Akkad AA (1990) Water Conservation in Arabian Gulf Countries. Journal of the American Water Works Association 82: 40–50.
6. Al-Tamimi, Izz El Din (1991) Religion as a Power for Protection of the Environment (in Arabic). Environmental Research and Studies, Jordan Environment Society, Jordan.
7. Farshad A, Zinck JA (1995) The fate of agriculture in the Semi-arid Regions of Western Iran - A Case Study of the Hamadan Region. Annals of Arid Zone 34: 235-242.
8. Abdulrazzak MJ, Khan MZA (1990) Domestic Water Conservation Potential in Saudi Arabia. Journal of Environmental Management 14: 167–178.
9. Al-Sodi, Abdul Mahdi (1993) Attitudes of Jordanian Citizens towards Environmental Protection in the Sweileh and Naser Mountains Areas (in Arabic). Environmental Research and Studies, Jordan Environment Society, Jordan.
10. Madani, Ismail (1989) Islam and Environment. For Environmental Awareness in the Gulf Countries, Ministry of Information, Bahrain.
11. Hamdan M, Toukan Ali, Shaniek M, Abu Zaki M, Abu Sharar T, et al. (1997) Environment and Islamic Education. International Conference on Role of Islam in Environmental Conservation and Protection, Al-Najah University, Palestine.

CHAPTER 5

Drinking Water Quality and Occurrence of *Giardia* in Finnish Small Groundwater Supplies

Tarja Pitkänen [1,*], Tiina Juselius [2], Eija Isomäki [3], Ilkka T. Miettinen [1], Matti Valve [3], Anna-Liisa Kivimäki [3,4], Kirsti Lahti [4] and Marja-Liisa Hänninen [2]

[1]*National Institute for Health and Welfare, P.O.BOX 95, Kuopio 70701, Finland*

[2]*Department of Food and Environmental Hygiene, Faculty of Veterinary Medicine, University of Helsinki, P.O.Box 66, Helsinki 00014, Finland*

[3]*Finnish Environment Institute, P.O.Box 140, Helsinki 00251, Finland*

[4]*The Water Protection Association of the River Vantaa and Helsinki Region, Asemapäällikönkatu 12 B, Helsinki 00520, Finland*

ABSTRACT

The microbiological and chemical drinking water quality of 20 vulnerable Finnish small groundwater supplies was studied in relation to environmental risk factors associated with potential sources of contamination. The microbiological parameters analyzed included the following enteric pathogens: *Giardia* and *Cryptosporidium*, *Campylobacter* species, noroviruses, as well as indicator microbes (*Escherichia coli*, intestinal enterococci, coliform bacteria, *Clostridium perfringens*, *Aeromonas* spp. and heterotrophic bacteria). Chemical analyses included the determination of pH, conductivity, TOC, color, turbidity, and phosphorus, nitrate and nitrite nitrogen, iron, and manganese concentrations. *Giardia intestinalis* was detected from four of the water supplies, all of which had wastewater treatment activities in the neighborhood. Mesophilic *Aeromonas salmonicida*, coliform bacteria and *E. coli* were also detected. None of the samples were positive for both coliforms and *Giardia*. Low pH and high iron and manganese concentrations in some samples compromised the water quality. *Giardia intestinalis* was isolated for the first time in Finland in groundwater wells of public water works. In Europe, small water supplies are of great importance since they serve a significant sector of the population. In our study, the presence of fecal indicator bacteria, *Aeromonas* and *Giardia* revealed surface water access to the wells and health risks associated with small water supplies.

Keywords: *Giardia*; groundwater quality; small water supply; surface water contamination

1. INTRODUCTION

Small drinking water supplies that provide water to small communities are an important public health issue because they are often vulnerable and may cause microbiological or chemical quality-associated health risks to the water consumers [1,2]. Even if the number of users in a supply is low, the total number of these supplies is high, and they often constitute the major water supplies, especially in rural areas. In 2001, there were 1359 water supplies in Finland, of which 61% distributed groundwater [3]. Estimates indicate that approximately 500 of these supplies serve fewer than 500 consumers in their distribution area and are, therefore, defined as small water supplies [4]. Most small water supplies are owned by cooperatives with operators who are often community members who work part-time, usually with no required professional training in the management of a water supply.

Small Finnish water supplies with fewer than 50 users (or with a water distribution of less than 10 m^3 per day) are controlled by local health authorities according to the Decree of the Finnish Ministry of Social Affairs and Health [5] governing small water supplies. Supplies with more than 50 consumers are controlled on the basis of Finnish regulations governing the quality demands and surveillance of household water based on criteria stated in European Union drinking water directive 98/83/EC [6,7]. The microbiological and chemical quality of tap water is monitored in certain periods according to the number of consumers or the quantity of water distributed, but at least once every three years.

Many small European supplies have been found to have occasional contamination by surface water associated with increased coliform counts, indicating that they are at risk for fecal contamination [2,8]. However, usually the microbiological quality of groundwater supplies in Finland meets the quality criteria of the legislation, and most of the supplies distribute drinking water without disinfection treatment. Moreover, the chemical quality generally meets the criteria, but some chemical parameters, such as low pH and high iron and manganese concentrations, degrade the technical quality of the water in small private wells, as well as in some large water abstraction plants [9,10].

Occasional waterborne outbreaks have been reported annually in Finland. The National Institute for Health and Welfare has reported in 1998–2009 a total of 67 waterborne outbreaks, which caused illness for more than 27,000 people [11]. Most of these outbreaks originated from contaminated small groundwater supplies [12]. The most important microbial agents in the outbreaks have been noroviruses and *Campylobacter jejuni* [13]. No outbreaks associated with protozoan parasites *Giardia* or *Cryptosporidium* were reported in Finland prior to 2007 when sewage contaminated tap water in the town of Nokia contained *Giardia* among other pathogens [14]. In a small number of the waterborne outbreaks, the causative agent remained unidentified [11].

The bedrock in Finland in dominated by Precambrian igneous rocks, the crust consisting mainly of plutonic and metamorphic rocks [15]. Although drilled wells utilizing deep groundwater flowing in the bedrock fractures are widely used in household water supply, the majority of municipal water supply relies on groundwater resources in shallow aquifers, *i.e.*, glaciofluvial and glacigenic deposits. Our aim was to study the microbiological quality, including the enteric pathogens (*Giardia* spp., *Cryptosporidium* spp., *Campylobacter* spp. and noroviruses), of small groundwater supplies and the association of microbiological quality with environmental risk factors that could increase the possibility of fecal contamination. In addition, we studied indicator bacteria and some chemical quality parameters of the small groundwater supplies. The supplies were selected on the basis of a recent history of coliform contamination or reported river or lake bank filtration, agricultural fields, household septic tanks or gravel mining pits close to the supply, potentially affecting groundwater quality in the studied shallow aquifers. The distance between the identified risk factors and the production wells ranged from less than 50 m to 200 m.

2. RESULTS

2.1. Water Supplies

The characteristics of the 20 water supplies selected for the study are presented in Table 1. All studied water supplies were considered as small supplies since they served less than 500 inhabitants. The number of inhabitants in the area served by the water supplies varied from 38 to 450 (median 120 inhabitants). The average daily water intake was from 4 to 125 m^3/d (median 30 m^3/d) and the maximum water intake from 12 to 700 m^3/d (median 138 m^3/d). The majority of the studied wells derive groundwater from a shallow aquifer, the average thickness of a vadose zone being, in most cases, less than 5 m (Table 1). Only limited data on aquifer characteristics were available. These small groundwater supplies in Finland are prone to the surface water influence due to the insufficient depth of protective layers above the water table [2].

The water supplies were located around Finland and were regulated by the national Decree of the Finnish Ministry of Social Affairs and Health [5] when having fewer than 50 users or a water distribution of less than 10 m^3 per day or by Finnish regulations based on European Union drinking water directive [6] when having more than 50 consumers or a water distribution of more than 10 m^3 per day. Problems mentioned in the previous monitoring by the local public health protection authorities include the occasional detection of coliform bacteria (14 of the 19 water plants), low pH (11/19), and excess iron or manganese (7/19), or both. In six of the water supplies, the water was alkalized with either lye (NaOH), soda (Na_2CO_3), or limestone ($CaCO_3$), and in three supplies, excess iron or manganese was removed prior to distribution to consumers (Table 1). In three of the supplies, water was continuously disinfected with either UV light or sodium hypochlorite, in two supplies water was occasionally disinfected as reactive action to a contamination episode. In the majority of the water supplies (14/19), no disinfection procedure was performed. We have no information on the details of one supply.

The well maintenance category presented in the Table 1 is based on information gathered from the personnel of the water supplies through the questionnaire. The well maintenance category is based on estimated technical state of well structures, piping, as well as general groundwater protection measures in the surroundings of the production wells. Instead of exact locations of the drinking water supplies, we report, herein, the results using the anonymous water supply number codes 1–20. All the results exceeding the drinking water quality standards have been announced to the corresponding water supply operators and health protection authorities to enable their corrective actions.

Table 1. Characteristics of the studied water supplies.

Water Supply	Aquifer Type	Average Thickness of Vadose Zone (m)	Well Type [1]	Well Depth (m)	Well Maintenance [2]	Water Intake Constructed	Main Risk Factor	Water Treatment [3]
1	Glaciofluvial esker, unconfined, sand and gravel	7	Driven well	8	3	1998	Bank filtration	1
2	Glaciofluvial esker, semi-confined, silty till and sand	<5	Dug well	NA	3	1987	Bank filtration	1
3	Ablation moraine, semi-confined, sandy till	<5	Spring well	2	3	1963	Bank filtration	0
4	Glaciofluvial esker, unconfined, till and gravel	<5	Dug well	6	2	1978	Bank filtration	0
5	Deep bedrock aquifer, confined	NA	Drilled well	NA	1	1979	Agriculture, sewage	0
6	Glaciofluvial esker, confined, clay and sand	NA	Spring well	3	2	1960	Agriculture	0
7	Ablation moraine, semi-confined, sandy till	<5	Driven + dug well	185	3	NA	Agriculture	2
8	Glaciofluvial sand formation, confined, clay and sand	NA	Dug well	4	2	1988	Agriculture	3
9	Moraine, confined, clay and silty till	NA	Dug well	10	1	1983	Agriculture	3
10	Littoral sand, semi-confined, clay and till	NA	Dug + spring wells	3	3	1940	Agriculture	0
11	Glaciofluvial esker, unconfined, sand and silt	3	Driven well	7	2	1988	Agriculture	1
12	Glaciofluvial ice-marginal formation, semi-confined, sand and silt	3	Driven well	10	3	1986	Sewage	3

13	Glaciofluvial interlobate formation, semi-confined, sand, gravel and till	NA	Dug well	3	1–2	1962	Sewage	0
14	Littoral sand, semi-confined, till and sand	2	Dug well	4	1	1978	Sewage	NA
15	Moraine, semiconfined, till and sand	3	Spring wells	5–6	3	1962	Sewage	0
16	Littoral sand, semi-confined, till and sand	1	Dug wells	2–4	1	1948	Gravel mining	0
17	Deep bedrock aquifer and moraine, confined, till and gravel	NA	Dug well	NA	3	1993	Gravel mining	1
18	Ablation moraine, semi-confined, sandy till and gravel	6	Spring well	4 *	1	1992	Gravel mining	1
19	Littoral sand, semi-confined, till and sand	<5	Dug wells	NA	1–2	NA	Surface water runoff	0
20	Moraine, confined, till and sand	NA	Spring well	NA	1–2	NA	Flooding	0

Notes: [1] Well type "Spring well" is a shallow dug well installed in close vicinity to a spring, the groundwater level reaching the ground surface; [2] Well maintenance categories based on information on well structures and piping plus the surroundings of the production wells: 1 = poor, 2 = moderate, 3 = good; [3] Treatment after sampling point before water distribution: 0 = no treatment, 1 = alkalization, 2 = reverse osmosis, 3 = disinfection. NA = not available. * Well is located in a pit where gravel layers have been removed during previous gravel mining.

2.2. Microbiological Quality

The results of the microbiological analyses appear in Table 2. *Giardia* cysts were detected in autumn in samples from four small groundwater supplies (Table 2). All were later identified as *Giardia intestinalis*. No *Cryptosporidium* oocycts, *Campylobacter* spp. or noroviruses were detected in any of the samples.

Low CFUs of coliform bacteria (10–150 CFUs/1000 mL) were found in five samples and a relatively high CFU (830 CFUs/1000 mL) in one sample. In the spring sampling, no coliforms were detected in any of the plants and the difference between the coliform counts at spring and autumn can be considered statistically significant (p = 0.027, Wilcoxon signed rank test). Low CFUs of *Escherichia coli* were found in two samples in autumn (2/1000 mL and 40/1000 mL), but in none of the samples in spring. Intestinal enterococci were found in three 1000-mL samples (2–5 CFUs/1000 mL), and no *Clostridium perfringens* was detected in any of the samples.

Table 2. Microbiological quality parameters of 20 small water supplies in spring and autumn 2005.

Water Supply	Risk Factor Category	Coliforms CFUs/1000 mL		*E. coli* CFUs/1000 mL		Intestinal Enterococci CFUs/1000 mL		HPC CFUs/mL		*Aeromonas* CFUs/1000 mL		*Giardia* cysts/100 L	
		Spring	*Autumn*	*Spring*	*Autumn*	*Spring*	*Autumn*	*Spring*	*Autumn*	*Spring*	*Autumn*	*Spring*	*Autumn*
1	1	-	-	-	-	-	-	<1	<1	-	-	-	1
2	1	-	-	-	-	-	-	2	0	-	-	-	-
3	1	-	-	-	-	-	-	<1	<1	-	-	-	-
4	1	-	830	-	-	-	-	<1	18	-	9	-	-
5	2	-	-	-	-	-	-	3	14	-	12	-	1
6	2	-	10	-	-	-	-	<1	<1	1	-	-	-
7	2	-	-	-	-	-	-	1	<1	-	-	-	-
8	2	-	-	-	-	-	-	83	1	33	5	-	-
9	2	-	10	-	2	-	-	3	5	-	2	-	-
10	2	-	-	-	-	-	-	6	<1	-	-	-	-
11	2	-	-	-	-	-	-	<1	1	-	-	-	-
12	3	-	50	-	40	-	-	<1	8	-	-	-	-
13	3	-	-	-	-	-	-	3	0	15	3	-	1
14	3	-	-	-	-	-	4	296	10	-	-	-	2
15	3	-	70	-	-	-	5	30	25	5	-	-	-
16	4	-	-	-	-	-	-	267	1	-	-	-	-
17	4	-	-	-	-	-	-	14	8	-	-	-	-
18	4	-	-	-	-	-	2	200	214	-	-	-	-
19	5	-	-	-	-	-	-	4	8	-	-	-	-
20	5	-	150	-	-	-	-	<1	45	-	-	-	-

Notes: Recognized environmental risk factors: 1 = bank filtration, river or lake; 2 = agricultural load; 3 = wastewater treatment plant/household septic tanks <100 m; 4 = sand/gravel mining; 5 = multiple threats to contamination. HPC = heterotrophic plate count. - not detected/below the detection limit.

The CFUs of heterotrophic aerobic bacterial counts varied in the range of <1 to 300 CFUs/mL in spring and <1 to 200 CFUs/mL in autumn. No significant differences (p = 0.795, Wilcoxon signed rank test) in heterotrophic CFUs were detected between the spring and autumn samples.

Aeromonas spp. was detected in spring samples from four supplies and in autumn samples from five supplies; two of the positive wells were the same (Table 2). Further identification at the genospecies level showed that all *Aeromonas* species were mesophilic *A. salmonicida* [16].

2.3. Chemical Parameters

The chemical parameters appear in Table 3. Water temperature was significantly lower at the spring samples compared to the autumn samples ($p < 0.001$, Wilcoxon signed rank test). The pH of the wells studied varied from 5 to 7.6, the TOC varied from 0.5 to 11 mg/L, and no significant differences in the levels between the spring and autumn samples were detected (pH; $p = 0.657$ and TOC; $p = 0.159$, Wilcoxon signed rank test). Color, which is supposed to be about 5 mg Pt/mL, was high (20–40 mg Pt/mL) in two water supplies (14, 15) independent of the season (Table 3). In one of those supplies (14), turbidity was also significantly higher than in other samples. The iron concentrations were relatively high, ranging from 20 to 780 μg/L. The high concentrations were found mostly in the same wells both in spring and autumn. Manganese concentrations were clearly above the recommended limits in two wells (up to 180 μg/L), and the same wells had high concentrations in both seasons (Table 3). The concentration of nitrogen compounds was low in all samples: nitrate nitrogen varied from below 1 to 8.7 mg/L and nitrite nitrogen from below 0.01 to 0.02 mg/L.

2.4. Association of Recognized Risk Factors with Microbiological and Chemical Parameters

At one groundwater supply (15), where the septic tank of a household and a subterranean sand filter were located in close proximity (<100 and <500 m) to the well, TOC and color were high, and intestinal enterococci, coliform bacteria and *Aeromonas* spp. were detected. Intestinal enterococci and coliform bacteria were detected in autumn and *Aeromonas* spp. in spring. Conductivity was high in both seasons, and TOC, total phosphorus, color, and the nitrate nitrogen content were the highest of all the water supplies studied.

In one water supply (14), located in the middle of a small rural village and close to a common road, some of the study parameters showed increased values. A septic tank and a subsurface sand filter were located <100 m from the water supply. Intestinal enterococci and *Giardia* were detected in autumn, while TOC, color and turbidity values were high.

Giardia intestinalis was also detected in a water supply (1) in which no indicator bacteria were detected. Of the chemical parameters, only iron was high. The recognized risk factor in the supply was bank filtration, floodwater also had access to the well, and a wastewater treatment system was situated at a distance of 200–500 m.

Table 3. Chemical quality parameters of 20 small water supplies in spring and autumn 2005.

Water Supply	T (°C)		pH		EC (µS/cm)		TOC (mg/L)		Color mgPt/mL		Turbidity (FTU)		P (µg/L)		Fe (µg/L)		Mn (µg/L)	
	Spring	*Autumn*	*Spring*	*Autumn*	*Spring*	*Autumn*	*Spring*	*Autumn*	*Spring*	*Autumn*	*Spring*	*Autumn*	*Spring*	*Autumn*	*Spring*	*Autumn*	*Spring*	*Autumn*
1	7.1	10.2	5.6	6.2	92	93	1.7	2.0	5	5	0.50	1.10	7	11	440	780	25	32
2	2.2	8.4	6.1	6.5	48	53	5.7	1.0	5	<5	1.05	1.33	16	18	120	170	11	8.2
3	4.8	6.5	6.2	6.3	41	46	1.2	1.0	<5	<5	0.16	0.15	28	27	<20	<20	<5	<5
4	0.7	17.8	6.9	7.6	44	95	1.8	4.1	15	10	0.15	0.15	8	12	55	110	<5	56
5	7.5	7.3	7.5	6.9	328	337	1.1	1.2	<5	5	0.22	0.08	8	13	26	23	25	28
6	7.9	8.1	7.5	7.6	391	421	0.9	1.0	<5	<5	0.06	0.52	23	24	<20	<20	130	180
7	6.6	8.0	6.5	7.0	259 *	ns	1.2	1.3	<5	<5	0.06	0.04	9	11	<20	53	12	12
8	7.8	6.5	6.6	ns	178 *	ns	3.7	3.3	5	<5	0.69	0.35	11	12	150	83	79	46
9	6.8	5.7	6.5	ns	215 *	ns	1.3	1.2	<5	<5	0.80	0.70	14	9	71	126	<5	7
10	3.5	7.9	5.6	6.0	ns	ns	1.1	0.8	<5	<5	0.04	0.02	15	16	<20	<20	<5	<5
11	5.6	5.7	6.0	5.9	46	153	1.4	1.4	<5	<5	0.94	0.30	13	7	390	240	59	51
12	5.0	7.0	ns	6.5	148 *	150	0.8	0.9	<5	<5	0.52	0.92	5	11	38	220	<5	<5
13	5.2	8.0	7.0	6.9	214	196	1.0	1.0	5	<5	0.39	0.26	10	15	35	31	<5	<5
14	5.4	8.5	ns	5.4	12	202	5.3	5.2	20	20	3.7	3.46	20	21	440	120	<5	<5
15	4.3	8.5	6.1	6.0	231	479	11.0	10.8	40	40	0.73	0.79	182	108	180	150	100	110
16	1.5	9.7	5.5	5.9	ns	30	3.5	2.8	10	5	0.15	0.12	8	9	<20	<20	<5	<5
17	3.3	7.4	7.2	6.3	150	148	1.7	1.2	<5	<5	0.20	0.24	15	15	<20	<20	<5	<5
18	4.7	6.9	6.2	6.3	81	79	0.5	0.6	<5	<5	0.16	0.16	21	12	<20	<20	<5	<5
19	4.0	6.1	6.3	5.1	173	ns	1.8	1.5	<5	<5	0.18	0.16	16	15	31	29	<5	<5
20	3.5	7.1	5.1	5.0	ns	ns	1.8	1.8	<5	<5	0.12	0.15	12	13	<20	21	<5	<5

Notes: EC = Electrical Conductivity; TOC = Total Organic Carbon; ns = not defined; * = measured only in laboratory.

The *Giardia*-positive well (5) also tested positive for *A. salmonicida* in autumn. Of the chemical parameters, conductivity was high (337 µs/cm). The identified potentially contaminating risk factor was agricultural load, but a wastewater treatment system was also located 200–500 m, and a wastewater drain 100–200 m, from the well.

Giardia intestinalis was also detected in a well (13), where *Aeromonas* spp. was detected in both spring and autumn samples, though no other indicators or pathogens were detected. The principal recognized threat to water safety was waste water treatment, and indeed the septic tank of a household was situated 50–100 m from the plant. Surface water also had access to the well.

Coliforms (830/1000 mL) were detected in an autumn sample from a well (4) located on a narrow isthmus between two lakes where bank filtration is substantial. The difference in the water temperature of the well between spring and autumn samplings was marked and suggests the close impact of large water bodies of the nearby lakes.

In one water supply (12), coliforms (50/1000 mL) and *E. coli* (40/1000 mL) were detected in autumn, though UV light disinfection was in use after the point of sampling. A septic tank as well as an oil tank, gravel pit, and a cultivated field were situated <50 m from the well.

Some of the water quality parameters associated with the recognized environmental risk factor categories. Heterotrophic CFUs were significantly higher in the supplies at gravel mining category compared to the bank filtration category supplies ($p = 0.020$, Kruskal-Wallis test). Electrical conductivity and manganese concentrations were more elevated at supplies close to agricultural fields than supplies with other risk factors ($p = 0.005$ and 0.009, respectively, Kruskal-Wallis test). Elevated water color and turbidity associated with supplies in close proximity of septic tanks ($p = 0.022$–0.023, Kruskal-Wallis test). The iron concentration was lower at the supplies associated with gravel mining than supplies in other risk factor categories ($p = 0.005$, Kruskal-Wallis test).

2.5. Associations within Microbiological and Chemical Parameter Results at Spring and Autumn

Coliform bacteria, *E. coli*, intestinal enterococci and *Giardia* findings were occasional and originated solely from the autumn samples (Table 2). Physico-chemical parameters pH, TOC, color, turbidity, and the concentrations of phosphorus, iron, and manganese were more stable and showed significant relation between the spring and autumn samples in the correlation analysis (Spearman correlation coefficient, $r_S > 0.719$, $p = 0.001$).

As regards associations between the microbiological parameters, intestinal enterococci counts associated with heterotrophic CFUs ($r_S = 0.319$, $p = 0.045$) when all samples were considered ($n = 40$) in the Spearman correlation analysis. The associations between the chemical parameters were strong: TOC associated with color ($r_S = 0.693$, $p < 0.001$) and iron concentration ($r_S = 0.418$, $p = 0.007$) and the iron concentration was also associated with color ($r_S = 0.413$, $p = 0.008$) and turbidity of the water ($r_S = 0.727$, $p < 0.001$). Manganese concentration was associated with the electrical conductivity of the water ($r_S = 0.468$, $p = 0.008$).

Overall, the associations between microbiological and chemical parameters were rare. When analyzing the autumn samples ($n = 20$), there was a weak association between the water color and the detection of *Giardia* ($r_S = 0.463$, $p = 0.040$) and when all samples were considered, water temperature associated with the *Giardia* counts ($r_S = 0.377$, $p = 0.016$). Furthermore, *Aeromonas* counts were associated with the water pH ($r_S = 0.420$, $p = 0.011$) and electrical conductivity ($r_S = 0.438$, $p = 0.014$).

3. DISCUSSION

The definition of a small drinking water supply varies in Europe and in the United States. In the USA, a small supply is one that supplies drinking water to fewer than 10,000 people. In the EU the DWD (Drinking Water Directive 98/83/EC) [7] does not require reporting from the supplies which either provide water to fewer than 50 users or which distribute less than 10 m^3/d. The EU research program WEKNOW collected data from European small and very small water supplies and defined a small supply as one which provides water to fewer than 5000, but more than 50 people, and distributes 10–1000 m^3/d [8]. The small water supplies in our study had 38 to 450 users and were thus not totally within any of the EU definitions mentioned above. Of the 20 Finnish small groundwater supplies studied, 4 contained *Giardia intestinalis* in autumn samples. The detected *Giardia* cyst counts were low (1–2 cysts/100 L). Due to the lack of exact recovery rates of the method used, we cannot rule out the possibility of false negative results in the *Giardia* and *Cryptosporidium* analysis. However, the detection of *Giardia* in these samples was proved and can be considered as a reliable result. No other studied enteric pathogens, *Campylobacter* spp. or noroviruses were detected. In addition, intestinal enterococci, coliforms and *E. coli* were detected in 8 of the 20 wells. All these results indicate that these water supplies are at increased risk for fecal contamination, thus extending the results of the studies of Corkal *et al.* [17] from Canada and of Rutter *et al.* [18] as well as those of Richardson *et al.* [1] from England and Wales. In our study, a trend emerged in which indicator organisms were detected more often in autumn samples than in spring samples, suggesting the impact of weather. Spring 2005 was dry, and occasional rain was common in autumn.

Giardiasis, as an endemic disease in Finland, has most likely been underestimated. Hörman *et al.* [19] estimated on the basis of their meta-analysis study that there could be as many as 4664 unregistered symptomatic *Giardia* cases per 100,000 general population compared to 5.38 registered symptomatic cases per 100,000 general population (ratio 1:867). In our previous studies, we identified *Giardia* cysts in Finnish surface water samples [20] as well as in municipal wastewater samples [21] and in tap water contaminated with sewage [14]. Our present study is the first Finnish study to identify *Giardia* cysts in a small drinking water supply. No data were available on potential asymptomatic carriers or symptomatic *Giardia* cases among the water users. If contaminated water caused the human illness, the number of patients was most likely so low that they remained undetected. The impact of small water supplies as source of *Giardia* infections in humans requires further study. *Giardia intestinalis* is a zoonotic pathogen, thus the contamination source of the supplies can be of either human or animal origin.

Only three of the water supplies distributed disinfected tap water, revealing that in most cases there were no preventive barriers between the aquifer and consumers' taps. The *Giardia* positive wells (Supplies 1, 5, 13, and 14) seem to be unprepared for the microbiological risk as there was no advanced water treatment or disinfection in place after the sampling point. The *E. coli* positive wells were better prepared presumably due to the earlier noncompliance with the microbiological water quality standards as the supplies 9 and 12 had disinfection in use. Fecal contamination of drinking water increases the risk for enteric illness among users even if no such documented data are available from our study region. Small supplies have been shown to be frequently prone to fecal contamination in Finland [2,11] and elsewhere. Richardson *et al.* [1] analyzed the microbial quality of 11,233 private drinking water supplies within England and found that *E. coli* was detected in at least one sample from 32.4% of the water supplies. In accordance with the increased risk for fecal contamination of small water supplies, analysis of the distribution of waterborne outbreaks in England and Wales showed that small supplies were associated with 36% of all drinking water outbreaks even though they serve only 0.5% of the entire population [22]. The impact of small supplies on the disease burden of their users remains unknown. The life-long consumption of drinking water from a contaminated source could, on the other hand, also lead to acquired immunity [23,24].

In our study, all *Giardia*-positive wells were located near wastewater treatment activities; either a wastewater treatment plant for municipal human sewage was located at a distance of 200–500 m or the septic tank of a household was located within 50–100 m. One of the *Giardia*-positive wells (14) also had increased levels of multiple indicator parameters (intestinal enterococci, TOC, electrical conductivity, and turbidity), indicating surface water access to the well. However, neither *E. coli* nor *C. perfringens*, suggested as suitable indicators for *Giardia* and *Cryptosporidium* [25,26], were detected in the 100-mL samples. This well was known to be located in the middle of a rural village and close to a road.

One *Giardia*-positive well (1) had no increased levels of indicator bacteria or chemical indicators of surface water contamination even though a wastewater treatment system was located at a distance of 200–500 m and surface water was known to have access to the well. In two of the *Giardia*-positive wells, *A. salmonicida* was also detected, but not coliform bacteria. In another study, protozoa and total coliform levels were clearly correlated [27]. High turbidity (>1.0 FTU), which indicated surface water access to the groundwater source, was observed in three autumn samples, two of which were *Giardia*-positive.

The wells studied were known to be located where contamination sources such as roads, sewage treatment/treatment plants, and habitations were known to be rather close to the aquifer. Contamination by surface water after snow thawing or rainfalls was possible in most of the wells. Most of the water supplies were opened decades ago, when habitation and human activities around the wells were most likely much less than today.

Ten of the wells were located close to a river or lake, a common location for small aquifers in Finland because this kind of location will guarantee a consistent supply of water. The location may allow bank filtration of river or lake water into the well during the dry season, when the groundwater table is low. After a heavy rain, floodwater may also contaminate the well. Studies of past Finnish waterborne outbreaks have shown that some outbreaks were associated with groundwater wells located close to a river or lake [11]. All of the four plants with *Giardia* findings were located less than 50 m from a lake, river or ditch.

A. salmonicida was found in four samples in spring and in five samples in autumn. *A. salmonicida* is a common bacterium in natural waters as well as in well water, and has been previously isolated in Finnish groundwater wells [28,29] and elsewhere [30]. Its isolation in groundwater well may indicate surface water contamination. *A. salmonicida* was not connected to fecal contamination in a study by Hirotani *et al.* [31]. The pathogenicity of *Aeromonas* spp. as an enteric pathogen is not confirmed, and mesophilic *A. salmonicida* in particular has been regarded as an environmental organism with low pathogenic potential for humans. In addition to indicating surface water contamination, its presence could be associated with increased heterotrophic plate counts [32].

In 14 of the water supplies examined, coliform bacteria had been detected in previous samplings, and *E. coli* had occasionally been detected in five of the wells. These wells placed consumers at increased risk for acquiring waterborne illness, especially because only four of them had undergone disinfection treatment with either UV light or hypochlorite solution before distribution. Two of these wells tested positive for *E. coli*, indicating fecal contamination. *E. coli* was detected in the 1000-mL sample, but not in the 100-mL, suggesting that the use of volumes larger than 100 mL would be more accurate in monitoring fecal contamination. Larger volumes have proved useful in tracking contamination sources associated with waterborne epidemics [33]. Both *E. coli*-positive wells were located in either a pit or flat ground. The other was located near (under 50 m) the septic tank of a household, and both were near a cultivated field and ditch.

Intestinal enterococci were found in three water supplies, and all of these wells were located within 100 m of the septic tank of a household. They were found in samples of either 500 mL or 1000 mL, but not 100 mL. These results again suggest that using larger volumes of water may more often facilitate the detection of indicators of contamination than do the 100-mL samples used in the EU Directive [7].

Our study supports previous findings showing that the pH of Finnish groundwater is typically low (<6.5) [9]. In approximately half of the groundwater wells monitored, pH was lower than recommended (pH 6.7). Low pH may cause the corrosion of iron water pipes. In the project questionnaire, 11 of 19 water supplies reported low pH as their problem. In six of the water supplies, the water was alkalinized prior to distribution to consumers.

Another common problem in Finnish groundwater is its high content of iron and manganese [9]. Iron and manganese in excess decrease the usability of water as drinking water or for household use. In our survey, 20% of the water supplies contained excess iron (up to 780 μg/L, with a recommended maximum of 200 μg/L), and 25% of them contained excess manganese (up to 180 μg/L, with a recommended maximum of 50 μg/L). In the questionnaire, seven water supplies reported the problem.

In only three of them, the iron and manganese were removed before distribution to consumers.

4. EXPERIMENTAL SECTION

4.1. Selection and Characterization of the Water Supplies

Water samples were collected from 20 small groundwater supplies around Finland in April and September–October 2005. The sampled supplies were selected based on a questionnaire answered for 248 small groundwater supplies owned mostly by cooperatives and operated by part-time working persons in the preceding year. Data on the water supply characteristics, microbiological quality and factors considered as potential fecal contamination threats of each water supply were obtained from the questionnaire and analyzed. The sites were selected according to these evaluations and the location of a water supply. The main selection criteria were the presence of an existing potential fecal contamination source in the neighborhood and the occasional detection of coliforms in the water quality compliance monitoring of the local health authorities.

4.2. Sampling and Analysis

Water samples were taken at a water supply from a tap or from a tap and well (if the tap was located after a collection tank), or only from a well with a submersible pump in the absence of a tap at the water plant. One sample was taken from the overflow of a well. The samples were taken prior to any potential treatments. For the microbiological analyses, all the equipment was disinfected before the sampling. The filters were removed from the taps, which were sterilized by flaming. The submersible pump was disinfected by submerging it in 10 mg/L hypochlorite solution for a minimum of 30 min. Each sample was taken with a new hose. Sterile containers were rinsed twice with sample water before sampling. For the *Giardia* and *Cryptosporidium* analyses, 100 L of water were filtered through an Envirochek® HV filter capsule (PALL Life Sciences, Port Washington, NY, USA). Other samples were taken with disinfected plastic containers and with a sterile glass bottle. Temperature, pH, and electrical conductivity were measured at the sampling site, and the evaluation of sensory qualities, such as odor, color, and turbidity, was performed repeatedly in a laboratory.

The water samples were refrigerated and transported to laboratories in Helsinki and Kuopio within 24 h and stored at refrigerated temperature prior to examination.

4.3. Detection of Enteric Pathogens

For the detection of *Giardia* spp. and *Cryptosporidium* spp., Envirochek® - concentrated samples were further treated according to the USEPA Method 1623 [34] as described by Rimhanen-Finne *et al.* [35]. In brief, for further concentration, the sample was first filtrated through a polycarbonate filter, which was then spooled with 10 mL of PBS-Tween20. The cysts and oocysts in the suspension were captured by using the immunomagnetic separation technique (Dynabeads® GC-Combo, Dynal Biotech ASA, Oslo, Norway). The final concentrate of 100 μL was divided into two portions, one of which was stored frozen at −20 °C for further molecular PCR and sequencing analyses. The other portion was immunostained using an Aqua-Glo G/C Direct Comprehensive Kit (Waterborne Inc., New Orleans, LA, USA). (Oo) cysts were counted under an epifluorescense microscope by using positive controls, *i.e.*, the cysts of *Giardia intestinalis* (H3 isolate, Waterborne Inc. New Orleans, LA, USA) and the oocysts of *Cryptosporidium parvum* (Iowa isolate, Waterborne Inc. New Orleans, USA) as described by Rimhanen-Finne *et al.* [35]. The positive control cysts and oocysts were stored at 4 °C. The numbers of purified cysts and oocysts was enumerated from five stock solution aliquots in hemocytometer resulting in a mean concentration of 1.1×10^6 cysts mL^{-1} and 11.8×10^6 oocysts mL^{-1} [35]. DNA from frozen water concentrate that tested positive for *Giardia* in microscopy was isolated through five rounds of the frozen-thaw procedure following DNA isolation [35]. *Giardia*-specific PCR was performed with glutamate dehydrogenase gene-targeted PCR using the primers GDH1 and GDH4 [36].

Thermophilic campylobacters were identified in 4000-mL samples using the ISO 17995 method [37] with Bolton enrichment (LabM, Lancashire, UK) and modified Charcoal Cefoperazone Deoxycholate Agar (Oxoid, Cambridge, UK) incubated in a microaerobic atmosphere.

Noroviruses were analyzed from a 1000-mL water sample that was concentrated by filtering it through a positively charged nylon membrane [38]. Norovirus detection was carried out using the RT-PCR method, and the result was confirmed with microplate hybridization [39].

4.4. Analyses of Indicator Bacteria

CFUs of *Escherichia coli* and coliform bacteria were analyzed from both 100- and 1000-mL samples according to the SFS 3016 standard [40] using m-Endo LES (Merck KGaA, Darmstadt, Germany) agar plates and membrane filtration. The heterotrophic plate count was determined according to the ISO 6222 standard [41] on tryptone-yeast agar (Oxoid) incubated at (22 ± 1) °C for three days. The CFUs of intestinal enterococci were analyzed from 100-mL, 500-mL and 1000-ml samples according to the ISO 7899-2 standard [42] with membrane filtration and a Slanetz-Bartley medium (Oxoid).

Clostridium perfringens CFUs were counted in 100-mL and 1000-mL samples with the membrane filtration method on Tryptose-Sulphite-Cyclocerine agar (Difco) using the ISO/CD 6461-2 method.

The CFUs of *Aeromonas* spp. were counted in 100-mL and 1000-mL samples by using the membrane filtration technique on ADA (Ampicillin-Dextrin Agar) plates with ampicillin as a selective substance [43]. After 24 h of incubation at 30 °C, typical yellow colonies were subcultivated on blood agar for further identification. A total of 28 colonies from five *Aeromonas*-positive water supplies from autumn 2005 were further identified as *Aeromonas* spp. with API20 NE (bioMérieux, Marcy l'Etoile, France). For genospecies identification, the DNA of the colonies was isolated, and a fragment of the 16S rRNA gene was amplified with PCR according to the method described by Borrell *et al.* [16]. PCR products were digested with *AluI* and *Mbo1*. The pattern of fragments on agarose gels was compared with the results of Borrell *et al.* [16], and genospecies identification was performed on the basis of an RFLP pattern.

4.5. Chemical Analyses

Temperature, pH, and conductivity were measured on site using a YSI 556 MPS multiple parameter instrument or a pH/Cond 340i WTW-meter. Color was measured according to the ISO 7887-4 standard [44], and turbidity according to the ISO 7027 standard [45]. Iron and manganese concentrations were measured according to the SFS 5502 standard [46]. Nitrite nitrogen and nitrate nitrogen was measured using ISO 13395 standard method [47].

4.6. Statistical Analyses

Statistical analyses were performed using SPSS Statistics 22. For the results below the detection limit, half of the detection limit was used as a numerical value. The normality of microbiological and chemical parameter results from spring and autumn samples was tested using Shapiro-Wilk test and by visual evaluation of frequency distributions. Non-parametric methods were used, because normal distributions of the variables could not be obtained. Wilcoxon signed rank test was used to determine if there were statistically significant differences between results obtained from spring and autumn samples. The variation of the water quality results in the recognized environmental risk factor categories were tested with non-parametric Kruskal-Wallis one-way analysis of variance (ANOVA) test. Spearman's correlation coefficients were used to display the relationships between the water quality parameters. Differences and correlations were considered statistically significant when $p < 0.05$.

5. CONCLUSIONS

We detected *Giardia intestinalis* for the first time in the Finnish groundwater supplies. *E. coli* and coliform bacteria, as well as intestinal enterococci were detected in some wells, which together with *Aeromonas* findings indicate surface water access and possible contamination from the surroundings to the wells. These findings suggest an increased health risk associated with small drinking water supplies, even though among public they are usually considered safe. In addition, high iron and manganese concentrations, and low pH, which have also been detected previously in Finnish groundwater, degraded the quality of drinking water.

ACKNOWLEDGMENTS

We would like to acknowledge the water supply operators for their responses to the questionnaire and participation in sampling.

REFERENCES

1. Richardson, H.Y.; Nichols, G.; Lane, C.; Lake, I.R.; Hunter, P.R. Microbiological surveillance of private water supplies in England—The impact of environmental and climate factors on water quality. *Water Res.* **2009**, *43*, 2159–2168.
2. Pitkänen, T.; Karinen, P.; Miettinen, I.T.; Lettojärvi, H.; Heikkilä, A.; Maunula, R.; Aula, V.; Kuronen, H.; Vepsäläinen, A.; Nousiainen, L.L.; *et al.* Microbial contamination of groundwater at small community water supplies in Finland. *Ambio* **2011**, *40*, 377–390.

3. Lapinlampi, T.; Raassina, S. *SY541 Vesihuoltolaitokset 1998–2000*; Finnish Environment Centre (SYKE): Helsinki, Finland, 2002. (In Finnish)
4. Isomäki, E. Pienet pohjavesilaitokset Suomessa. *Vesitalous* **2006**, *3*, 11–16. (In Finnish).
5. Decree of the Ministry of Social Affairs and Health Relating to the Quality and Monitoring of Water Produced by Small Water Supplies 401/2001, Finlex Data Bank; Edita Publishing Oy: Helsinki, Finland, 2001. Available online: http://www.finlex.fi/fi/laki/alkup/2001/20010401 (accessed on 20 August 2015). (In Finnish).
6. Decree of the Ministry of Social Affairs and Health Relating to the Quality Requirements and Monitoring of the Household Water 461/2000, Finlex Data Bank, Edita Publishing Oy: Helsinki, Finland, 2000. Available online: http://www.finlex.fi/fi/laki/alkup/2000/20000461 (accessed on 20 August 2015). (In Finnish).
7. European Union. Council directive 98/83/EC of 3 November 1998 on the quality of water intended for human consumption. *Off. J. Eur. Communities* **1998**, *L330*, 32–54.
8. Hulsmann, A. *Small Systems Large Problems—A European Inventory of Small Water Systems and Associated Problems*; Report of Web-based European Knowledge Network on Water WEKNOW/ENDWARE, European Commission; KWR Watercycle Research Institute: Nieuwegein, The Netherlands, 2005; p. 41.
9. Lahermo, P.; Tarvainen, T.; Hatakka, T.; Backman, B.; Juntunen, R.; Kortelainen, N.; Lakomaa, T.; Nikkarinen, M.; Vesterbacka, P.; Väisänen, U.; *et al. Tuhat kaivoa—Suomen kaivovesien fysikaalis-kemiallinen laatu vuonna 1999; (Summary: One thousand wells—The physical-chemical quality of Finnish well waters in 1999)*; Report of Investigation 155; Geological Survey of Finland: Espoo, Finland, 2002; p. 92. Available online: http://tupa.gtk.fi/julkaisu/tutkimusraportti/tr_155.pdf (accessed on 20 August 2015).
10. Zacheus, O. *Talousveden valvonta ja laatu vuonna 2008. Yhteenveto viranomaisvalvonnan tuloksista*; Avauksia 18/2010; Terveyden ja hyvinvoinnin laitos (THL): Helsinki, Finland, 2010; p. 67. Available online: http://urn.fi/URN:NBN:fi-fe201205085417 (accessed on 20 August 2015). (In Finnish)
11. Zacheus, O.; Miettinen, I.T. Increased information on waterborne outbreaks through efficient notification system enforces actions towards safe drinking water. *J. Water Health* **2011**, *9*, 763–772.
12. Miettinen, I.T.; Zacheus, O.; von Bonsdorff, C.H.; Vartiainen, T. Waterborne epidemics in Finland in 1998–1999. *Water Sci. Technol.* **2001**, *43*, 67–71.
13. Pitkänen, T. Review of *Campylobacter* spp. in drinking and environmental waters. *J. Microbiol. Methods* **2013**, *95*, 39–47.
14. Rimhanen-Finne, R.; Hänninen, M.L.; Vuento, R.; Laine, J.; Jokiranta, T.S.; Snellman, M.; Pitkänen, T.; Miettinen, I.; Kuusi, M. Contaminated water caused the first outbreak of giardiasis in Finland, 2007: A descriptive study. *Scand. J. Infect. Dis.* **2010**, *42*, 613–619.
15. Simonen, A. *The Precambrian in Finland*; Bulletin 304; Geological Survey of Finland: Espoo, Finland, 1980; p. 58. Available online: http://tupa.gtk.fi/ julkaisu/ bulletin/ bt_304.pdf (accessed on 20 August 2015).
16. Borrell, N.; Acinas, S.G.; Figueras, M.J.; Martinez-Murcia, A.J. Identification of *Aeromonas.* clinical isolates by restriction fragment length polymorphism of PCR-amplified 16S rRNA genes. *J. Clin. Microbiol.* **1997**, *35*, 1671–1674.
17. Corkal, D.; Schutzman, W.C.; Hilliard, C. Rural water safety from the source to the on-farm tap. *J. Toxicol. Environ. Health A* **2004**, *67*, 1619–1642.
18. Rutter, M.; Nichols, G.L.; Swan, A.; de Louvois, J. A survey of the microbiological quality of private water supplies in England. *Epidemiol. Infect.* **2000**, *124*, 417–425.
19. Hörman, A.; Korpela, H.; Sutinen, J.; Wedel, H.; Hänninen, M.L. Meta-analysis in assessment of the prevalence and annual incidence of *Giardia* spp. and *Cryptosporidium* spp. infections in humans in the Nordic countries. *Int. J. Parasitol.* **2004**, *34*, 1337–1346.
20. Hörman, A.; Rimhanen-Finne, R.; Maunula, L.; von Bonsdorff, C.H.; Torvela, N.; Heikinheimo, A.; Hänninen, M.L. *Campylobacter* spp., *Giardia* spp., *Cryptosporidium*, noroviruses and indicator organisms in surface water in southwestern Finland, 2000–2001. *Appl. Environ. Microbiol.* **2004**, *70*, 87–95.

21. Rimhanen-Finne, R.; Vuorinen, A.; Marmo, S.; Malmberg, S.; Hanninen, M.L. Comparative analysis of *Cryptosporidium*, *Giardia* and indicator bacteria during sewage sludge hygienization in various composting processes. *Lett. Appl. Microbiol.* **2004**, *38*, 301–305.
22. Said, B.; Wright, F.; Nichols, G.L.; Reacher, M.; Rutter, M. Outbreaks of infectious disease associated with private drinking water supplies in England and Wales 1970–2000. *Epidemiol. Infect.* **2003**, *130*, 469–479.
23. Von Hertzen, L.; Laatikainen, T.; Pitkänen, T.; Vlasoff, T.; Makela, M.J.; Vartiainen, E.; Haahtela, T. Microbial content of drinking water in Finnish and Russian Karelia—Implications for atopy prevalence. *Allergy* **2007**, *62*, 288–292.
24. Casemore, D. Towards a US national estimate of the risk of endemic waterborne disease—Sero-epidemiologic studies. *J. Water Health* **2006**, *4*, 121–163.
25. Ferguson, C.M.; Coote, B.G.; Ashbolt, N.J.; Stevenson, I.M. Relationships between indicators, pathogens and water quality in an estuarine system. *Water Res.* **1996**, *30*, 2045–2054.
26. Cizek, A.R.; Characklis, G.W.; Krometis, L.A.; Hayes, J.A.; Simmons, O.D., III; di Lonardo, S.; Alderisio, K.A.; Sobsey, M.D. Comparing the partitioning behavior of *Giardia* and *Cryptosporidium* with that of indicator organisms in stormwater runoff. *Water Res.* **2008**, *42*, 4421–4438.
27. LeChevallier, M.W.; Norton, W.D.; Lee, R.G. Occurrence of *Giardia* and *Cryptosporidium* spp. in surface water supplies. *Appl. Environ. Microbiol.* **1991**, *57*, 2610–2616.
28. Hänninen, M.L.; Siitonen, A. Distribution of *Aeromonas.* phenospecies and genospecies among strains isolated from water, foods or from human clinical samples. *Epidemiol. Infect.* **1995**, *115*, 39–50.
29. Hänninen, M.L.; Oivanen, P.; Hirvelä-Koski, V. *Aeromonas.* speciesin fish, fish-eggs, shrimp and freshwater. *Int. J. Food Microbiol.* **1997**, *34*, 17–26.
30. Altwegg, M.; Steigerwalt, A.G.; Altwegg-Bissig, R.; Luthy-Hottenstein, J.; Brenner, D.J. Biochemical identification of *Aeromonas.* genospecies isolated from humans. *J. Clin. Microbiol.* **1990**, *28*, 258–264.
31. Hirotani, H.; Sese, C.; Kagawa, H. Correlations of *Aeromonas. hydrophila* with Indicator Bacteria of Water Qualityand Environmental Factorsin a Mountain Stream. *Water Environ. Res.* **1999**, *7*, 132–138.
32. Kersters, I.; van Vooren, L.; Huys, G.; Janssen, P.; Kersters, K.; Verstraet, W. Influence of temperature and process technology on the occurrence of *Aeromonas.* species and hygienic indicator organisms in drinking water production plants. *Microb. Ecol.* **1995**, *30*, 203–218.
33. Hänninen, M.L.; Haajanen, H.; Pummi, T.; Wermundsen, K.; Katila, M.L.; Sarkkinen, H.; Miettinen, I.; Rautelin, H. Detection and typing of *Campylobacter jejuni* and *Campylobacter coli* and analysis of indicator organisms in three waterborne outbreaks in Finland. *Appl. Environ. Microbiol.* **2003**, *69*, 1391–1396.
34. United States Environmental Protection Agency. *Method 1623: Cryptosporidium and Giardia in Water by Filtration/IMS/FA*; EPA-821-R-99–006; United States Environmental Protection Agency: North Chelmsford, MA, USA, 1999.
35. Rimhanen-Finne, R.; Hörman, A.; Ronkainen, P.; Hänninen, M.L. An IC-PCR method for detection of *Cryptosporidium* and *Giardia* in natural surface waters in Finland. *J. Microbiol. Methods* **2002**, *50*, 299–303.
36. Homan, W.L.; Gilsing, M.; Bentala, H.; Limper, L.; van Knapen, F. Characterization of *Giardia duodenalis* by polymerase-chain-reaction fingerprinting. *Parasitol. Res.* **1998**, *84*, 707–714.
37. International Organization for Stardardization. *Water Quality—Detection and Enumeration of Thermotolerant Campylobacter Species*; ISO 17995; International Organization for Stardardization: London, UK, 2005.
38. Gilgen, M.; Germann, D.; Luthy, J.; Hubner, P. Three-step isolation method for sensitive detection of enterovirus, rotavirus, hepatitis A virus, and small round structured viruses in water samples. *Int. J. Food Microbiol.* **1997**, *37*, 189–199.

39. Maunula, L.; Piiparinen, H.; von Bonsdorff, C.H. Confirmation of Norwalk-like virus amplicons after RT-PCR by microplate hybridization and direct sequencing. *J. Virol. Methods* **1999**, *83*, 125–134.
40. Finnish Standard Association. *Water Quality—Membrane Filter Technique for the Enumeration of Total Coliform Bacteria*; SFS 3016; Finnish Standard Association: Helsinki, Finland, 2001.
41. International Organization for Stardardization. *Water Quality—Enumeration of Culturable Micro-Organisms. Colony Count by the Inoculation in a Nutrient Agar Culture Medium*; ISO 6222; International Organization for Stardardization: London, UK, 1999.
42. International Organization for Stardardization. *Water Quality—Detection and Enumeration of Intestinal Enterococci. Part 2: Membrane Filtration Method*; ISO 7899-2; International Organization for Stardardization: London, UK, 2000.
43. Havelaar, A.H.; During, M.; Versteegh, J.F. Ampicillin-dextrin agar medium for the enumeration of *Aeromonas.* species in water by membrane filtration. *J. Appl. Bacteriol.* **1987**, *62*, 279–287.
44. International Organization for Standardization. *Water Quality—Examination and Determination of Colour*; ISO 7887-4; International Organization for Standardization: London, UK, 1994.
45. International Organization for Standardization. *Water Quality—Determination of Turbidity*; ISO 7027; International Organization for Standardization: London, UK, 1999.
46. Finnish Standards Association. *Metal Content of Water, Sludge and Sediment Determined by Flameless Atomic Absorption Spectrometry. Atomization in a Graphite Furnace. Special Guidelines for Aluminium, Cadmium, Cobolt, Chromium, Copper, Lead, Manganese, Nickel and Iron*; SFS Method 5502; Finnish Standards Association: Helsinki, Finland, 1990.
47. International Organization for Standardization. *Water Quality—Determination of Nitrite Nitrogen and Nitrate Nitrogen and the Sum of both by Flow Analysis (CFA and FIA) and Spectrometric Detection*; ISO 13395; International Organization for Standardization: London, UK, 1996.

CHAPTER 6

Saline Intrusion Response to Sea Level Rise and Its Implications on Water and Coastal Management: A Case Study in Brazil

Raquel Toste[1], Paulo C. C. Rosman[2], Marcos A. V. de Freitas[3]

[1]*Laboratory for Computational Methods in Engineering, Federal University of Rio de Janeiro, Rio de Janeiro, Brazil*

[2]*Department of Oceanic Engineering, Federal University of Rio de Janeiro, Rio de Janeiro, Brazil*

[3]*Department of Energetic Planning, Federal University of Rio de Janeiro, Rio de Janeiro, Brazil*

ABSTRACT

Global temperature is predicted to increase in the end of the century and one of the primary consequences of this warming is the sea level rise. Considering the vulnerabilities on coastal systems and water resources, it is important to evaluate the potential effects of this rising in coastal areas, since the saline intrusion on rivers would be intensified, leading to problems related to water quality. In this context, the present work aimed to verify saline intrusion changes along an important river, São Francisco Canal, located in Rio de Janeiro State, Brazil. For this purpose, a hydrodynamic modeling was performed using SisBaHiA, considering different sea levels and tide conditions. According to the results, it was verified the intensification on saline intrusion and higher salinity values due to a sea level rise of 0.5 m. These results show that new licenses for water withdrawals must be carefully analyzed as the fluvial flow plays an important role to contain the saltwater intrusion on the studied river. Accordingly, it is recommended the evaluation of climate change effects in order to choose best strategies to reduce coastal vulnerability, and the use of this theme on environmental licensing and territorial planning, integrating water planning with coastal management.

Keywords: Saline Intrusion, Sea Level Rise, Water Resource Management

1. INTRODUCTION

Climate change and its effects on the environment and human life have been increasingly discussed and considered in the present society. Despite the increasing on greenhouse gas (GHG) emissions after Industrial Revolution, the causes of the global warming are not straightforward. Nevertheless, even natural or hu- man-caused, the climate change effects should be considered and investigated in different scales and spheres both for global and regional planning.

This increasing on global surface temperature has already been observed in the present days. According to IPCC (Intergovernmental Panel on Climate Change), the global temperature increased 0.74°C along the 20th century and its effects could also be observed on the global hydrological cycle [1]. It includes the global mean sea level (SL) rise observed in a rate of 1.7 mm per year [2]. However, for the next century, the SL is predicted to rise on speeder rates [1].

These projections are an important issue, mainly considering the coastal vulnerability to sea level rise in addition to the effects of potential changes on winds, waves and on the frequency of climate extremes. Among others, the SL rise leads to problems associated to coastal flooding and contamination of coastal aquifers. However, these effects vary according to regional characteristics being important to study different environments and regions [3].

In Brazil, just few studies have been realized considering climate change effects on coastal zones. Despite of this, in some regions, comprising the coastal zone of the state of Rio de Janeiro, losses associated to extreme events are being experienced. These current evidences of SL rising effects on coastal environments demonstrate that further investigation is a necessary and strategic factor [4]. However, due to the absence of historical information about the SL on the Brazilian coast, the coastal vulnerability in the country is also due to these uncertainties about past SL changes and their scales [5].

According to global climate models, the average prediction for global SL rise along the 21st century is about 0.47 m in relation to the mean SL for 1986-2005. For 2100, this change is projected in 0.53 m to the same reference period [6]. These projections were realized considering the stabilization scenario RCP4.5 reported on the Fifth Assessment Report by IPCC (Intergovernmental Panel on Climate Changes). In a regional scale, Belém [7] estimated SL oscillations along the Brazilian coast considering local salinity and temperature data and predicted an average rise of 0.50 m for the next 50 - 100 years.

This projected rise would deteriorate water quality, since saline wedge reaches inshore regions of the coastal aquifers. Saline intrusion is a serious environmental condition as 80% of the world's population live on coastal zones and use local aquifers for their water supply [8]. Thus, these long-term changes should be evaluated, since just few centimeters of SL rise can cause significant damages. Additionally, the combined effect of withdrawals and the SL rise on saline intrusion requires management actions to protect water resources, as the water conditions appropriated for human use may be affected [9]. Hence, many studies are necessary to understand those effects on ecosystems and human well-being.

In some regions, saline intrusion on coastal aquifers may be intensified by lower rainfall rates, what decreases the water availability for human supply, as previously observed in Senegal [10]. In Brazil, some studies were realized but there is still a lack of investigations linking saline intrusion to climate change. Thus, the present research is important to understand and predict further potential stresses on water resources. In this sense, the present work aimed to estimate the influence of SL rise on the saline intrusion in São Francisco Canal (Brazil), using numerical modeling, in order to verify possible implications on water management.

2. MATERIAL AND METHODS

São Francisco Canal is one of the main aquifers located in Rio de Janeiro State, Brazil (Figure 1), and it plays an important role on the regional economical development supplying great industries. This river corresponds to the downstream section of Guandu River and it flows to Sepetiba bay, being influenced by tide action. Its fluvial flow is controlled by the water transferred from Paraíba do Sul River to Guandu river basin, what results in an average flow of 108 m³/s [11]. However, the actual flow depends on the waters released by an hydroelectric power plant (HPP), called Pereira Passos HPP, which controls the total volume flowing to the bay, being important to control the saline wedge penetration.

To verify the saline intrusion on the studied region, hydrodynamic and dispersion modeling were performed using SisBaHia (Sistema Base de Hidrodinâmica Ambiental) system.

This system corresponds to a finite element hydrodynamic numerical model governed by barotropic forcing and optimized for natural water bodies where the effects of variable density can be neglected. This model is based on principle of conservation of momentum what, in conjunction with the continuity equation, the equation of state and the respective constituent transport equations, constitute the fundamental mathematical model for any water body [12].

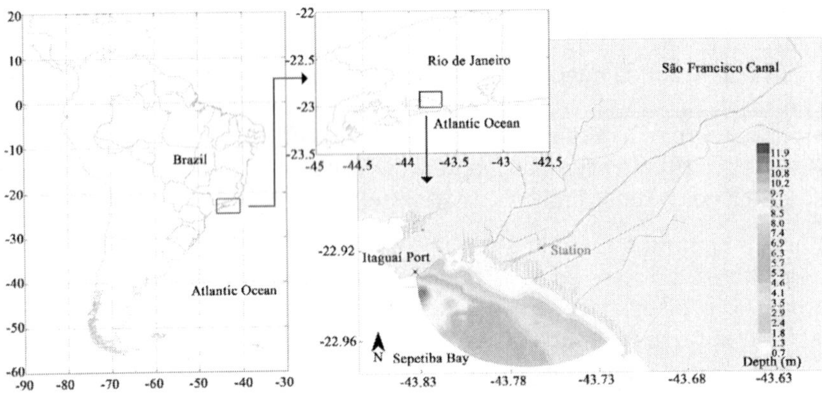

Figure 1. Representation of the study area and the bathymetry of the grid domain, in meters, used on the hydrodynamic model. The position of the grid point where modeled salinity and water level time series were extracted is marked in red (station).

Eulerian transport models are widely used on environmental analysis and are suitable to simulate the transport of dissolved scalars in the water. They simulate advective and diffusive transports and represent decay or increasing rates of the target property. In the present study, this model was coupled to the hydrodynamic model in order to verify the salinity dispersion in the study area.

This coupling is necessary since the salinity is an active tracer and it can interfere on the hydrodynamics [13].

The salinity dispersion model is depth integrated and defined as in:

$$\frac{\partial \hat{C}}{\partial t} + \hat{u}_i \frac{\partial \hat{C}}{\partial x_i} = \frac{1}{H}\frac{\partial}{\partial x_i}\left(H\hat{D}_{ij}^T \frac{\partial \hat{C}}{\partial x_j} \right) + \sum R_{P\&C} + \frac{1}{H}\left(q_p\left(C_p - \hat{C}\right) - q_e\left(C_e - \hat{C}\right)\right)$$
$$+ \frac{1}{H}\left(q_{in}\left(C_{in} - \hat{C}\right) - q_{ex}\left(C_{ex} - \hat{C}\right)\right) \quad (1)$$

Where $\hat{C}$ is the average salinity along the water column, $\hat{u}_i$ is the zonal barotropic velocity component, H is the total depth, $\hat{D}_{ij}^T$ is turbulent diffusivity tensor, $R_{p\&c}$ is the kinetic reactions of tracer production and consumption, q_p is the precipitation, q_e is the evaporation, q_{in} is infiltration and q_{ex} is the water exfiltration.

The hydrodynamic and dispersion simulations were realized in an unstructured grid with a minimum resolution of 280 m on the open boundary. The local bathymetry was interpolated in the numerical grid using information from the Brazilian Navy nautical charts. The numerical grid and the local bathymetry are presented in Figure 1.

Twelve hydrodynamic scenarios were used to evaluate the saline intrusion in the studied river. These scenarios comprise different combination of astronomical tide conditions, storm surge amplitudes and the rise of 0.5 m on the SL, as shown in Table 1. This value was used based on the predictions for global SL and Brazilian coast [6] [7].

Tides were included as boundary conditions based on the predictions for the tidal constituents from Itaguaí Port tidal station. Along the grid boundaries, this forcing presented a spatial variation by the use of a phase lag that was considered in function of the distance and the average depth.

In addition to the tidal forcing and the particular conditions for each scenario, the model was forced by hourly surface winds, measured in a meteorological station from INMET (Instituto Nacional de Meteorologia) located in Angra dos Reis (at 23.00°S and 44.32°W). The salinity values used as initial and boundary conditions were measured by the State Environmental Agency (INEA) during its monitoring campaigns along Sepetiba Bay.

The fluvial flow of some rivers including Saco do Engenho (0.5 m^3/s), Mazomba River (1.4 m^3/s), Itaguaí River (0.1 m^3/s), Guarda River (5.5 m^3/s), Santo Agostinho Canal (28.4 m^3/s), Guandú Canal (2.8 m^3/s) and Itá Canal (3.2 m^3/s) were used in the modeling. The average river flows were considered as permanent and used as boundary conditions.

Table 1. Scenarios and conditions of tides, storm surges and relative sea level (SL) used on hydrodynamic simulations.

Scenario	Astronomical tide	Storm surge (m)	SL (m)
1	Spring	-	-
2	Neap	-	-
3	Spring	0.40	-
4	Neap	0.40	-
5	Spring	0.80	-
6	Neap	0.80	-
7	Spring	-	0.5
8	Neap	-	0.5
9	Spring	0.40	0.5
10	Neap	0.40	0.5
11	Spring	0.80	0.5
12	Neap	0.80	0.5

Water volume releases and uptakes by the main water users, mainly corresponding to energy and siderurgy sectors, were also included as sinks and sources of water volume. The sum of all uptakes in the modeled region of São Francisco Canal corresponds to 53.9 m^3/s and the total release of water is realized to Guandú Canal corresponding to 20.1 m^3/s.

Each experiment was performed using an integration time of 15 days. The experiment results were used to compare the differences between scenarios with SL rise and with the present SL.

3. RESULTS

In this section, the hydrodynamic and the salinity dispersion results are presented aiming to verify the main changes observed after the consideration of a SL rise due to a future climate scenario.

3.1. Hydrodynamic Results

The Sepetiba bay is a semi-enclosed body and its hydrodynamics basically depends on the action of tides, winds and river flows. Because of this, they are the same factors influencing the water level on São Francisco Canal. The variations on the water level observed in the canal were verified with model results. In Figure 2, it is possible to verify that the free-surface displacement computed in a grid point located inside the canal, as showed in Figure 1, varies according to the occurrence of storm surges, which are observed up to eight days on average. In Figure 3, the results considering the SL rise are presented and it is possible to observe the same pattern in spite of the high levels due to the SL rise of 0.5 m.

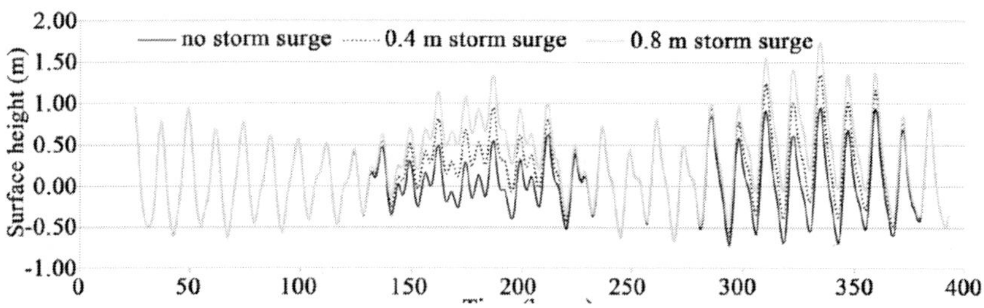

Figure 2. Surface heights computed for the current SL at the station indicated in Figure 1 considering scenarios with no storm surge (scenarios 1 and 2-black line); with 0.4 m (Scenarios 3 and 4-dotted line); and with 0.8 m of amplitude due to a storm surge event (Scenarios 5 and 6-gray line).

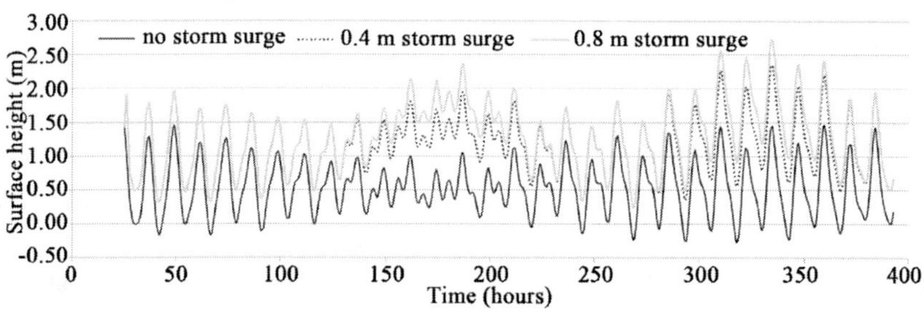

Figure 3. Surface heights computed for the future SL at the station indicated in Figure 1 considering scenarios with no storm surge (Scenarios 7 and 8-black line); with 0.4 m (Scenarios 9 and 10-dotted line); and with 0.8 m of amplitude due to a storm surge effect (Scenarios 11 and 12-gray line).

Changes on the velocity fields in response to the increasing sea level due to climate change, tide condition and storm surges were observed along the canal. In Figure 4 it is presented the extreme position where inward currents were observed for each scenario. It is noted that the maximum intrusion occurs on Scenario 11, followed by Scenarios 9 and 5. These three scenarios correspond to spring tide situations. Scenario 11 corresponds to the combined effect of mean SL rise and the presence of a storm surge event with 0.8 m of amplitude, and the second greatest intrusion was observed on the scenario that considers the mean SL rise in conjunction with the occurrence of storm surge event with 0.4 m of amplitude.

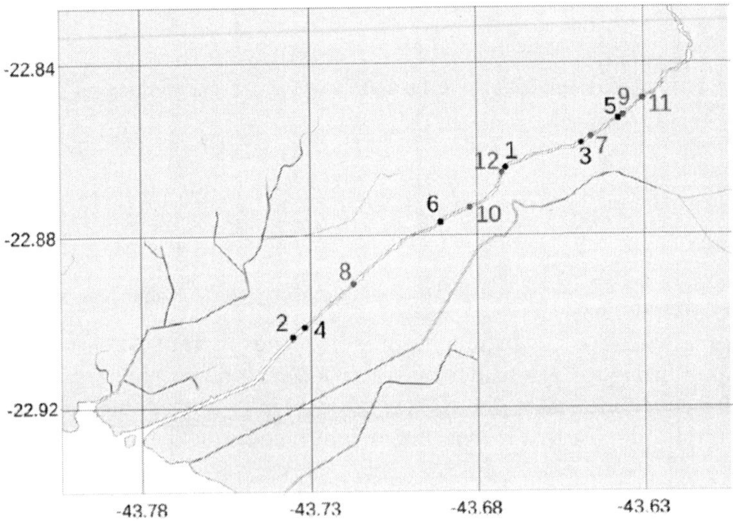

Figure 4. Position of flux inversion limits on spring tide for all scenarios (presented in Table 1) in São Francisco Canal. Red numbers correspond to sea level rise scenarios.

However, the third maximum intrusion was observed in the results obtained on Scenario 5, not on Scenario 7, showing, in this case, that the effect of the storm surge (0.80 m) during spring tide periods suppresses the effect of the mean SL change. Nevertheless, the impact of SL rise (Scenario 7) was greater than the impact with the scenario which just considers the effect of a storm surge with 0.4 m of amplitude (Scenario 3).

With the Scenario 1, that is the mildest of spring tide scenarios, the position of the upstream currents is near to the position verified on Scenario 12 results that represent a condition of neap tide with SL rise and 0.8 m of storm surge. Among the other neap tide scenarios, it was maintained the same relationships observed between spring tide scenarios.

In general, during neap tides, smaller impacts are observed on water levels and on upstream currents than during spring tide periods. However, considering worsening factors, as the SL rise and the presence of storm surge with higher amplitudes, the sea water from Sepetiba bay advances along the canal.

3.2. Saline Intrusion

The differences on the salinity concentrations in São Francisco Canal are presented by the average and maximum values computed in each scenario. Considering the extreme values obtained with the model results, the displacement of the mean SL was not main responsible for the higher values, as can be observed in Figure 5. Comparing scenario 1 and 7, it is possible to verify that during a spring period with no storm surges, higher salinity values are not correlated to the SL rise. The same is observed during neap tide periods (red bars in Figure 5). Nevertheless, considering the combined effect of storm surges and spring tides, extreme values are observed with the SL rise (Figure 5).

Considering the average salinity obtained in each scenario, in general higher salinity values were computed on SL scenarios (Figure 6), both during neap and spring tide periods. The highest averaged salinity was computed in scenario 11, corresponding to a SL rise scenario with the effect of storm surges and spring tides.

On scenarios with the effect of storm surges with 0.8 m of height and the SL rise (Scenarios 11 and 12), concentrations above the national normative limits (0.5) for this type of water body were observed for longer periods (Figure 7).

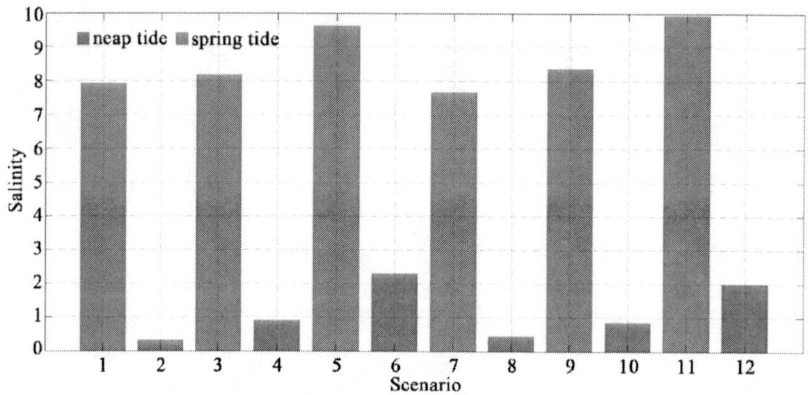

Figure 5. Maximum salinity values computed in each modeling scenario presented in Table 1 at the station located inside the São Francisco canal as shown in Figure 1. The blue and red bars represent spring and neap tide scenarios, respectively.

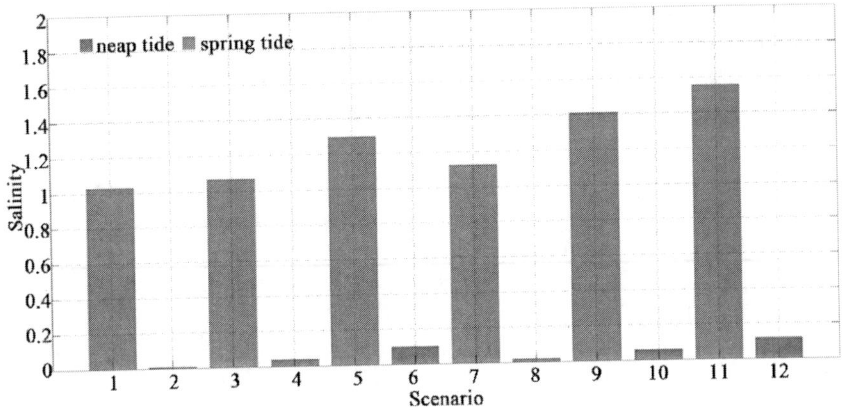

Figure 6. Average salinity values computed in each modeling scenario presented in Table 1 at the station located inside the São Francisco canal as shown in Figure 1. The blue and red bars represent spring and neap tide scenarios, respectively.

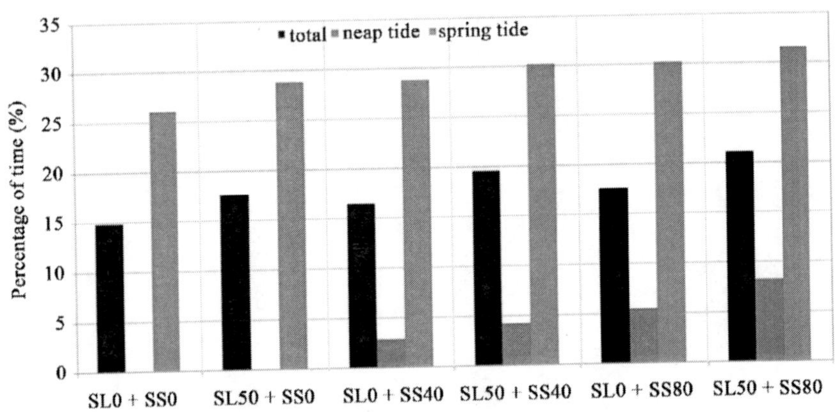

Figure 7. Percentage of simulation time with salinity values greater than 0.5 at the station located inside São Francisco Canal, considering neap and spring periods (black bars), only spring periods (blue bars) and only neap periods (red bars). SL0 indicates no sea level rise and SL50 represents 0.5 m of rise. SS0 represents conditions with no storm surge occurrence, SS40 represents the occurrence of storm surges with 0.4 m of height and SS80 with 0.8 m.

The average salinity values for these two scenarios also indicate this pattern (Figure 6).

Both average salinity values and the maximum inward position of the salt wedge indicate that the worse water quality condition is obtained in response to the advance of the surface currents along the river. Therefore, higher saline intrusions are observed with the SL rise and during the spring tide, being more intense in accordance to the amplitude of storm surges.

4. DISCUSSION

There are three strategies to reduce coastal vulnerability. These strategies are: protection, when the risk is minimized due to the reduction on its occurrence probability; retreat, when the risk is reduced by the limitation on its potential effects; and accommodation, when the capacity of the society to recover itself from the effects is increased [14].

In the case of SL rise, protection would mean the reduction on GHG emissions, but, even if the emissions are stabilized in the next decades, a significant increase on SL should occur due to the temperature differences between oceanic layers [15]. Furthermore, currently there is no conclusive knowledge that relates climate changes to GHG emissions. Then, this strategy would not be suitable for this problem. Other interpretation would be through engineering interventions in order to preserve the present human occupation on coastal zones, as by the construction of barriers and dykes, but it is usually an expensive measure.

Retreat means the abandonment of present occupations and improvements, and the migration toward the continent. It is suitable for new occupations, when the new land uses would be projected according to the SL rising. However, considering intensely occupied waterfront areas, the alternative would be the combination of retreat and accommodation strategies, adapting the coastal zone to the new SL [5].

According to Titus (1991), in established coastal areas, engineering interventions should be used, and, when these protection measurements are expensive, they should be replaced by accommodation ones. Retreat strategies, according to Titus, are only appropriate in underdeveloped areas, with few investments in infrastructure, or where is possible to occur a natural adaptation [16]. Despite the interest on accommodation and retreat strategies in the environmental management of coastal zones, an integrated approach with coastal management is still necessary [14].

In Brazil, the National Policy on Climate Change (PNMC, from Portuguese), recognized by the Federal Law n. 12,187/2009, establishes principles to avoid, predict and minimize climate change due to anthropogenic causes, aiming to adopt adaptation measures and integrated mitigation and adaptation strategies in local, regional and national spheres.

In Rio de Janeiro, the Law n. 5690/2010 establishes principles, aims, guidelines and instruments to adapt and prevent climate change negative effects, similar, in general, to those on National Policy. Among these instruments, economic-eco- logical zoning is important in light of the SL rise. However, environmental licensing, as contained in the Law, is just related to the control of GHG emissions by license applicants and it does not state about the local vulnerabilities to climate changes in installation site and the possible effect of this licenses on this vulnerability. Both in National and State policies, these issues are not addressed.

The state of Rio de Janeiro has been frequently affected by extreme rainfall events. Even though the frequency of these events is not related to the global climate change, they show the possible consequences of these changes [17]. The coastal area of Rio de Janeiro has been reported as vulnerable to SL rise and to the occurrence of extreme events, including storm surge events, worsening runoff problems and coastal flooding [17]. The continental margin of the Sepetiba Bay is considered one of the most vulnerable areas to SL rise in Rio de Janeiro [17] and the results of the present work exemplify that.

Observing the changes in response to the future SL rise, the flood currents reached more than 19 km along the São Francisco Canal considering the most extreme scenario. Higher water levels were also observed along the canal, of about 2.72 m, while under normal conditions this value only reaches 0.96 m during spring tides.

Comparing the results from hydrodynamic and transport models, the highest salinity concentrations are not observed in scenarios with higher water elevations, but in scenarios where the stream inversion limits occurred upstream on the canal.

The most extreme scenario (storm surge, spring tide and SL rise) presented an average increase of 52% on salinity values in relation to the milder scenario and of 22% in relation to the corresponding scenario (storm surge and spring tide) with no SL rise. Thus, it is possible to verify that an increase of 0.5 m in the local SL would increase the salinity in the San Francisco Canal, what would deteriorate water quality. This increase is mainly evidenced by the further advancement of the salt wedge, affecting upstream areas.

According to the results, the SL rise would lead the salt wedge 450 m upstream along the river, comparing the results between Scenarios 1 and 7. Considering the effect of storm surges with amplitudes of 0.40 and 0.80 m, during the periods of spring tide, the differences in response to the SL rise are of 380 and 267 m, respectively.

The saline intrusion obtained in scenario 11 that corresponds to the most extreme situation, is positioned 4.4 km and 0.8 km upstream to the positions obtained with the milder scenarios, representing the period of neap and spring tides, respectively. The maximum intrusion was observed at 5.4 km from the river mouth.

This same type of analysis was performed by [18] in the Gorai River basin in Bangladesh. Considering a SL rise of 0.59 m, the authors verified an increase of 0.9 in salinity values, 80 km upstream of the mouth, representing a salinity increase of 1.5 per meter of elevation. Their results demonstrated for the region that this intrusion would result in socioeconomic problems, especially in agriculture and aquaculture activities.

Despite the saline wedge intrusion in São Francisco Canal is not as extensive as that of the Gorai River, salinity concentrations ranged approximately in accord to the proportion cited by the authors. Near to the mouth of São Francisco Canal, the salinity increased on the average in a rate of 1.4 per meter of SL rise.

It is necessary to note that other changes besides SL rise were not considered in the present case study. However, it is easy to assume that in a future scenario, environmental conditions will be different from the current scenario, especially considering the effects from climate changes. Thus, future bathymetric and morphological conditions, as well as winds and rainfall rates, and therefore, river flows, could cause significant changes in the results presented in this work. These considerations may lead to the worsening or the attenuation of the results.

Thus, comprehensive studies on the local conditions are necessary to allow the strategic planning of water resources in the studied river basin, considering the imminent interference of climate change on water availability. Moreover, previous studies on the effects of climate change show that the water management in coastal areas should be made considering the potential saline wedge intrusions.

However, such planning is a complicated achievement as it is needed to consider the best water extraction locations and to control the water quality [19]. An efficient system can be obtained by considering all flow extractions, in order to previously predict the amount of river flow or reservoir volume that will decrease. However, this type of analysis is complex, given the degree of uncertainty associated with climate predictions, and due to disagreements regarding the responsibilities and investment needs, as climate change time cycle and the political cycle occur in very different time scales [7]. Thus, the choice of proactive attitudes is often left out, delaying the planning and the consideration of these atmospheric and environmental variables on the land use management, even if they are part of government policies.

The sustainable exploitation of aquifers, according to some researchers, may be the solution to prevent the degradation by saltwater intrusion. However, it does not seem feasible for aquifers in broad expansion. São Francisco Canal, because of its location, becomes strategic for industrial use, especially because it is located nearby Itaguaí Port, which attracts the establishment of a range of economic sectors in the region.

According to the strategic plan for the region of Guandu river basin, some criteria must be considered in the implementation of new raw water users [20]. Currently, the fluvial flow downstream of Pereira Passos HPP is 120 m^3/s, but after current water uptakes only 7.3% returns to the system and reach the Sepetiba Bay.

In this context, new water transfers must be carefully evaluated. As Guandu river flow depends on the present transfer from Paraíba do Sul River, even possible new transfers of water from this source could pose risks in maintaining the required flow to control the saltwater intrusion along the São Francisco Canal. Additionally, new water uptakes in Paraíba do Sul River could possibly exacerbate adverse conditions considering future climate change scenarios, such as drought extremes in river headwaters.

Thus, water resource and coastal managements must be integrated and in accordance with the principles stated in climate change policies to ensure that the environmental quality will be maintained. In addition, to turn possible the maintenance of the current water withdrawals and the establishment of new grants, it is necessary a thorough study of climate variables and considering them in the licensing process, in order to do not compromise the water supply and to choose strategies to reduce the vulnerability of the coastal zone.

5. CONCLUSIONS

In the present work, the effect of the sea level rise on water resources was evaluated in a river located in the region of Sepetiba Bay, Rio de Janeiro, Brazil. Using numerical modeling, changes on water levels and on canal hydrodynamics were observed. The upstream currents during rising tides would be more intrusive due to the intensification of pressure gradients in response to future SL rise.

This is an important result as climate change on the numerical modeling was just included as the SL displacement. Then, the modeled changes correspond just to the response to the modification on the SL and do not comprise the combined effects in consequence of this rise, as the increasing on wave heights near to the shore, due to the decreased bottom friction, and the morphological changes, due to the intensification on transport processes.

Resulting from these modeled hydrodynamical changes, the water quality along the river would be injured, in which more areas influenced by saline waters would be observed so as the increasing on salinity values at the river points previously affected by saline intrusion. The rise of 0.5 m on the mean SL would lead to a salt wedge intrusion up to 5.4 km away from the river mouth, affecting water uptakes of some industries established in the region.

The present results illustrate how this coastal environment would respond to sea level changes, however, further investigations are needed to better understand the water availability in a future climate scenario. Nevertheless, the potential future degradation on São Francisco Canal water quality indicates that future water grants should be evaluated by the local environmental agencies after considering climate scenarios. The decreasing on fluvial flow in the studied river would lead to more intense saline intrusion.

Finally, it is recommended the consideration of climate change effects along the environmental licensing, in order to establish suited strategies for each type of project and for each implementation area, and the integration between water resources and coastal management. Thus, it would be possible to determine the most vulnerable areas where major investments are needed and to manage the coastal environment according to charts of climate change vulnerability.

ACKNOWLEDGEMENTS

The authors would like to thank the CNPq (National Council of Technological and Scientific Development) for the financial support.

REFERENCES

1. IPCC (2007) Climate Change 2007: The Physical Science Basis. Contribution of Working Group I to the Fourth Assessment Report of the Intergovernmental Panel on Climate Change. Cambridge University Press, Cambridge.
2. Church, J.A. and White, N.J. (2006) A 20th Century Acceleration in Global Sea-Level Rise. Geophysical Research Letters, 33, 4 p. https://doi.org/10.1029/2005GL024826

3. Santos, R.F. and Caldeyro, V.S. (2007) Paisagens, Condicionantes e Mudanças. In: Vulnerabilidade Ambiental—Desastres Naturais ou Fenômenos Induzidos? Santos, R. F. Org., Brasília, 192 p.
4. Tagliani, C.R., Calliari, L.J., Tagliani, P.R. and Antiqueira, J.A. (2010) Vulnerability to Sea Level Rise of an Estuarine Island in Southern Brazil. Quaternary and Environmental Geosciences, 1, 18-24. https://doi.org/10.5380/abequa.v2i1-2.12821
5. Neves, C.F. and Muehe, D. (2008) Vulnerabilidade, Impactos e Adaptação às Mudanças do Clima: A Zona Costeira. In: Parcerias Estratégicas: Mudança do Clima No Brasil: Vulnerabilidade, Impactos e Adaptação, Brasília.
6. Church, J.A., Clark, P.U., Cazenave, A., et al. (2013) Sea Level Change. In: Stocker, T.F., Qin, D., Plattner, G.K., et al., Eds., Climate Change 2013: The Physical Science Basis, Cambridge University Press, Cambridge.
7. Belém, A.L. (2007). Impactos das Mudanças Climáticas Globais No Risco de Inundaçoes em Zonas Costeiras. Jornadas Internacionales sobre Gestión del Riesgo de Inundaciones y Deslizamientos de Laderas. Brazil, 13 p.
8. Chang, S.W., Clement, T.P., Simpson, M.J. and Lee, K. (2011) Does Sea-Level Rise Have an Impact on Saltwater Intrusion? Advances in Water Resources, 34, 1283-1291. https://doi.org/10.1016/j.advwatres.2011.06.006
9. Abd-Elhamid, H.F. and Javadi, A.A. (2011) Impact of Sea Level and Over-Pumping on Seawater Intrusion in Coastal Aquifers. Journal of Water and Climate Change, 2, 19-28. https://doi.org/10.2166/wcc.2011.053
10. Niang, I., Dansokho, M., Faye, S., Gueye, K. and Ndiaye, P. (2010) Impacts of Climate Change on the Senegalese Coastal Zones: Examples of the Cap Vert Peninsula and Saloum Estuary. Global and Planetary Change, 72, 294-301. https://doi.org/ 10.1016/ j.gloplacha.2010.01.005
11. Molisani, M.M., Kjerfve, B., Silva, A.P. and Lacerda, L.D. (2006) Water Discharge and Sediment Load to Sepetiba Bay from an Anthropogenically-Altered Drainage Basin, SE Brazil. Journal of Hidrology, 331, 425-433.
12. Rosman, P.C.C. (2013) Referência Técnica do SisBaHia. 249 p. http://www. sisbahia. coppe.ufrj.br/SisBAHIA_RefTec_V92.pdf
13. Rosman, P.C.C. (1997) Subsídios Para Modelagem de Sistemas Estuarinos. In: Rosman, P.C.C., Almeida, A.B., Eiger, S., Eds., Métodos Numéricos em Recursos Hídricos, 3rd Edition, ABRH, Porto Alegre, 238-348.
14. Klein, R.J.T., Nicholls, R.J., Ragoonaden, S., Capobianco, M., Aston, J. and Buckley, E.N. (2001) Technological Options for Adaptation to Climate Change in Coastal Zones. Journal of Coastal Research, 17, 591-543.
15. Wigley, T.M.L. (1995) Global-Mean Temperature and Sea Level Consequences of Greenhouse Gas Concentration Stabilization. Geophysical Research Letters, 22, 45-48. https://doi.org/10.1029/94GL01011
16. Titus, J.G. (1991) Greenhouse Effect and Sea Level Rise: The cost of Holding Back the Sea. Coastal Management, 19, 171-204. https://doi.org/10.1080/08920759109362138
17. Gusmão, P.P. (2010) Vulnerabilidade das Megacidades Brasileiras às Mudanças Climáticas: Região Metropolitana do Rio de Janeiro. In: Nobre, C. and Hogan, D.J., Eds., Projeto Megacidades, Vulnerabilidade e Mudanças Climáticas. INPE/UNI-CAMP/UFRJ, 32 p.
18. Bhuiyan, M.J.A.N. and Dutta, D. (2012) Assessing Impacts of Sea Level Rise on River Salinity in the Gorai River Network, Bangladesh. Estuarine, Coastal and Shelf Science, 96, 219-227. https://doi.org/10.1016/j.ecss.2011.11.005
19. Ferreira Da Silva, J. and Haie, N. (2007) Optimal Locations of Groundwater Extractions in Coastal Aquifers. Water Resources Management, 21, 1299-1311. https://doi.org/ 10.1007/s11269-006-9082-7
20. ANA (2007) Water National Agency (Agência Nacional de Aguas). Plano Estratégico de Recursos Hídricos das Bacias Hidrográficas dos Rios Guandu, da Guarda e Guandu Mirim: Relatório Gerencial. Brasília, 63 p.

CHAPTER 7

The Environmental Dimension of Groundwater in Brazil: Conflicts between Mineral Water and Water Resource Management

Ana Lucia Desenzi Gesicki[1], Francesco Sindico[2]

[1]*National Department of Mineral Production (DNPM), Superintendence of São Paulo, São Paulo, Brazil*

[2]*Strathclyde Centre for Environmental Law and Governance (SCELG), University of Strathclyde, Glasgow, UK*

ABSTRACT

There are three legal categories of groundwater in Brazil. Mineral water and potable table water are considered mineral resources, are part of the Federative Union's assets, and follow the legal regime applicable for the mining sector. "Normal" groundwater, on the other hand, falls under State's jurisdiction and is dealt within the Brazilian System of Water Resource Management, which promotes a decentralized and participatory management of water resources on surface or stored underground. This has led to conflicts of competences between federal (mining regulation) and state agencies (water resource management) because different concepts and styles of manage- ment are involved. This article argues for the urgent need to reopen the discussion aimed at a possible major revision of

Brazilian mineral water legislation, in order to duly take into account the environmental dimension of groundwater as a public good for common use. An open-minded and transparent discussion involving the government and different sectors of society with com- peting interests in mineralized groundwater would therefore be highly recommended.

Keywords: Mineral Water, Water Resources, Mining Law, Water Policy, Brazil

1. INTRODUCTION

In spite of Brazil's continental land area and abundance of fresh water, the country faces challenging tasks in managing water resources due to significant regional diversity and inequity of water distribution within its terri- tory. While the sparsely populated northern portion of the country encompasses the largest river basin in the world (the Amazon river basin), the Northeast suffers from periodic droughts and the Southeast has to tackle with the issues of water resources having an unbalanced supply-demand relationship and increasing pollution due to urbanization and industrial activities [1] [2] .

Regarding specifically Brazilian groundwater resources, these are subject to competing interests from various categories of use, mainly in the regions where the availability of surface fresh water of good quality is under threat from growing risk of pollution and scarcity due to critical climatic events. An overall estimate reveals that there must be about 400,000 wells across the country pumping groundwater for several purposes, at a likely total abstraction rate of 4000 m^3/s [3]. This estimate still gives a wide margin for safe water use considering that re- newable reserves of groundwater in Brazil have been estimated at about 42,000 m^3/s [4].

The Brazilian Water Reform, carried out in the late 1990s, includes several principles and instruments inter- nationally accepted for dealing with water resources, such as integrated and decentralized management and mul- tiple uses of water. The water legislation, however, is silent on economic uses of mineralized groundwater. This is not surprising considering the aged legislation for mineral waters, namely Code of Mineral Waters (CMW), was enacted in 1945 and has never been updated. The CMW considers mineral water as a mineral substance with economic value and deals with it separately from water resource management. By doing so, mineral water falls under the same legal regime applicable for the mining sector. This has led mineral water and groundwater to be covered by different legal frameworks that embody distinct concepts and styles of management, which re- sults in conflicts of competences between federal (mining regulation) and state agencies (water resource man- agement).

Against this background, this article aims to critically assess the likely sources of conflicts between mineral water and water resource management, by providing an overview of both legal frameworks. This article is di- vided in six sections. The first three are devoted to provide an overview of the relevant aspects of the water, mining and environmental legislations in Brazil. The next two sections examine in detail the reasons why min- eral water legislation is in disagreement with the principles of the mining sector and of water resource manage- ment, including a reported case of conflict of interests between federal and local institutions. The last section fi- nally addresses key legal standpoints, which we consider important enough to justify a major revision of the current mineral water legislation.

2. WATER RESOURCES—PRINCIPLES AND POLICIES

Since the first decades of the 1900s, rivers have been extremely important in Brazil because of their huge poten- tial to generate energy. As a result, Brazil's energy matrix relies heavily on hydroelectricity. The Water Code of 1934 represents the early efforts to discipline the use of water resources in a country with an increasing demand for energy due to the growth of an urban-industrialized society [1]. During more than 60 years, the Water Code played an important role in governing the use of surface water resources for hydropower generation purposes, but several provisions of the law related to other matters (for example, groundwater use) have never been prop- erly regulated because the law did not take into account the complexity of different water uses. Although the Water Code is still in force, several provisions were implicitly revoked by the 1988 Federal Constitution. This has been the case for example, for water bodies previously ascribed as private property and municipal rivers.

The awareness for the need for environmental protection that emerged in the 1970s, combined with the further widespread understanding of the need to promote sustainable development, culminated in a change of paradigm from an ancient culture of disregard for available water to a new concept of water as a scarce resource [2]. The new environmental mind-set when dealing with water has led to a review of governmental strategies and actions for the implementation of an integrated management programme for water resources in Brazil [3]. The first step towards this achievement was the promulgation of the 1988 Federal Constitution itself, which assigned all wa- ters to the public domain and mandated the implementation of the national system for water resource manage- ment.

International legal instruments about water and the environment, such as the Dublin Declaration of 1992, reinforced the discussion about principles for water resource management and played an important role in the process of reviewing laws and policies at federal and state levels[1]. This led to the enactment of the National Water Act in 1997 (Law 9.433/1997), also known as "Water Law", which laid the basis for the National Policy for Water Resources and created the National System of Water Resource Management (SINGREH), launching new principles, objectives and administrative instruments for managing water resources in Brazil.

The National Policy for Water Resources is based on six principles that underpin the Brazilian system of wa- ter resource management. First, water is considered to be a public good, following constitutional mandatory provisions. Second, it is recognized that water is a limited resource with economic value. Third, in the case of water shortage, the law establishes priority water uses for human and animal consumption. Fourth, the multiple uses of water must be considered within the management of water resources. Fifth, it establishes the river basin as the basic territorial unit for the implementation of the National Policy for Water Resources and the National System of Water Resource Management. Sixth, the law promotes the decentralized and democratic management of water resources, counting on the participation of governmental institutions, stakeholders and communities involved in the river basins.

An important advance in Water Reform was the introduction of a permit system, at a national level, to grant rights for any use of water resources stored underground. As groundwater is under the exclusive jurisdiction of the States[2], groundwater users must follow the rules established by the water permit system defined at State lev- el[3]. For granting water rights, some relevant issues are observed by the State bodies, such as the water availabil- ity of the aquifer in the river basin, its expected recharge rate and the stipulation of a maximum yield for groundwater abstraction that is considered "sustainable", i.e. the volume of water in a rate of abstraction that guarantees the availability of groundwater for other uses [5]. The authorization for water use issued by the State institution is time limited, a time limit that varies from State to State[4].

The dual jurisdiction over water bodies has been identified as a challenge for the integrated management of water resources. That is the case of surface waters, which can be under the jurisdiction of the Union or the States depending on their geographic localization[5]. It is not different for groundwater. Even though the Constitution defines the jurisdiction of groundwater to the States, there are not yet legal provisions to integrate mineralized groundwater management to the National Policy for Water Resources and so providing a desirable integration between federal and State agencies. For a country of great regional diversity, the governance of water resources depends primarily on the cooperation between the different levels of the government, which is not simple to achieve [3] considering that Union, States and Municipalities all have administrative autonomy to define rules and standards to deal with protection of the environment and other issues of their own competences.

3. PROTECTION OF THE ENVIRONMENT

At the time when the legislation for mineral water was drafted, Brazil had great potential for development and was starting to face a significant increase in the rate of industrialization and also undergoing a population explo- sion, mainly in the southern urban centres. Paradoxically, this accelerated economic growth was founded on in- creasing regional economic discrepancies, and occurred at the expense of individual freedoms and furthermore imposed environmental costs because of the uncontrolled use of the country's natural resources [6] -[8] .

The model of development at all costs started to be questioned in the 1970s, mainly as a result both of a movement of civil society organizations and of several serious environmental damages in Brazil[6]. Environmen- tal regulation at federal level in Brazil started with the promulgation of Law 6.938/1981, which established the National Environmental Policy, and set up the National System for the Environment (SISNAMA) multi-institu- tional framework. Within SISNAMA there are institutions and entities at Federal, State and Municipal levels that grant environmental licenses, carry out inspections and enact supplementary rules and policies within the ambit of their own competences in matters related to environmental protection and pollution control.

Environmental regulation of economic activities in Brazil stems on two internationally widespread principles: the precautionary and the polluter-pays principles. The precautionary principle requires a proactive approach towards the protection of the environment, using two main instruments of the National Environmental Policy —environmental impact assessment (EIA) and licensing of economic activities that cause potential or effective hazards to the environment[7]. The environmental viability of a project is analyzed by SISNAMA bodies at Fed- eral, State or Municipal level, depending on the type of activity and the scale and localization of the project. The polluter-pays principle is expressed in the penalties, sanctions and requirements for reparations to which offend- ers are subject if they cause harm to the environment. Note that mining is an activity that admittedly implies damages to the environment; hence, the miner has the constitutional obligation to restore what has been de- graded.

The environmental legislation, combined with the environmental principles of the Federal Constitution and of the Brazilian Civil Code[8], establishes water as a public good for common use, describing it generally as an environmental resource, together with other elements of ecosystems. Environmental resources are for general use of the whole community, being essential to acquire and maintain a healthy quality of life and to reach the ideal ecological equilibrium. On the other hand, the private use of water resources through the permit system of the National Policy for Water Resources does not denote possession of the water itself [9] , considering the premise of inalienability of public goods for common use [10] . Rather it implies the right to use water for economic or individual purposes, observing the restrictions imposed by law and by the collective interests. Integrating the public domain with other environmental resources, water in general cannot be privately or individually owned. Rather the opposite since the public bodies with legal competence for managing public goods are bound by law to guard them (include water) for the overall benefit of the community [10] .

Against this background, groundwater acquires a strong environmental dimension. As an environmental re- source, it constitutes as an important reserve of drinking water for future generations and must subject to protective and conservative policies for keeping an ecologically balanced environment. The economic use of ground- water resources promotes either private or collective interests, supporting economic activities or supplying the population with essential public services. Nevertheless, the economic dimension of groundwater must be com- patible with the intergenerational obligation of guaranteeing groundwater reserves in adequate quality and quan- tity for the future.

4. MINING AND MINERAL WATERS—REGULATORY FRAMEWORK

The Code of Mineral Waters (CMW) was enacted while the Constitution of 1937 and the former Code of Mines of 1940 were in force. The Constitution of 1937 reiterated most of the principles for mineral resources exploitation introduced in 1934. According to it, the property of mines and other subsoil wealth was considered distinct from the ownership of the land, and their economic exploitation relied upon federal grants. These main principles were later reinforced by the Federal Constitution of 1988, currently in force, which extends the Federative Union's domain over mineral resources even for those undiscovered, wherever they are settled (on surface or in the subsoil), and also defines that mining activity must be carried out in the national interest, following policies and rules for protection of the environment and for the recovery of degraded areas by the mining company.

The CMW was promulgated in 1945 after the work of a Hydrology Commission nominated by the Ministry of Agriculture in 1940. The law clarifies that the economic exploitation of mineral waters is governed by the legal regime for mineral resources based on the provisions of the Code of Mines of 1940, subsequently updated by the Code of Mining (CM) of 1967.

Since 1967, the CM has governed all aspects of the mining sector in Brazil[9], from exploration of mineral re- sources to production and commerce of mineral commodities. Mineral rights are established and extinguished by successive administrative proceedings, mostly of a binding nature [11] , issued by both the Ministry of Mines and Energy (MME) and the National Department of Mineral Production (DNPM).

Authorization and concession are legal regimes defined by the CM that represent the two main phases of the mining activity, respectively the exploration and exploitation stages. The authorization regime comprises a time-limited title (2 to 3 years, renewable) that allows either individuals or companies to conduct, at their own expense, all prospecting works needed for the definition of an ore deposit, evaluation of mineral reserves and grades, as well as the determination of their economic feasibility. The concession title is issued by the MME, has unlimited duration and allows private companies to carry out mining operation projects that have been pre- viously submitted to an environmental licensing process.

The Constitution empowers the Federative Union to dispose of mineral wealth, be it on the surface or under- ground, without any legal restriction. The first competent applicant to request a survey grant for a particular available area can obtain exploration priority rights. The right of priority, also known as the "first come, first served" rule, is one of the mining activity principles that guarantees democratic access to mineral resources, as the Union must grant mineral rights to whoever first legitimately applies for it.

The mining sector has a peculiar dynamic as economic activity. Despite the public nature of mineral resources, mining is not a public service. All investments (normally of a long term nature) and usually high risks are at the expense of private parties. Once a mining title has been issued, the miner begins to have a right/duty to exploit mineral deposit until its depletion. The property of the mined product is guaranteed to him, as is the right to use, enjoy, sell or rent the mine and to use the mineral rights as collateral. Thus, the exploitation concession is the ad- ministrative act that allows the shift of the ownership of a public asset (in situ ore deposit) from the Union to private hands (mined ore deposit), incorporating it into their assets [12] [13].

Mining is constrained by the immovable nature of the ore deposits. Mineral resources can be found anywhere, outcropping the surface or deep inside the soil. Considering the random and unpredictable natural occurrence of mineral resources, mining is considered to be of public utility or of social interest and direct exploitation in some protected areas can be allowed only under specific environmental law provisions[10].

4.1 The Code of Mineral Waters (CMW)

Mineral water is legally defined as groundwater with a distinct chemical composition or physical-chemical properties which confer a medicinal action. The CMW recognizes two categories of groundwater: mineral waters and potable table water. Both categories of groundwater are considered mineral substances and are governed by the mining legal framework. The CMW regulates two only possible end uses, for bottling process and for spas purposes.

Mineral waters are defined as having a medicinal action depending on their composition. They are characte- rized by a wide range and defined amounts of some chemical compounds (radium, sodium and calcium bicar- bonate, sulphate, sulphur, nitrate, sodium chloride and iron), by the presence of dissolved gases (carbon dioxide, hydrogen sulphide, and short half-life radioactive gases) and by the contents of noteworthy trace elements (iodine, arsenic, lithium, etc.). Together with chemical composition, some properties of the water at the source, such as temperature and temporary radioactivity, are also used to classify a mineral water source.

Potable table water, on the other hand, has nothing special at all, i.e. comprises groundwater which does not reach the chemical or physical parameters defined by the CMW, but only potability. Potable table water theoreti- cally is not associated with any sort of presumable beneficial health effect, but is suitable for bottling purposes.

According to the CMW, water with particular chemical and physical parameters is defined as having thera- peutic benefits, even though no therapeutic effects may have been shown empirically by medical experts. The medicinal action is the main legal criterion that defines a mineral water, but the procedure of testing for any fa- vorable effect on health is not a legal requirement[11], because the law sets down the chemical and physical para- meters for mineral waters (Art. 35 and Art. 36 of CMW).

The CMW was enacted at a time when several spas and hydromineral resorts were designated as important healing centres in Brazil. Mineral waters were normally known as “mineral-medicinal” or “thermal-medicinal” waters, because Crenotherapy was experiencing great popularity as a traditional practice in medicine [5] [14] . After the Second

World War, however, the discovery of penicillin and the spread of the pharmaceutical industry have been identified as the main reason for the decline of those medicinal practices, not only in Brazil but also worldwide [15] . Bottled mineral waters have their origin in the spas and resorts, retaining a connexion to their sources and the brands that carried the prestige of some known effects on health, but slowly they became more and more separated and diverged from the resorts [15] , splitting up completely and developing as independent companies in a fast growing sector. The production of bottled mineral water before 1970s were lower than 100 million litres/year, increasing to around 800 million litres/year in 1990 and then boosting to more than 3000 million litres/year from 2000 forward [16] . By the end of the 1950s, bottled mineral water was no longer sold in pharmacies but had begun to be marketed in supermarkets, restaurants and bars [15] , and was seen not anymore as a medicine, but also as a domestic consumer good. The mineral water sector changed its focus from healing purposes, leaving the medical scope and approaching the bottled beverage industry, to dealing with the primary purpose of the water: thirst quenching.

Considering the extinction of the Crenotherapy specialty or subspecialty in medicine, the confirmation of some therapeutic use of a mineral water has not been made since the 1960s.

In that way, the classification of mineral water has been based only on the chemical or physical properties defined by law. On the other hand, the legally defined parameters used to classify a mineral water cover such a wide range (for example the tempera- ture in the source) that practically any grade of mineralization or even waters containing low amounts of trace elements (for example fluoride) are enough to categorize them as mineral water. In practice, today, any naturally potable groundwater is likely to be classified either as mineral water or as potable table water [15] [17] if suita- ble for bottling or for spa purposes.

5. MINERAL WATERS VERSUS MINERALS—A COMPARISON

Mineral water seems an alien mineral substance within the mining sector and remarkable differences between mineral water and other minerals can be pointed out as follows: 1) mineral water is neither a mineral nor a fossil substance; 2) mineral water is a renewable resource, in theory; 3) mineral water is produced directly for human consumption, and because of that, the sector must follow standards and rules promoted by health, sanitation and environmental institutions; 4) inspection procedures for mineral water are more similar to the ones taking place in the sanitation and health sectors than in the mining one; 5) exploitation of mineral water implies a minimal alteration of the surface and minimized visual impacts at the point of extraction (punctual extraction); 6) mineral water exploitation is conservative because it involves a delimitation of a protection area around the source with restriction to any underground work or intervention; and 7) mineral water extraction and processing cause desirable minimal change in the original water composition.

The first basic difference between minerals and mineral waters refers to their nature. Water is not considered a mineral because it is liquid and does not show a crystalline structure [18] . Mineral water is not the only exception within other mineral substances though. For example, mercury occurs as a liquid and opal is amorphous, so both are considered mineral-like substances, named mineraloid, but are not formally accepted as minerals. Some chemical compounds induced by biological processes, such as marine phosphorites, are not truly minerals either. In fact, being a mineral is not a sine qua non for being regarded as a mineral resource.

Another key difference regards renewability. Mineral water is a drinkable groundwater which, theoretically, is a renewable resource connected directly or indirectly to surface waters, taking part in the hydrological cycle. According to Rebouças, 97% of all groundwater is meteoric in origin and was recharged by infiltration of sur- face water into the ground at some moment in geological time [19].

The concept of renewability for groundwa- ter, however, is relative to time and can be divided into renewable and non-renewable. Renewable aquifers re- ceive natural recharge and have the ability to replenish their reserves depending on the level of connectivity with surface waters, in a period of time comparable to a human lifespan. On the other hand, aquifers that need a very long period (hundreds to thousands of years) to replenish their stocks are known as non-renewable or fossil [20]. Non-renewable groundwater is normally found in unconfined aquifers where contemporary recharge amounts are small or infrequent due to severe climatic conditions, as well as in confined sections of some aquifers where active recharge is negligible [20] [21] . "Groundwater mining" is the convenient term often used for groundwater developments that involve the abstraction from fossil or non-renewable water reservoirs. In other words, ground- water mining may be defined as the exploitation of groundwater at a rate that is much greater than the rate of replenishment of the aquifer storage [22] [23].

Nevertheless, the principles that underpin the mining activity and water resource management are quite dif- ferent. One of the problems of mineral water management is the premise that mineral resources can be exploited until ores deposits are exhausted, because this principle implies some environmental concerns related to the pro- tection and conservation of groundwater as an environmental resource, which are not properly covered by the CMW or the CM. Because the CMW and the CM antedate the Brazilian environmental legislation, most of the policies and rules concerning environmental protection affect the mining sector directly without disrupting the mining legal framework.

The environmental legislation introduced another step for the mining activity, by sub- mitting mining developments to environmental licensing process regarding its installation and operation. Consi- dering that mining is legally recognized as a potentially hazardous activity to the environment, mineral exploita- tion (including mineral water) is not authorized unless previous environmental impact assessment, or another similar instrument, has been carried out. The theoretical depletion of a mineral water "deposit" can be legally authorized[12], but this is unlikely to happen because the exploitation would not be licensed by the environmental authority in the first place.

The second problem in considering mineral water as a mineral resource refers to the ownership transference of a public asset to a private party. Legally, the miner owns the mined product as part of his private assets. This means that the abstracted mineral water belongs to the miner, differently than users of water resources. As seen before, according to Brazilian Water Law's principles, water rights refer to the private use of a public good with the public good remaining in public hands.

Then the mining principle conflicts directly with the constitutional assumption that water is a public good for common use by the people and is inalienable. Under Brazilian Ad- ministrative Law, public assets for common use are not susceptible to being ascribed to individual ownership, since they belong to everyone and the government is in charge of managing them for the benefit of the society [9].

From an economic perspective, the exclusive use of a water resource by a private party depends on an ad- ministrative permit act as stated by a national and regional set of policies and rules. The management of mineral water under the mining exploitation regime seems, therefore, to be incompatible with the premise of the inalie- nability of water.

Finally, some inconsistency between the CM and the CMW rests on the principle of management of mineral water for end use. Mineral water is managed exclusively for bottling process or for spa purposes. Any other use (for example, water supply or irrigation) is not regulated by the CMW and is out of the federal jurisdiction of DNPM. Mining activity, however, is deeply dependent on the fixed location of ore deposits and on its non-renew- ability rather than on the end uses of extracted mineral substances.

6. CONFLICTS BETWEEN MINERAL AND GROUNDWATER SECTORS

Mineral water management can potentially conflict with other sectors, such as health and sanitation, environ- mental protection, land use planning, and groundwater resource management [24] . Regarding the health sector, there are concurrent administrative levels of the government, such as the Ministry of Health and the Ministry of Mines and Energy, which enact conflicting norms. This is the case for sanitation inspection procedures for min- eral water industries and labelling of bottled mineral water[13]. Another source of conflict lies in the necessary in- tervention in preserved wooded areas (named Permanent Protection Areas defined in the Brazilian Forest Code) to construct the source protection building around mineral water springs, that would be theoretically authorized only under special circumstances defined by law[14]. While environmental law states that Permanent Protected Areas must be preserved, the mineral water law determines deforestation can be carried out in localized portions of those protected wooded areas [24] , which can result in fines by the environmental body.

Nevertheless, the mineral water sector collides directly and more frequently with groundwater sector. Official statistics demonstrate that almost 190,000 permits for water use had been granted by July 2011, representing a total authorized yield of 6800 m^3/s, where groundwater still represents a minor part, of about 8400 permits and a total yield of 520 m^3/s (or 1,872,000 m^3/h) [25] . Compared with the official statistics for the water resource management sector, the production of mineral water represents a minimal, if not negligible, proportion of the end use of groundwater in Brazil, which is estimated at 1024 exploitation concessions granted by the end of 2011, representing a total yield of 0.28 m^3/s (or 1023 m^3/h) [26] . Although there is such a quantity difference, the potential conflicts between the groundwater and mineral water use still occur, but are rather a localized issue, especially in certain regions where mineral and thermal waters play an important role in production of bottled beverages or for spa tourism.

A well-reported case of conflict takes place in the Caldas Novas region, in the central part of Brazil, an im- portant tourist destination in Goias State, where the presence of thermal water sources has been attracting tour- ists from all over the country for leisure tourism and balneotherapy since the 1960s. A large number of wells (more than 400 wells) had been drilled and groundwater abstraction was almost uncontrolled, resulting in a drastic decrease in the regional water level of the thermal aquifer, with a decrease of more than 50 m from its original level [27] . From 1996, the National

Department of Mineral Production (DNPM) has been adopting a set of control measures for the thermal aquifer in the Caldas Novas region, aiming to recover of its falling potentio- metric surface and to maintain the thermal characteristics of the aquifer that was about to be depleted because of groundwater overexploitation [27] . The set of procedures comprised the interdiction of 180 illegal wells, the prohibition of use of thermal water from 155 wells for domestic purposes, the restraint and regular monitoring of the production yield in the 89 authorized thermal water wells and the suspension of granting of new mineral rights in the region[15]. Recently, the control measures successfully led the aquifer to recover about 30 m above the lowest water level registered in the region in the middle 1990s, but some concerns still remain because mon- itoring data have shown a progressive tendency for thermal aquifer drawdown that has not adequately been ad- dressed[16]. On the other hand, the supremacy of thermal groundwater use for mineral water spas has also con- flicted directly with the local public water supply service, because the municipality had their thermal water wells restricted by the federal regulatory body, which ultimately led the matter to be brought before a court of justice to be settled[17]. Only 30 shallow wells are authorized by the State body for abstraction of cold groundwater for domestic uses inside the restricted area [27].

Another reported case of conflict refers to the Nestlé industry in São Lourenço city, a traditional hydromineral resort in Southeast Brazil. The local population organised protests from 1999 to 2004 against the multinational company, which was suspected of overexploiting the aquifer by intensive pumping of wells for bottled water production [28] . Protests focused on the distrust of the multinational company being granted with mining rights to exploit mineralized groundwater from wells enclosed in a resort with reputed medicinal springs. The supposed- ly high rate of groundwater pumping for bottling purposes would have caused the falling water level of the aquifer, with apparent changes in water composition and the drying of some secular springs. The Nestlé industry was involved in a trial, with great negative repercussions shown both in national and international media. In the end, some irregularities were found in the industry's operation, but there was no direct evidence of aquifer overexploitation due to the production of bottled water. Nestlé had their activities adjusted to the Public Minis- try's requirements by signing an agreement in 2006 that restricted one of the pumping wells, and also limited its total production to 13 million litres of mineral water per year [28].

Aware of the conflicts of jurisdiction between the federal regulatory body (management of mineral water) and State institutions (management of water resources), the National Water Resource Council (CNRH) in 2002 started a long discussion on this issue, conducted within its Groundwater

Technical Chamber (CTAS). The dis- cussion about the integration between the management of mineral water and groundwater were particularly lively in 2004, when a seminar was held focusing conceptual and legal issues in each sector. The discussion was somewhat heated and opinions were clearly polarized into two divergent points of view: on the one side representatives of the National Department of Mineral Production (DNPM) and of the National Confederation of In- dustry were against any attempt to integrate mineral water developments with the permit system of the National System of Water Resource Management (SINGREH). On the other side, representatives of Federal and State bodies for the environment and water resource management argued the opposite [16] [29] [30] . In the end, the Council published Resolution 76/2007 focusing mainly on guidelines to promote the dialogue between federal and states administrative institutions, as well as the sharing of technical information of each permit system. This resolution seeks, to stimulate the "integration and coordinated action" of the organisations involved (federal and state agencies) without formally stating that SINGREH should be an additional management instrument for mineral water developments[18].

A likely reason for this sort of conflict is that mineral water today is not so far distinct from "ordinary" groundwater. According to Queiroz, 88% of all mineral water sources in Brazil produce very lightly to lightly mineralized water, showing a total concentration of dissolved chemical compounds no higher than 300 mg/L [31] . About 10% of the sources are classified as potable table water and 27% of the sources are classified as mineral water exclusively due to insignificant amounts of fluoride as a noteworthy component [31] . This study also shows that 43% of all Brazilian mineral water sources are characterized exclusively by intermittent physical and chemical parameters, such as temperature and/or temporary radioactivity, which means they are truly min- eral waters only at, or near to, their sources. Away from their sources, they are only potable table waters (or common potable groundwater) because their temperature has changed and their dissolved gases have been lost at the time they are available for consumption. While those intermittent physical-chemical characteristics are sig- nificant for spa purposes, they are irrelevant for the consumer of bottled mineral water.

Only a small part of all mineral water sources in Brazil, about 20%, is characterized by greater diversity of chemical compounds and/or high temperature and/or natural carbonation, characteristics that are strictly in ac- cordance with the legal chemical limits established in Article 35 of the CMW and with the legally stated defini- tion of a supposed medicinal action for mineral water. This percentage is roughly the proportion of mineral water sources (14%) that historically were devoted to some medicinal purpose or aimed at leisure tourism in spas and hydromineral resorts [31].

Considering only bottled mineral water, one third are very lightly mineralized (total dissolved solids < 50 mg/L), associated with fast water circulation through shallow aquifers (pH < 6, acidic), theoretically highly vul- nerable to surface-driven contamination, and furthermore one quarter of the sources shows some signs of anth- ropic impact indicated by the presence of nitrate[19], even though they are potable [32] . There are some sources in the North and Northeast of Brazil that produce mineral water compositionally similar to rainwater, as they show total dissolved solid contents below 5 mg/L and pH between 4 and 5 [17] [32].

In fact, the CMW currently fails to sort out conflicts between the water and mining sectors because the law does not provide reasonable well-defined limits, based on intrinsic parameters, which can without doubt differ- entiate mineral water, potable table water and common groundwater. With some exceptions, in practice, they are almost the same, because the CMW allows a loose interpretation of the legal parameters used to classify any potable groundwater either as mineral water or as potable table water.

In the 1940s, when the CMW was drafted, there was no stated policy at federal level to deal with other eco- nomic uses of groundwater apart those regulated by the CMW. Furthermore, the limited knowledge in Hydroge- ology in the 1940s and the lack of technology in underground works have possibly led to the misconception that mineralized groundwater was disconnected from surface waters, causing mineral water management to be regu- lated more closely to the mining sector, completely separated from surface water resource management [15] . The CMW's provisions worked without major conflicts for years, until the Federal Constitution came into force in 1988, which introduced an environmental dimension for water, previewed a national-level integrated water resource management, and empowered the States with jurisdiction over groundwater.

With groundwater being defined by the Constitution as a natural resource under States' jurisdiction, there is a risk that potable table water would in fact be interpreted as being "common" groundwater, since they are technically the same. If this were the case, the CMW's provisions designating potable table water as a mineral re- source would conflict directly with the 1988 Federal Constitution. Accordingly, neither potable table water nor mineral waters would be regarded as assets belonging to the Union, because the Federal Constitution makes no distinction between different types of water, but includes all groundwater under the States' domain. As vehe- mently argued by Serra, if no exception is made for groundwater, the CMW as a whole would be implicitly re- voked by the 1988 Federal Constitution and mineral water would fall under States' competence as a water re- source [15] . This is a controversial point of view, however, and is not shared by the Ministry of Mines and Energy's advisory board of Federal Attorneys (Advogado Geral da União). They argue that the

CMW is in ac- cordance with the Federal Constitution, because mineralized groundwater is solidly defined as a mineral re- source at infra-constitutional level [29] .

7. THE NEED FOR UPDATING THE BRAZILIAN LEGISLATION FOR MINERAL WATER

The concepts used by the CMW for defining mineral water lie in the sparse scientific and medical knowledge about the therapeutic uses of mineralized, radioactive and thermal waters that was available in the 1940s. Some properties of the water that are directly linked to the source, such as high temperature and dissolved gas content, are relevant for therapeutic purposes and have been extensively reported as beneficial to health in both recent and older international literature on spa therapy [33] -[35] . Some therapeutic effects of mineral waters, however, are currently asserted in law, but are unsubstantiated by science. The actions of chemicals are now much better understood than they were when the law was drafted, and it is now clear that the law defining the therapeutic benefits of some of the chemical compounds found in mineral water seems to be misleading.

For example, nitrate and arsenic are defined by the CMW as likely to be associated with some medicinal ac- tion, but they in fact have been found to have some harmful effects in humans, so that their concentration in drinking water is now determined by rigid standards[20]. On the other hand, the presence of radioactivity (perma- nent or temporary) in water, which for a long time Crenotherapy had credit with the mild therapeutic effects of lightly mineralized water [36] , has had its beneficial action on health questioned, and it has even been described as hazardous to humans [17] [37] . If there are controversies about the benefits or possible harms to humans re- lated to certain components in drinking water, precautionary concerns should be taken into account in using these parameters for mineral water classification and its use for human consumption.

The concept of mineral water carrying some medicinal effect is currently in disuse in Brazil, mainly because medicinal practices using mineral and thermal waters are virtually extinct due to the lack of experts in Creno-therapy [14] [15] . As the linkage between medicinal therapy and mineral water has been lost, the principles that underpin the Brazilian legal framework for dealing with mineralized groundwater should be revised. As pointed out by the reputed Brazilian jurist, Miguel Reale, "the legal rule must be formally valid and socially effective" [38] . In other words, the effectiveness of the law is dictated by its engagement with the social experience, to the point that "the disuse [of a law] can occur because the legal rule has never been applied or, at a certain time, has ceased to be applied, since the obedience to a diverse customary norm started to prevail within the community, with the oblivion of the legal rule" [38] . The same author also argues that: "the public authorities have a duty to avoid the divorce between social reality and certain legal rules, which are not, or never have been effective, be- cause it [the law] is in conflict with trends and authentic dominant interests within the community" [39] .

The current political climate in Brazil is particularly favorable to making public authorities aware of the ur- gent need for reopening the discussion for updating mineral water legislation, because the government has been making efforts to modernize outdated laws. In this direction, the federal government has released the bill that will replace the 1967 Code of Mining and will improve the institutional and legal framework of the mining activity in Brazil. The bill number 5.807/2013 is before the National Congress since June of 2012 and several changes in principles and management instruments for dealing with mineral resources are likely to happen by the end of 2014. The expected changes, however, will not affect the current management style of mineral waters.

A desirable scenario would be for the government to open a public consultation process followed by a trans- parent discussion through public meetings, involving different sectors of society with competing interests in the mineral water issue, such as the mining, groundwater, health and food sectors. Academic knowledge and the experience of other countries would be of great value in supporting the discussion and the law-making process. The government should provide an open-minded debate for answering the big question: which is the model for mineral water management that society really wants and should pursue? The multidisciplinary debate should lead to answers regarding the establishment of (new) principles and concepts for mineral water, the definition of the management instrument, a decision about the legal regime of exploitation and the determination of well-defin- ed technical parameters that unquestionably differentiate mineral water from normal groundwater.

The current design of the mineral water sector has evolved so that most of the mineral water sources today are barely distinguishable from normal groundwater. The connection between mineral water and medicine has been lost for a long time and so is the linkage with the principles that underpin the mining legal framework. Keeping mineral water management in the mining sector can only be justified if the principles governing mineral water legislation are in accordance with the non-renewability of aquifers and with the fixed location of the mineral water sources[21]. Otherwise, mineral water will continue to be treated like an alien substance with an alien model of management within the mining sector.

Nevertheless, keeping mineral water management in the mining sector does not sort out concerns about the compatibility of the mining legal regime for mineral water exploitation, because the mining alienability principle conflicts directly with the environmental dimension of groundwater.

8. CONCLUSIONS

For reasons that have little to do with hydrogeology, but much to do with currently disused traditional spa prac- tices, mineral water in Brazil is considered to be a mineral substance and even today it is treated differently to water resources. The CMW is a curious case in which the law did promote a change in the nature of a natural resource, namely, groundwater. The legal ascription of mineral water as a mineral substance should never ex- clude its natural condition as a groundwater resource. In fact, it is almost physically impossible to separate mi- neralized groundwater from "ordinary" groundwater because they are interconnected in an aquifer, taking part of a dynamic system composed of flowing water and rock. Furthermore, mineral water has now acquired a water resource status in its own right, taking into account how legal instruments for water management have evolved in Brazil in the last decade.

The cases of conflict reported in this article highlight the inversion of priorities stated by the Constitution, which was supposed to have the supremacy of public over private interests. In the Central Brazil conflict case, spa tourism is favoured in the end use of mineralized water for private interests rather than for the public water supply service. The regime of mineral water production under the mining legal framework sometimes denies some principles and guidelines of environmental and water resource legislations that should not happen today.

There will be inevitable conflicts of jurisdiction on mineralized groundwater while the CMW remains in force, because it is founded on outdated definitions and disused concepts regarding water intrinsic parameters and end uses. The CMW has ceased to be effective and barely reflects the current design of the mineral water market, since it has been drafted since the 1960s primarily for the production of bottled beverages rather than for medi- cinal purposes. Bottled mineral water today favours characteristics agreeable to consumer tastes (lightly minera- lized bottled water), with the oblivion or the principles originally stated by law (mineral water as a remedy). More than 80% of mineral water sources in Brazil are currently devoted to production of bottled water. For the consumers of bottled mineral water, the only thing that really matters is the quality of groundwater for drinking and not its presumable medicinal effects on humans.

There is an urgent need to reopen the discussion aimed at a possible revision of the CMW. This piece of leg- islation has been frozen in time and does not adequately take into account the deep changes in the way society has been dealing with water, in general, and with the environment, in particular. For the discussion process, some aspects should be taken into account, such as serious concerns about the compatibility of the CMW's prin- ciples with the environmental dimension of groundwater. The ascription of mineral water as a mineral substance belonging to the Union's assets and the adoption of the legal regime of mining are frankly in disagreement with the constitutional premises of groundwater under the States' jurisdiction and with the inalienability of water as a public good for common use.

ACKNOWLEDGEMENTS

Ana Lucia Desenzi Gesicki, Francesco Sindico (Brazilian Council of Technological and Scientific Development) for the post-doctoral scholarship provided during 2012-2013 period.

REFERENCES

1. Porto, M. and Kelman, J. (2000) Water Resources Policy in Brazil. Rivers—Studies in the Science Environmental Policy and Law of Instream Flow, 7, 250-259. http://www.kelman.com.br/pdf/Water_Resources_Policy_In_Brazil_2.pdf
2. Benjamin, A.H., Marques, C.L. and Tinker, C. (2005) The Water Giant Awakes: An Overview of Water Law in Brazil. Texas Law Review, 83, 2185-2244.
3. ANA, Brazilian National Water Agency (2007) GEO Brazil Tematic Series, Water Resources Component of a Series of Reports on the Status and Prospects for the Environment in Brazil. ANA/PNUMA,Brasília. http://arquivos.ana.gov.br/institucional/sge/CEDOC/Catalogo/2010/GEOBrasilResumoExecutivo_Ingles.pdf
4. Hirata, R., Zoby, J.L.G. and Oliveira, F.R. (2010) Groundwater: Strategic or Emergency Reserve. In: Bicudo, C.E., Tundisi, J.G. and Scheuenstuhl, M.C.B., Org., Waters of Brazil: Strategic Analysis, Instituto de Botânica, São Paulo, 149-161. (InPortuguese) http://www.ianas.org/books/aguas_do_brasil_Final_02_opt.pdf
5. ANA, Brazilian National Water Agency (2011) Permit System for Water Resources Use. ANA, Brasília. (Cadernos de Capacitação em Recursos Hídricos, Vol. 6) (In Portuguese) http://arquivos.ana.gov.br/institucional/sge/CEDOC/Catalogo/2012/OutorgaDeDireitoDeUsoDeRecursosHidricos.pdf
6. Drummond, J. and Barros-Platiau, A.F. (2006) Brazilian Environmental Laws and Policies, 1934-2002: A Critical Over- view. Law & Policy, 28, 83-108. http://dx.doi.org/10.1111/j.1467-9930.2005.00218.x
7. Daibert, A. (2009) Historical Views on Environment and Environmental Law in Brazil. The George Washington International Law Review, 40, 779-840.
8. McAllister, L.K. (2008) Revisiting a "Promising Institution": Public Law Litigation in the Civil Law World. Georgia StateUniversity Law Review, 24, 693-734.

9. Pompeu, C.T. (2010) Water Rights in Brazil. 2nd Edition, Editora Revista dos Tribunais, São Paulo. (In Portuguese)
10. Di Pietro, M.S.Z. (2010) Private Use of a Public Good by a Private Party. 2nd Rdition, Editora Atlas S/A, São Paulo. (In Portuguese)
11. Freire, W. (2010) Fundamentals of Mining Law. Jurídica Editora, Belo Horizonte.
12. Trindade, A.D.C. (2009) Mining Law Principles. In: Souza, M.M.G., Coord., Mining Law in Evolution, Editora Mandamentos, Belo Horizonte, 47-76. (In Portuguese)
13. Trindade, A.D.C. (2011) Perspectives for a Mining Law Reform. In: Martins, J., Lima, P.C.R., Queiroz Filho, A.P., Schüller, L.C. and Pontes, R.C.M., Coord., Mineral Sector: On Route to the New Legal Framework, Câmara dos Deputados, Brasília, Série Cadernos de Altos Estudos 8, 205-220. (In Portuguese) http://www2.camara.leg.br/a-camara/altosestudos/arquivos/setor-mineral-rumo-a-um-novo-marco-legal/setor-mineral-rumo-a-um-novo-marco-legal
14. Quintela, M.M. (2004) Thermal Knowledge and Practices: A Compared Perspective in Portugal (São Pedro do Sul Thermal Spa) and in Brazil (Caldas da Imperatriz Hot Springs). História, Ciências, Saúde, 11, 239-260. (In Portuguese) http://www.scielo.br/scielo.php?script=sci_pdf&pid=S0104-59702004000400012&lng=en&nrm=iso&tlng=pt
15. Serra, S.H. (2009) Mineral Waters of Brazil. Millennium Editora, Campinas. (In Portuguese)
16. Caetano, L.C. (2005) Mineral Water Policy: An Integrated Proposal for the Rio de Janeiro State. Ph.D. Thesis, State University of Campinas, Campinas. (In Portuguese)
17. Bertolo, R.A. (2006) Reflections on the Mineral Water Classification and Chemical Characteristics of Bottled Mineral Water in Brazil. Proceedings of the 14th Brazilian Groundwater Congress, Curitiba, 7-18 November. (Revista Águas Subterrâneas, Supplementary Issue) (In Portuguese) http://aguassubterraneas.abas. org/ asubterraneas/article/view/23114/15229
18. Nickel, E.H. (1995) The Definition of a Mineral. The Canadian Mineralogist, 33, 689-690.
19. Rebouças, A.C. (2006) Groudwaters. In: Rebouças, A.C., Braga, B. and Tundisi, J.G., Org., Freshwaters in Brazil, 3rd Edition, Editora Escrituras, São Paulo, 111-144. (In Portuguese)
20. Margat, J., Foster, S. and Droubi, A. (2006) Concept and Importance of Non-Renewable Resources. In: Foster, S. and Louks, D.P., Eds., Non-Renewable Groundwater Resources, A Guidebook on Socially-Sustainable Management for Water-Policy Makers, UNESCO, IHP-VI, Series on Groundwater 10, 13-24. (IHP/2006/GW-10) http://unesdoc.unesco.org/images/0014/001469/146997e.pdf
21. Foster, S., Nanni, M., Kemper, K., Garduño, H. and Tuinhof, A. (2005) Utilization of Non-Renewable Groundwater: A Socially-Sustainable Approach to Resource Management. The World Bank, Washington DC. (GWMTE Briefing Note Series, Note 11). http://documents.worldbank.org/curated/en/2003/01/5161421/utilization-non-renewable-groundwater-socially-sustainable-approach-resource-management
22. Custodio, E. (2002) Aquifer Overexploitation: What Does It Mean? Hydrogeology Journal, 10, 254-277. http://dx.doi.org/10.1007/s10040-002-0188-6
23. Llamas, M.R. and Custodio, E. (2003) Intensive Use of Groundwater: A New Situation Which Demands Proactive Action. In: Llamas, M.R. and Custodio, E., Ed., Intensive Use of Groundwater, Challenges and Opportunities, A. A. Balkema Publishers, Lisse, 13-34.
24. Caetano, L.C., Pereira, S.Y. and Dourado, F. (2012) Conflicts of Mineral Water Management in Brazil—Study Case: Rio de Janeiro State. Holos Environment, 12, 132-146. (In Portuguese) http://www.periodicos.rc.biblioteca.unesp.br/index.php/ holos/article/view/2080/4935
25. ANA, Brazilian National Water Agency (2012) Water Resources Conjuncture in Brazil. 2012 Report. ANA, Brasília. (In Portuguese) http://arquivos.ana.gov.br/imprensa/arquivos/Conjuntura2012.pdf
26. DNPM, National Department of Mineral Production (2012) 2012 Mineral Summary. DNPM, Brasília, Vol. 32, 3-4. (In Portuguese) https://sistemas.dnpm.gov.br/ publicacao/mostra_imagem.asp?IDBancoArquivoArquivo=7368

27. Andrade, A.M.A. and Almeida, L. (2012) Potenciometric Level Performance of the Thermal Caldas Novas Aquifer ? Goias State and Control Measures Imposed by the National Department of Mineral Production (DNPM). Águas Sub- terrâneas, 26, 99-112. (In Portuguese)
28. Guimarães, B.C. (2009) Environmental Collective Law and the (un)Sustainable Exploitation of Mineral Waters. Mandamentos, Belo Horizonte. (In Portuguese)
29. Scalon, M.G.B. (2011) Mineral Waters and Water Resources: A Perspective of Integrated Management. Revista de Direito, Estado e Recursos Minerais, 1, 131-160. (In Portuguese) http://seer.bce.unb.br/index.php/rdern/article/view/5173
30. Reis, A.M. (2005) Searching for an Integrated Management of Mineral Waters in Brazilian Legislation. In: Freitas, W.P., Ed., Environmental Law in Evolution, Vol. 4, Editora Juruá, Curitiba, 15-39. (In Portuguese)
31. Queiroz, E.T. (2004) Mineral Waters of Brazil: Distribution, Classification and Economic Importance. DNPM, Brasília. (In Portuguese) http://www.dnpm.gov.br/ mostra_arquivo.asp?IDBancoArquivoArquivo=377
32. Bertolo, R., Hirata, R. and Fernandes, A. (2007) Hydrogeochemistry of Bottled Mineral Waters in Brazil. Revista Brasileira de Geociências, 37, 515-529. (In Portuguese)
33. Lopes, R.S. (1956) Mineral Waters of Brazil, Value and Therapeutic Uses. Ministério da Agricultura, Rio de Janeiro. (In Portuguese)
34. Mourão, B.M. (1992) Hydrological Medicine—Modern Therapy of Mineral Waters and Healing Centres. Secretaria Municipal de Educação, Poços de Caldas. (In Portuguese)
35. Bender, T., Karagülle, Z., Bálint, G.P., Gutenbrunner, C., Bálint, P.V. and Sukenik, S. (2005) Hydrotherapy, Balneotherapy, and Spa Treatment in Pain Management. Rheumatology International, 25, 220-224. http://dx.doi.org/10.1007/s00296-004-0487-4
36. Nasermoaddeli, A. and Kagamimori, S. (2005) Balneotherapy in Medicine: A Review. Environmental Health and Preventive Medicine, 10, 171-179. http://dx.doi.org/10.1007/BF02897707
37. WHO, World Health Organization (2008) Guidelines for Drinking-Water Quality. http://www.who.int/entity/water_sanitation_health/dwq/GDW9rev1and2.pdf
38. Reale, M. (2001) Preliminary Lessons of Law. 25th Edition, Saraiva, São Paulo. (In Portuguese)
39. Reale, M. (2002) Law Philosophy. 19th Edition, Saraiva, São Paulo. (In Portuguese)

NOTES

[1]Porto and Kelman cite that for the law-making process, several workshops took place during the first half of the 1990s, involving a broad set of participants, such as politicians, water professionals, NGOs and local communities in order to reinforce the need for decentralization of water management and the need for counting on democratic participation by society [1]

[2]Article 26 of the 1988 Federal Constitution.

[3]Water permit systems vary in national range because States are allowed to promulgate their own legislation of water resource management, following the guidelines established in the National Policy for Water Resources. For this reason, there are States with long-term and well established water permit systems (e.g. São Paulo and Minas Gerais States) and others which

have just promulgated its own water legislation and still face difficulties with the enforcement of the law.

[4]The maximum period of validity of water rights is 35 years.

[5]Surface fresh water bodies (rivers and lakes) are under the States' domain except if they serve as boundaries with other countries or be- tween States, which are considered of federal jurisdiction.

[6]One of the most serious events of environmental damage occurred in Cubatão, an industrial city in São Paulo State, where uncontrolled pollution caused by 24 petrochemical and steel companies resulted in mountain landslides, contamination of soil, rivers and air, as well as grave health problems to local population. The companies were involved in a lawsuit filed under the Public Civil Action conducted by the State Public Ministry from 1985 and obliged to fully remediate the environment [9]

[7]The construction, installation, expansion and operation of any potentially hazardous activity are subject to a three-step licensing process (preliminary, installation and operation licences). The preliminary, installation and operation licences of a project are time-limited, not exceeding five years, six years and ten years, respectively. The environmental licensing process eventually involves the realisation of public meetings for divulgation of the project to the affected community.

[8]Article 99 of Brazilian Civil Code (Law 10.046 of 10/01/2002).

[9]By the time of drafting this article, 1967 Code of Mining was still in force. Nevertheless, the bill number 5.807/2013 has been in progress in the National Congress since 18/06/2013, which will update the cited law.

[10]Resolution 369/2006 of the National Environmental Council (CONAMA) reinforces the public utility and the social interest of activities of exploration and extraction of mineral substances, by permitting eventual intervention or removal of wood cover in Permanent Protected Areas defined in the Brazilian Forest Code.

[11]Groundwater classified as "oligomineral", "nitrated" and "chlorided" are kinds of mineral water that exceptionally must have its beneficial effect on health attested through empirical measurements carried out by specialized doctors, which results depending on approval of a board of experts in Crenotherapy. CMW, Art. 35, I together with Art. 1, §2 and §3.

[12]The concept of groundwater reservoir depletion is somewhat distinct from mineral ore deposit depletion. Depletion of groundwater storage does not mean the complete drainage of the aquifer because is rather unlikely to happen, since an important part of an aquifer is renewable and the stocks can be replenished by recharge. Nevertheless, the continuing abstraction of non-renewable aquifers represents, in fact, reduction in groundwater

reserves, triggering several negative side effects. According to Llamas & Custodio, the common undesirable side effects of intensive abstraction of groundwater are the drawdown of the hydraulic head locally and regionally, alteration in the river-aquifer relations (drying up of springs and reduction of river-base flow), alteration in the groundwater quality (normally degradation) and land collapse [23]

[13]Ordinance 470/1999 of the Ministry of Mines and Energy conflicts with Resolution RDC 259/2002 of the National Sanitary Surveillance Agency about labelling of bottled products. Likewise, Ordinance 374/2009 of the DNPM conflicts with Ordinance 326/1997 of the Ministry of Health regarding sanitary standards for bottling industries [24]

[14]See Note 10.

[15]DNPM Ordinance 52/1999 defined an area of restriction of 107.3 hectares in the Caldas Novas and Rio Quente regions in which 89 min- eral water wells have their yields monitored monthly, with a maximum total yield of 1800 m^3/hour, at a pumping rate of 14 hours/day.

[16]The drawdown tendency of the Araxá Thermal Aquifer System, in Caldas Novas region, can possibly be related to some interference be- tween the thermal water wells authorized by the DNPM and the municipal wells producing groundwater for the public water supply [28]

[17]According to Andrade & Almeida, the Caldas Novas' municipal water supply service agreed to replace the volume of groundwater ab- stracted from the Araxá Thermal Aquifer System for surface water catchment work carried out in Pirapitinga River. As the work was fin- ished at the end of 1996, the six municipal wells were restricted by the DNPM in 1997. Restrictions of the abstraction of groundwater were later released, from 2008 until the end of 2011 by a judicial sentence, using the justification that the use of surface water by the municipal water service did not adequately supply the local population [28].

[18]Although there was no consensus in the CNRH, Serra observes that some states actually provided legal means for the integration of mineral water claims into the SINGREH permit system (states of Bahia, Paraná, Pernambuco and Rio Grande do Norte). This occurs because States have legislative authority to enact administrative laws regarding environmental protection and water use permits, which vary from State to State [15]

[19]Bertolo et al. argue that groundwater with nitrate concentration above 3 mg/L has been somewhat impacted by surface derived anthropic contamination [32]

[20]According to Resolution 274/2005 of the National Health Surveillance Agency, the maximum amount of dissolved nitrate and arsenic in drinking water is, respectively, 50 mg/L and 0.01 mg/L.

[21]That would be the case of mineralized groundwater from fossil aquifers, for example.

CHAPTER 8

Collaborative Modelling for Informed Decision Making and Inclusive Water Development

Laura Basco-Carrera[1,2,3,4], Eelco van Beek[1,3], Andreja Jonoski[2], Camilo Benítez-Ávila[3] & FX PJ Guntoro[5]

[1] *Deltares, Boussinesqweg 1, P.O. Box 177, 2629 HVDelft, the Netherlands*

[2] *UNESCO-IHE Institute for Water Education, Westvest 7, 2601 DADelft, The Netherlands*

[3] *University of Twente, 7500 AE, Enschede, the Netherlands*

[4] *Present address: Unit InlandWater Systems, Department ofWater Resources and Delta Management,*

Deltares, Boussinesqweg 1, P.O. Box 177, 2629 Delft, HV, Netherlands

[5] *PT Virama Karya, Jl Durian Raya 70, Semarang, Central Java, Indonesia*

ABSTRACT

Ensuring availability and adequate management of water and sanitation for all is one of the key priorities for sustainable development. Integrated Water Resources Management is widely considered to be the best management process for securing water for all. However, implementing it remains a challenge for many decision makers and practitioners. In this article, we demonstrate an application of collaborative modelling to enhance and inform decisions and ensure an inclusive process. This collaborative modelling approach was applied to a river basin master plan for Pemali Comal River Basin Territory in Indonesia. Based on surveys of its development and the stakeholders and decision makers, collaborative modelling and traditional planning approaches for river basin planning were compared. The results of the evaluation demonstrate the benefits of collaborative modelling for structuring stakeholder participation in modelling processes, managing conflicts among competing water users and social learning.

Keywords: Water security, River basin planning, Collaborative modelling, Stakeholder, participation, Indonesia

1 INTRODUCTION

Water management is an important challenge that numerous countries must face in order to survive and improve their levels of economic development. Water is a key condition for generating rural livelihoods, growing food, producing energy, ensuring industrial and service sector growth, and guaranteeing the integrity of ecosystems and the goods and services they provide. However, the allocation of water resources can and likely will create disputes among water users. Moreover, population growth and its associated urban developments to accommodate it leads to increasing water demand and water pollution; major causes of the growing pressures on our water resources systems (UN-Water 2006).

Integrated Water Resources Management (IWRM) has been accepted worldwide as the best mechanism to deliver improved water management decisions. One of the central aims of IWRM is to promote increased coordination and integration across different water users and sectors as a means of achieving more holistic water management and improving water resource sustainability (GWP 2000; Muller and Lenton 2009). The Sustainable Development Goals (SDGs) consider implementing IWRM as one of the key targets to be achieved by 2030 (United Nations 2016). In practice this requires an analysis of the inter-connection between problems and a clear focus on the goals to be achieved (Biswas 2008; Serrat-Capdevila et al. 2011; Vennix 1999). Moreover, human well-being and socio-economic development cannot ensue by focusing only from within the 'water box'. It is crucial to understand that water is not the main factor; rather it is just one of the factors of interest (Medema et al. 2008). The concept of water security has emerged as an attempt to address these challenges (Van Beek and Arriens 2014).

Multiple definitions of the concept of water security exist (Cook and Bakker 2012; Grey and Sadoff 2007; GWP 2012; WEF Water Initiative 2010). In this article the definition of UN-Water (2013) is used: water security is the capacity of a population to safeguard sustainable access to adequate quantities of acceptable quality water for sustaining livelihoods, human well-being, and socio-economic development for ensuring protection against water-borne pollution and water-related disasters; and for preserving ecosystems in a climate of peace and political stability. In an attempt to quantify water security, the Asian Development Bank (ADB), in partnership with Asia-Pacific Water Forum (APWF), developed the Asian Water Development Outlook (AWDO) (ADB 2013). This framework provides a better understanding of the complexities and dimensions of water resources problems (i.e. household, urban, economic and environmental water security, and resilience to water-related disasters). The framework supports policy and decision making processes in terms of opportunities for investment, governance and capacity development.

Indonesia has a national water security index of 2 according to AWDO (2013), as it scores low in the key dimensions of household and urban water security and in resilience to water-related disasters. Despite significantly increasing the percentage of its population with access to safe drinking water from 70% (1990) to 85% (2012), only 20% of households in rural areas presently have access to piped water. In urban settings, such access only increases to 36%. Similarly, approximately half of the total population (54%) have access to sanitation. In terms of water quality, only 34% of the water is treated in waste water treatment plants in urban settings (ADB 2015). On Java, water security is exacerbated by the high risk of floods and droughts (ADB 2016; Deltares 2012).

Indonesia's policy and decision making processes for achieving water security are broadly based on an IWRM approach. This commenced in 1974 with the national river management water law (11/1974) and the establishment and formulation of River Basin Territories and master plans during the 1990s. However, these plans tended to follow a project-oriented development approach. In 2004, the legislation was renewed with Law No. 7/2004 on water resources and related governmental regulations (Peraturan Pemerintah), with the intention to shift to a more comprehensive and sustainable management approach. The integrated river basin master plan for Pemali Comal River Basin Territory was developed under this law. The new law aimed to protect, manage, rationalize usage, reduce waste and supervise the community utilization of water resources to guarantee water supply, quality and conservation. In line with the IWRM approach, the water resources law also recognized data and information, and stakeholder participation as supportive components for sustainable planning and implementation of water resources management policy (ADB 2016; Sukardi et al. 2013). However, in 2015, the Indonesian Constitutional Court revoked the 2004 water resources law due to disagreements surrounding the maintenance of clean water access for the whole population. The previous water law of 1974 was reinstated until a new legislation is adopted.

This study presents a collaborative modelling approach used in the preparation of the operational integrated river basin master plan in Pemali Comal River Basin Territory. This approach facilitates integration of data and information with decision making using modelling tools with stakeholder participation and negotiation for river basin planning. The approach was design considering the context and conditions of Pemali Comal River Basin Territory, specific use of the modelling tools, information handling, stakeholder involvement, modelling and organizing team, and means (Basco-Carrera et al. 2016; Henriksen et al. 2009). Secondly, the article presents an evaluation of the impacts of using collaborative modelling for river basin planning in comparison with traditional water resources planning approaches. To address this second objective, the Pemali Comal River Basin Territory was divided into two areas. In one area the collaborative modelling approach was applied for river basin planning, while in the other traditional planning methods were used. Both methods were then compared using two semi-structured questionnaires (i.e. pre-evaluation and post-evaluation forms) distributed to 29 stakeholder representatives.

The next section outlines the methods used to address both objectives. It provides a description of the study area and a comprehensive overview of the design and application of the collaborative modelling approach used for the preparation of the integrated river basin master plan in Pemali Comal River Basin Territory. Finally, the article reflects on the outcomes of using collaborative modelling in comparison with traditional planning approaches and provides lessons learnt for future applications of collaborative modelling in water resources management and planning and decision making processes.

2 METHODS

2.1 Study Area

Indonesia has been divided into River Basin Territories, which are areas with hydrological, and in some cases political boundaries, comprising one or more hydrological catchments (Peraturan Pemerintah RI 38/2011 Tentang Sungai) (ADB 2016). Pemali Juana covers two River Basin Territories: Pemali Comal and Jratunseluna. This study case focuses on Pemali Comal River Basin Territory. Pemali Comal is classified as cross-provincial River Basin and it is located in Central Java. It covers an area of about 4900 km^2 along the North Coast of Java and consists of 32 watersheds. The river discharge in Pemali Comal varies considerably from one river to the other. The lowest average annual discharge is found in the Waluh river (1m^3/s), and the highest in the Comal river (130 m^3/s). Average rainfall in the area is relatively high at 1700–5000 mm/year and there are considerable groundwater resources and springs in the region.

At administrative level, Pemali Comal is composed of 5 districts (Kabupaten) and 2 cities (Kota) (Fig. 1). Its population is approximately 6.5 million (BAPPENAS 2008), with Brebes and Pemalang the most densely populated districts. Agriculture is the major economic activity in the region. Around 50% of the area is used for highly productive food crops. Coastal areas are the preferred location for the construction of fish ponds, and the textile (e.g. batik) and ship-building industries (ADB 2016).

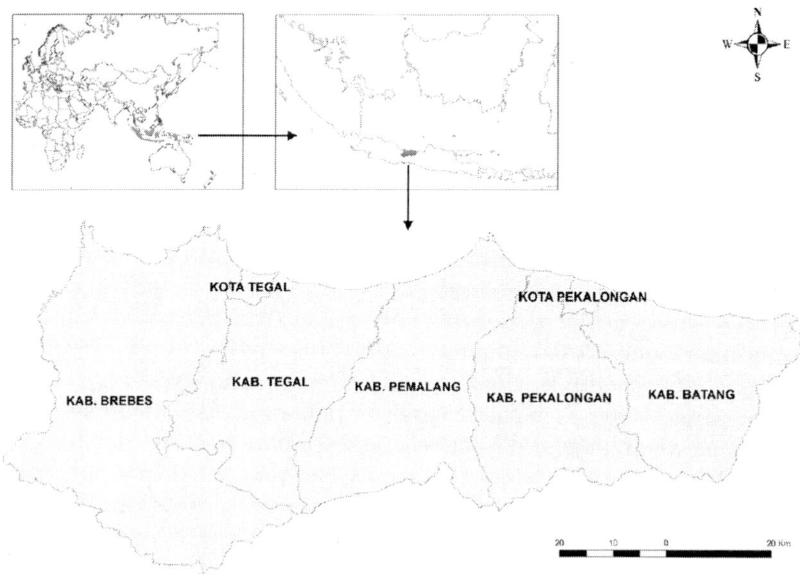

Figure. 1 The Pemali Comal River Basin Territory

2.2 Problem Context

Pemali Comal River Basin Territory has moderate vulnerability towards water security. Water stress is the main issue (ADB 2016; Deltares 2012), regardless of its relatively high rainfall. Water pollution (mainly from the high disposal of pollutants into the river by households and industries) and sea water intrusion have also become considerable issues in the coastal area. Moreover, the over-abstraction of groundwater for drinking water supply has caused significant ground subsidence (8–10 cm/year at the coast). As such, regional water supply authorities have been forced to use springs located upstream for supplying water to coastal areas. This directly affects the water available for irrigation, causing water stress during dry seasons. Conflicts between public water supply companies and farmer associations have arisen as a consequence of this policy and its impacts. The poor condition of secondary infrastructure, erosion and sedimentation (mainly caused by the mining industry) are other important issues in Pemali Comal. These have resulted in a considerable reduction of reservoir storage capacity and reduced flood protection (Wahyudi et al. 2012).

Limited knowledge about the water resources system and capacities aggravate this situation. Thus, conservation and land use, awareness, education, institutional and administrative settings are among the other related issues facing decision makers and stakeholders.

2.3 Collaborative Modelling Approach

The collaborative modelling approach was developed according to the complexity of the policy problem encountered in Pemali Comal, as well as project means (2 years duration, 450,000 Euros budget). Its high degree of knowledge uncertainty and levels of stakeholder disagreement meant that this case was considered to constitute an unstructured policy problem (Graaf and Hoppe 1996; Hommes 2008; Hurlbert and Gupta 2015). The developed approach was thus conceived as an interactive and adaptive planning process in which stakeholder participation was complemented by the use of computer-based models and communication tools (Basco-Carrera et al. 2016). It included high levels of participation for key stakeholders and reduced participation for other interested stakeholders. A specific goal of the collaborative modelling approach was to enhance cooperation between the stakeholders (Sadoff and Grey 2005), as this is a critical factor for the implementability and sustainability of integrated river basin master plans. The planning and decision making process was treated as an adaptive process in which decision makers and involved stakeholders could determine the way to proceed depending upon the situation, goals and outputs obtained.

2.3.1 River Basin Modelling Framework

The river basin modelling framework used for the preparation of the master plan was adapted from the analytical framework used by Deltares for water resources studies (Loucks et al. 2005).

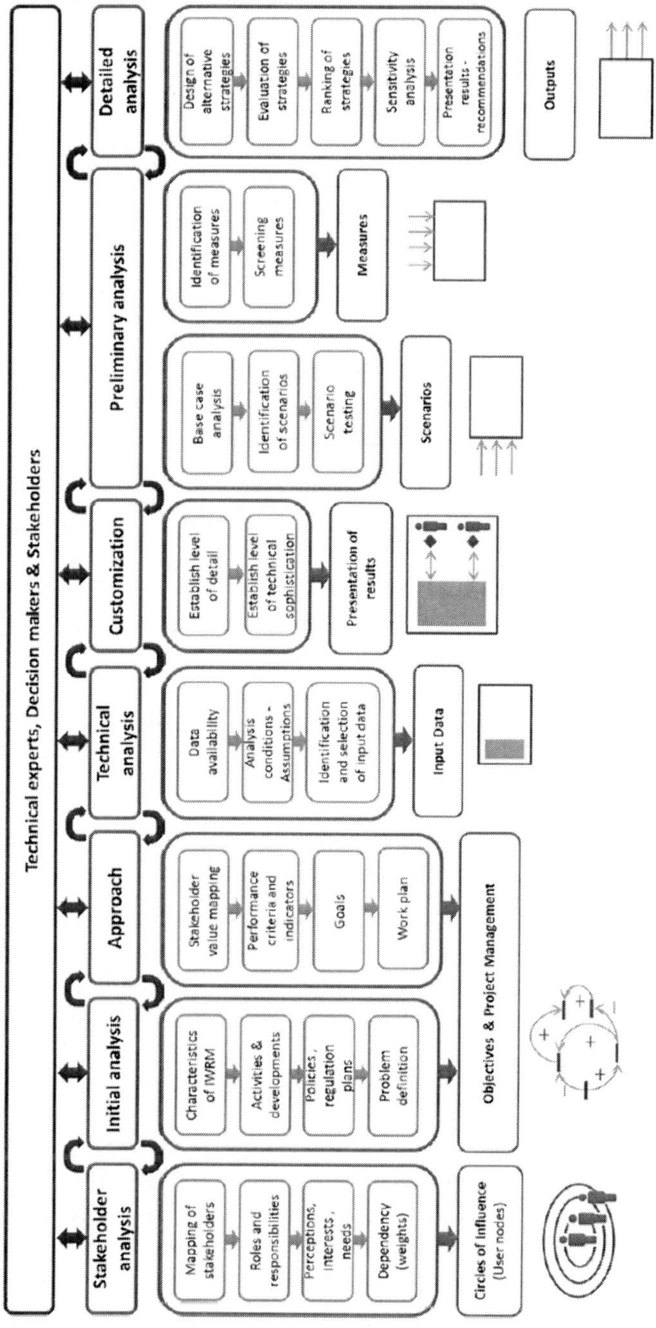

Figure. 2 River basin modelling framework

It consists of an iterative step-by-step framework for informed planning and decision making by making use of data and modelling tools. The river basin modelling framework included certain additional steps relevant for Collaborative Modelling. The framework was composed of seven main planning stages: Initial analysis, Stakeholder analysis, Approach, Technical analysis, Customization, Preliminary analysis, and Detailed analysis (Fig. 2). During each planning stage, a model and/or tool was developed or used jointly with stakeholders.

A system dynamics model was constructed to identify and analyse the main water-related issues in Pemali Comal. The Group Model Building approach was used for its construction (Vennix 1999; Videira et al. 2017). The model aimed at creating a collective, integrated overview of the current and future problems in the territory. The water balance and water allocation analyses for Pemali Comal River Basin Territory were conducted using a RIver BAsin SIMulation (RIBASIM) model. Its design started during the technical analysis phase (Fig. 2). The construction of the model finalized with the selection of scenario conditions and potential measures during the preliminary analysis phase. The identification of potential measures was carried out together with stakeholders using the system dynamics model. The process followed an adapted mediated modelling approach (Van den Belt 2004). The RIBASIM model was then used for analysing the impacts of selected measures and their possible combinations, and supported the selection of the preferred strategy that was then included in the integrated river basin master plan (Fig. 2; detailed analysis).

2.3.2 Participatory Engagement Structure

The participatory engagement structure followed an adapted Circles of Influence approach (Fig. 3) (Bourget (2011); Cardwell et al. 2008) and was defined based on the results of a stakeholder analysis (Grimble and Chan 1995). It included four circles: (i) Circle A: model development team, (ii) Circle B: model users and validation team, (iii) Circle C: consultation and information team, and (iv) Circle D: decision makers. Stakeholder engagement comprised three levels of participation: co-construction, discussion and consultation (Arnstein 1969; Bruns 2003; Mostert 2003).

An organizing and modelling team supported by a regional team was involved throughout the planning process. Members supported the stakeholder consultation process for the development of both models: the conceptual model using system dynamics and RIBASIM.

The entire planning process was coordinated by a process manager responsible for (i) managing the stakeholder processes, (ii) guiding the organizing and modelling team, (iii) aiding in all negotiations, consensus building and decision making processes, and (iv) acting as a focal point between decision makers and the modelling team.

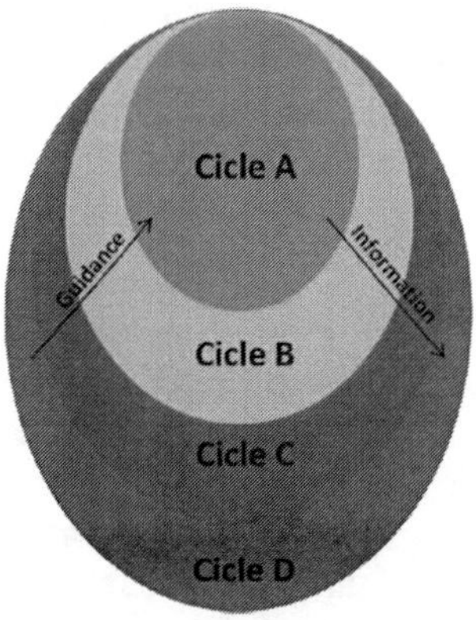

Circle	Involved stakeholders
A	1. Regional water resources management implementation centre 2. Public water supply authorities 3. Provincial office of water resources 4. Water users associations
B	1. Circle A 2. District water resources management authority 3. Provincial spatial planning authority 4. River basin water resources council
C	1. Non-governmental Organizations 2. Citizens organizations
D	1. Governor 2. Provincial water resources management authority 3. Provincial administrative water resources council

Figure. 3 Circles of Influence approach

2.3.3 Data and Simulation Model

The analysis of the performance of the water resources system in Pemali Comal was undertaken using Deltares' RIBASIM modelling package (van der Krogt and Boccalon 2013). This simulated the hydrological cycle under various hydrological conditions whilst giving consideration to existing water users in the basin. Simulations established water allocations according to their prioritization, and thereby supported impact assessments of possible measures to each simulated scenario.

Hydrological water inputs comprised rainfall data from 2003 to 2014. Data was obtained from the Tropical Rainfall Measuring Mission (TRMM). For the schematization of Pemali Comal river basin, variable inflow nodes were used to represent the hydrological water inputs. Two major water users were considered in the analysis: public water supply and irrigation water demands. The model also included existing, planned and proposed storage facilities and other infrastructural measures (e.g. reservoirs, weirs, etc.).

An adaptive planning approach was followed during modelling analyses to achieve flexibility in the allocation of water over time (Jeuken et al. 2015; Speed et al. 2013). Four scenario conditions (all in relation to spatial planning) were considered to analyse future system performance: population growth, economic development, climate change and institutional setting. Population was assumed to grow by 1.5% in urban areas and a reduction of 0.5% in rural areas. A moderate economic growth scenario (4.5%–6.5%) based on the GDP was agreed. Two scenarios conditions were considered for computing the effects of climate change in rainfall: ±3 mm/day (IPCC). Business as Usual and good governance were the two scenario conditions assumed for estimating the institutional setting in 2034. The good governance scenario assumed that decision makers and stakeholders would support integrated river basin master plan and follow its policies, and work jointly in the achievement of SDGs.

2.4 Collaborative Modelling Process

The collaborative modelling process was designed according to the river basin modelling framework. For the planning process Pemali Comal River Basin Territory was divided into two areas. In one area the collaborative modelling approach was applied (referred in this article as collaborative modelling area). In the other area, the traditional planning approach was followed (referred as conventional area). The collaborative modelling process comprised a public inception meeting, three collaborative modelling workshops and three public consultation events to discuss scope of the project, problem formulation, model design and construction, a set of rounds for model validation, water allocation and prioritization, model use for testing of measures and assessing their impacts, and finally the finalization of the plan. This process was combined with other participatory engagement methods: three consultation meetings and face-to-face interviews. In the conventional area, the stakeholder engagement process comprised a first meeting to present the project and three consultation meetings to present the results of the RIBASIM model, discuss possible measures and present the final integrated river basin master plan.

2.4.1 Inception Phase

The inception phase served to scope the project and collectively agree on specific goals and targets of the integrated river basin master plan. Collaborative modelling supported shared learning (Voinov and Bousquet 2010), consequently stakeholders understood that these could evolve over time and needed to be adaptive (Bousset et al. 2005; Medema et al. 2008; Mintzberg 1978). During this phase the stakeholder engagement structure and process, and the river basin modelling framework were also endorsed.

2.4.2 Problem Analysis, Model Construction and Validation

Information about system uncertainties and the dispute context was obtained through a set of individual consultation meetings. Two qualitative, concept models using system dynamics were constructed and validated jointly with stakeholders from Circle A to schematize systematic interactions and cause-effect relations between problems. The first model illustrated present water-related challenges. It showed cause-effect relationships regarding water withdrawal and water supply for agriculture and public water supply. The second model contained anticipated problems for the future horizon (2034). Outputs from the system dynamics model were used for constructing the schematization of the Pemali Comal water balance model using RIBASIM. Following this, data collection and model verification was performed by the modelling team in collaboration with stakeholders from Circle A.

2.4.3 Model Validation and Formulation of Measures

Due to persistent conflicts between water users in regards to the water withdrawals, uses, as well as cultural and hierarchical aspects (e.g. verbal dominance and freedom) (Akkermans and Vennix 1997) of the Indonesian policy and institutional setup, mediated modelling was applied during the model use phase. This commenced with the selection of the four scenario conditions (Section 2.3.3) with Circle A stakeholders followed by an interactive process of converting water resources issues into solutions using system dynamics. The RIBASIM model was run to quantify the cause-effect relations defined during the problem analysis phase. The complete concept system dynamics models used for problem analysis were transformed into incomplete models to determine suitable solutions to each specific problem. A total of 40 different measures were identified. They comprised infrastructural and soft measures, legal and institutional arrangements. These were then combined and evaluated based on the various scenario conditions using the RIBASIM model. After the prioritization of measures based on the outputs from the impact assessment, stakeholders and the modelling team selected the preferred strategy and incorporated it into the master plan. The master plan was then finalized and approved by the decision makers.

2.5 Evaluation Process

A comparative analysis was used to evaluate the impacts of using collaborative modelling for river basin planning (applied in the Eastern region) versus traditional water resources planning approaches (Western region). The collaborative modelling area comprised Pemalang, Pekalongan and Batang districts, and Pekalongan city. Similarly, Tegal and Brebes districts and Tegal city composed the conventional area (Fig. 1). Three mechanisms were used for data gathering: (i) project documents, (ii) semi-structured interviews with decision makers and local stakeholders, and (iii) semi-structured questionnaires with decision makers, local stakeholders, and the modelling and organizing team.

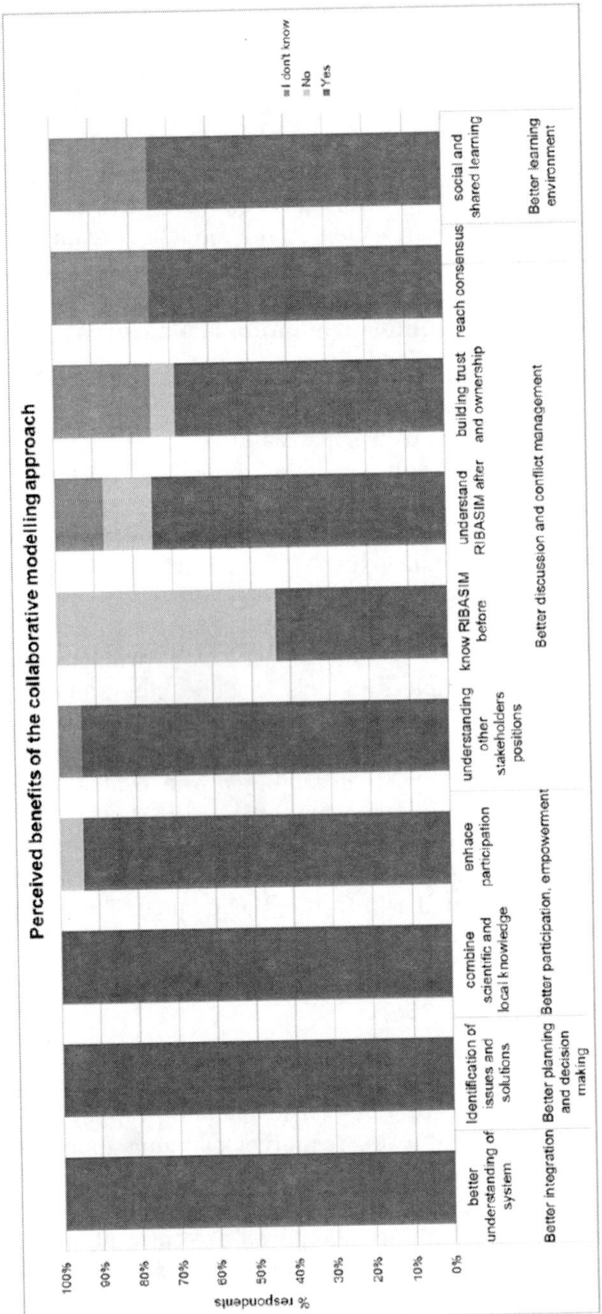

Figure. 4 Perceived benefits of the collaborative modelling approach

The semi-structured interviews were designed following the Protocol of Canberra (Jones et al. 2009). The outcomes served for adapting the collaborative modelling approach to the local context and conditions. Semi-structured questionnaires were used for evaluating the progress and experiences from participants during the modelling and planning processes. In particular, two semi-structured questionnaires (i.e. pre-evaluation and post-evaluation forms) were distributed to stakeholders from Circles A and B. Both forms were designed using some of the contingencies for Group Model Building interventions (Akkermans and Vennix 1997). In the collaborative modelling area, an additional evaluation of the collaborative modelling approach and outcomes was also carried out (Fig. 4). The questionnaires are available in Annex A.

3 RESULTS AND DISCUSSION

In this section, we present the results and stress practical lessons learnt from the collaborative modelling and traditional planning processes used for the preparation of the integrated river basin master plan for Pemali Comal River Basin Territory. Overall, the results of the three evaluation mechanisms suggest that the preparation of the master plan was effective, and it served to support informed and inclusive planning and decision making. Stakeholders and decision makers of the collaborative modelling area evaluated the general process and outcomes in a positive way (Fig. 4). Collaborative modelling has proved to be effective for structuring unstructured problems and the process has resulted in satisfactory outcomes.

3.1 Integrated Approach

An integrated river basin master plan needs to tackle the water-related issues in the basin in an integrated manner to secure water for all. The integrated approach is also conceived as a facilitating instrument to enhance group learning (Sterman 2001; Vennix 1996) and thus enhance cooperation among stakeholders (Bouwen and Taillieu 2004; Pahl-Wostl et al. 2007). The results of the comparative analysis corroborate these findings (Fig. 4). The collaborative modelling approach helped stakeholders gain an improved insight into the inter-linkages between issues. The results of the post-evaluation form in both areas show that more than 92% of respondents are aware of the inter-connections between problems. However, when asked to explain some of the (cause-effect) relations, stakeholders from the conventional area had difficulty in describing these. Less that 18% could actually describe more than two inter-connections. In the collaborative modelling area, stakeholders (even those stakeholders with less technical background) could explain not only the connections between the majorities of problems but also explain the problems faced by other stakeholders. Stakeholders and decision makers corroborated that the joint identification and analysis of future problems using system dynamics helped to create awareness of future collective problems.

3.2 Shared Learning: Better Understanding of the System and Evidence-Based Solutions

Engaging stakeholders in model development and technical analysis fostered shared learning (Kirono et al. 2014; Langsdale et al. 2013; Voinov and Bousquet 2010). 70% of the stakeholders and all the modelling team agreed that the collaborative modelling approach contributed towards shared learning. It helped decision makers and stakeholders having a good understanding of the functioning of the system and the impacts of possible interventions. Risks and costs could then be minimized. Collaborative modelling created a better learning environment for technicians to learn about the different systems by recognising the local knowledge and experience from the decision makers and stakeholders. Moreover, the joint and iterative construction of the RIBASIM model served to create awareness and consensus regarding the functioning of the different systems, water availability in the basin, water users, and present and future water-related problems. Moreover, construction a simple model and then improve it over time with input from stakeholders and experts (Langsdale et al. 2013) helped the process of getting familiar with system dynamics and RIBASIM.

Furthermore, collaborative modelling helps developing the technical capacity of stakeholders. Only 43,7% of the stakeholders had heard about RIBASIM before commencing the project (Fig. 4). Further, none of them were familiar with the application. This result is surprising as RIBASIM had been used for the preparation of the strategic integrated river basin master plan in Pemali Comal, prior to this project. This might imply that a pure traditional planning approach was used for its preparation. Most stakeholders had however a better understanding of the functioning of system dynamics and RIBASIM model thanks to the use of collaborative modelling. They felt confident to be able, with some technical support, to develop such models and use them for planning purposes.

The impacts of using collaborative modelling or a traditional planning approach are also reflected in the formulation of measures by decision makers and stakeholders. In line with the major problems identified, structural measures were most prominent in both regions. In the East,

O&M (40%), the use of Decision Support Systems (33,3%), regulations and licensing (40%) and awareness raising (13,3%) were most valued. By contrast, conservation and land use measures (41,7%), enhance cooperation (41,7%) and better water resources management (33,3%) were the most demanded measures in the West, where traditional methods were applied. This demonstrates that in the conventional area, stakeholders could perceive that the traditional planning approach did not enhance cooperation among stakeholders, so the disputes between stakeholders remained. Moreover, improving water resources management is a very generic solution. One could argue that decision makers and stakeholders did not have a good understanding of the physical, socio-economic and institutional system to find more detailed solutions. By contrast, in the collaborative modelling area, enhancing cooperation is not perceived as a major issue and the proposed solutions concerning the improvement of water resources management are much more concrete. Stakeholders and decision makers perceived more the benefits of soft measures and institutional mechanisms, rather than just grey infrastructure.

3.3 Enhanced use of Participatory and Modelling Tools

Collaborative modelling enhances the use of participatory, modelling and communication tools for inclusive and informed decision making. 93,3% of respondents in both areas considered the use of planning support tools as important for better planning and decision making processes (Geurts and Mayer 1996). Group discussions is the most demanded participatory technique, with 66,7% in the collaborative modelling area and 58,3% in the conventional area. The need of local knowledge from other stakeholders through group discussions and public consultation meetings was perceived as significantly important (46,7%). Not surprisingly, the use of Decision Support Systems and modelling tools was requested much more in the collaborative modelling area (60%) versus the conventional area (25%). Stakeholders in the collaborative modelling area provided also detailed information on the type of technical studies required (e.g. water quality, coastal zone management), scenarios and river basin simulations. These technical studies were not only requested by stakeholders with technical and non-technical backgrounds (e.g. water users associations). Both perceived the advantages of using computer-based models for obtaining a better understanding the system, the water-related issues and testing of possible measures (Kumar et al.; Loucks and da Costa 2013; Loucks et al. 2005).

3.4 More Efficient Modelling Process

Collaborative modelling accelerates data collection, model construction and validation. Consequently the construction of the model is carried out much faster than following a traditional modelling approach. The engagement of stakeholders in the early modelling phases not only boosted shared learning but it also facilitated data collection. Stakeholders gained a comprehensive understanding of the data required (quantity and quality) for the construction of the models. This resulted in data collection not becoming a "bottle neck" for the project. Gaining time in data collection implied more available time for model validation. This iterative modelling process concluded with a widely accepted version of the Pemali Comal water balance and allocation model.

3.5 Tackling Uncertainty to Build Trust and Ownership

Collaborative modelling helped in enhancing trust in the models, gaining acceptance for the model results and this built ownership of the models and tools, as well as for the measures proposed and the integrated river basin master plan for Pemali Comal. Uncertainty related to data accuracy and completeness was one of the key factors affecting trust (Voinov et al. 2016). Although the functioning of the model was explained in several occasions, for some stakeholders it was not sufficient to obtain a good understanding of the model. For others, the collaborative modelling approach allowed them to spend time comprehensively discussing the model and its results. The iterative process helped them building ownership. Building trust and ownership was thus directly related to the time spent in developing knowledge and learning.

3.6 Customization and use of Free-Available Modelling Tools is Critical

A best practice identified by Langsdale et al. (2013) is to select software that is easy to learn and can be made available to all. The use of system dynamics for problem identification and measures formulation supported this practice; however, we encountered difficulties in customizing RIBASIM due to its "rigid" user-interface. RIBASIM was developed as a decision making tool for technicians. Its configuration is not sufficiently adequate for engaging all stakeholder groups in its construction and use. Respondents considered that making RIBASIM more user-friendly considering the needs, skills and backgrounds of stakeholders and making it free-available would both increase and enhance its use in decision making processes.

3.7 Social Learning: Better Participation and Cooperation

The majority of stakeholders (70%) corroborated that collaborative modelling contributed towards social learning (Hurlbert and Gupta 2015; Mostert et al. 2007; Pahl-Wostl et al. 2007; Standa-Gunda et al. 2003; Voinov and Bousquet 2010) and had a significant effect in terms of consensus building. The comparison between the mental models and the quantitative model (i.e. RIBASIM) became a tipping point in creating a cooperative environment, as the information provided by the model was considered as being neutral during the discussions (Gaddis et al. 2010). Collaborative modelling facilitated that stakeholders and decision makers could learn from the concerns, needs and interests of other stakeholders. They could finally visualize and understand the problems faced by other stakeholder groups in regards to water withdrawal and supply. The process helped reaching the required level of consensus and trust for addressing the collective problems for the present and future situations and finding collective, sustainable solutions (Smajgl 2010).

3.8 Acceptance Requires of an Inclusive Well-Structured Stakeholder Engagement Process

Early and continuous engagement of stakeholders is of key importance in collaborative modelling (Korfmacher 2001; Voinov and Gaddis 2008). The selection of appropriate stakeholder group representatives to create an appropriate working group is often considered as trivial; however this study has demonstrated its relevance. Local NGOs participated in the study; however their involvement was restricted to Circle C. Not incorporating and considering sufficiently all forms of stakeholder knowledge and their concerns, especially from NGOs, resulted in the Indonesian Constitutional Court revoking the 2004 water resources law in 2015 due to disagreements arisen mainly by local NGOs on the maintenance of clean water access. The integrated river basin master plan can thus not be endorsed and implemented; instead it needs to be prepared again under a new legislation is adopted.

4 CONCLUSIONS

This article has demonstrated the benefits of using a collaborative modelling approach for the preparation of the integrated river basin master plan for Pemali Comal in comparison with traditional planning methodologies. Results from the study demonstrate the importance of the adaptive structure of a collaborative modelling approach.

In this case, the heterogeneity of stakeholder groups in terms of hierarchical diversity, verbal dominance, freedom, organizational practices, background, culture and existing conflicts required the adaptation of the approach from Group Model Building for model construction to mediated modelling for model use. The approach also shows the benefits of using system dynamics in combination with RIBASIM. Both analytical models were useful for characterizing and analysing the socio-physical system and their inter-relations. The use of the water security framework helped to structure and guide the formulation of potential measures, and led to the integrated planning approach required for sustainable IWRM. However, further analysis is necessary to transform the outputs from the IWRM planning process into specific inputs for evaluating water security in the region.

This study demonstrates that the collaborative modelling approach helped better planning and decision making. Collaborative modelling supported the process of problem identification and measures formulation. It ensured fruitful stakeholder participation during the overall planning process, particularly during the modelling phases. Collaborative learning and the neutral atmosphere created by the use of modelling tools helped change the mental models of stakeholders regarding the functioning of the water resources system and the problems faced by other stakeholders.

This was critical for transforming the initial disputes among competing water users into a more cooperative environment. All these factors enhanced stakeholder trust in the model and its results, and strengthened the sense of ownership of the integrated river basin master plan for Pemali Comal River Basin Territory. This case demonstrates the benefits of collaborative modelling for enhancing informed decision making, evidence-based solutions and ensuring inclusive planning for sustainable water resources development.

ACKNOWLEDGEMENTS

The funding and knowledge support for this research provided by Deltares and WISMP2-BWRMP project from the World Bank is thankfully acknowledged. We appreciate the support from Aart van Nes, Jos Dijkman, Daniel Tollenaar and Aris Setiawan in Indonesia and the guidance and contributions from Karen Meijer and Olivier Tuyishimire in the Netherlands. We are very thankful for the support of Andrew Warren as reviewer.

REFERENCES

1. ADB (2013) Asian water development outlook 2013. Measuring water security in Asia and the Pacific. Asian Development Bank, Mandaluyong City
2. ADB (2015) Making it happen: Technology, finance and statistics for sustainable development in Asia and the Pacific. Asia-Pacific Regional MDG Report 2014/15. Asian Development Bank (ADB), United Nations Economic and Social Commission for Asia and the Pacific (UN ESCAP) and the United Nations Development Programme (UNDP), Bangkok
3. ADB (2016) River Basin management planning in Indonesia: policy and practice. Asian Development Bank, Mandaluyong City
4. Akkermans HA, Vennix JA (1997) Clients' opinions on group model-building: an exploratory study. Syst Dyn Rev 13(1):3–31
5. Arnstein SR (1969) A ladder of citizen participation. J Am Inst Plann 35(4):216–224
6. BAPPENAS (2008) Rencana Pembangunan Jangka Panjang (RPJP) Nasional 2005-2015
7. Basco-Carrera L, Warren A, Van Beek E, Jonoski A, Giardino A (2016) Collaborative modelling or participatory modelling? A Framework for Water Resources Management. (In review Environmental Modelling & Software - copy on file with author)
8. Biswas AK (2008) Integrated water resources management: is it working? Water Resour Dev 24(1):5–22
9. Bourget L (ed) (2011) Converging waters: integrating collaborative modeling with participatory processes to make water resources decisions. Institute for Water Resources, U.S. Army Corps of Engineers
10. Bousset J-P, Macombe C, Taverne M (2005) Participatory methods, guidelines and good practice guidance to be applied throughout the project to enhance problem definition, co-learning, synthesis and dissemination. SEAMLESS
11. Bouwen R, Taillieu T (2004) Multi-party collaboration as social learning for interdependence: developing relational knowing for sustainable natural resource management. J Community Appl Soc Psychol 14(3):137–153
12. Bruns B (2003) Water tenure reform: developing an extended ladder of participation. Politics of the Commons: Articulating Development and Strengthening Local Practices, Chiang Mai
13. Cardwell H, Langsdale S, Stephenson K (2008) The shared vision planning primer: how to incorporate computer aided dispute resolution in water resources planning. Institute for Water Resources, Alexandria, IWR Report
14. Cook C, Bakker K (2012) Water security: debating an emerging paradigm. Glob Environ Chang 22(1):94–102
15. Deltares (2012) Water security atlas. Atlas Ketahanan Air. Java Water Resources Strategic Study. Deltares, DHV, MLD, Wiratman & Associates, Indonesia
16. Gaddis EJB, Falk HH, Ginger C, Voinov A (2010) Effectiveness of a participatory modeling effort to identify and advance community water resource goals in St. Albans, Vermont. Environ Model Softw 25(11):1428–1438
17. Geurts J, Mayer I (1996) Methods for participatory policy analysis: towards a conceptual model for research and development. WORC report 9612
18. Graaf H, Hoppe R (1996) Beleid en politiek. Een inleiding tot de beleidswetenschap en de beleidskunde
19. Grey D, Sadoff CW (2007) Sink or swim? Water security for growth and development. Water Policy 9(6):545–571
20. Grimble R, Chan MK (1995) Stakeholder analysis for natural resource management in developing countries, Natural resources forum Wiley Online Library pp 113–124
21. GWP (2000) Integrated water resources management. TAC Background Papers. Global Water Partnership Technical Advisory Committee 4
22. GWP (2012) Increasing water security - a development imperative, TAC Background Papers. Global Water Partnership (GWP)

23. Henriksen HJ, Refsgaard JC, Højberg AL, Ferrand N, Gijsbers P, Scholten H (2009) Harmonised principles for public participation in quality assurance of integrated water resources modelling. Water Resour Manag 23(12):2539–2554
24. Hommes S (2008) Conquering complexity: dealing with uncertainty and ambiguity in water management. University of Twente, The Netherlands
25. Hurlbert M, Gupta J (2015) The split ladder of participation: a diagnostic, strategic, and evaluation tool to assess when participation is necessary. Environ Sci Pol 50:100–113
26. Jeuken A, Haasnoot M, Reeder T, Ward P (2015) Lessons learnt from adaptation planning in four deltas and coastal cities. Journal of Water and Climate Change 6(4):711–728
27. Jones NA, Perez P, Measham TG, Kelly GJ, d'Aquino P, Daniell KA, Dray A, Ferrand N (2009) Evaluating participatory modeling: developing a framework for cross-case analysis. Environ Manag 44(6):1180–1195
28. Kirono DG, Larson S, Tjandraatmadja G, Leitch A, Neumann L, Maheepala S, Barkey R, Achmad A, Selintung M (2014) Adapting to climate change through urban water management: a participatory case study in Indonesia. Reg Environ Chang 14(1):355–367
29. Korfmacher KS (2001) The politics of participation in watershed modeling. Environ Manag 27(2):161–176
30. Kumar S, Godrej AN, Grizzard TJ (2015) A web-based environmental decision support system for legacy models. J Hydroinf 17(6):874–890
31. Langsdale S, Beall A, Bourget E, Hagen E, Kudlas S, Palmer R, Tate D, Werick W (2013) Collaborative modeling for decision support in water resources: principles and best practices. JAWRA Journal of the American Water Resources Association 49(3):629–638
32. Loucks DP, da Costa JR (2013) Decision support systems: water resources planning. Springer Science & Business Media
33. Loucks DP, Van Beek E, Stedinger JR, Dijkman JP, Villars MT (2005) Water resources systems planning and management: an introduction to methods, models and applications. UNESCO, Paris
34. Medema W, McIntosh BS, Jeffrey PJ (2008) From premise to practice: a critical assessment of integrated water resources management and adaptive management approaches in the water sector. Ecol Soc 13(2):29
35. Mintzberg H (1978) Beyond implementation: an analysis of the resistance to policy analysis. École des hautes études commerciales
36. Mostert E (2003) The challenge of public participation. Water Policy 5(2):179–197
37. Mostert E, Pahl-Wostl C, Rees Y, Searle B, Tàbara D, Tippett J (2007) Social learning in European river-basin management: barriers and fostering mechanisms from 10 river basins. Ecol Soc 12(1):2007
38. Muller M, Lenton R (2009) Integrated water resources Management in Practice. Earthscan, London
39. Pahl-Wostl C, Craps M, Dewulf A, Mostert E, Tabara D, Taillieu T (2007) Social learning and water resources management. Ecol Soc 12(2):2007
40. Sadoff CW, Grey D (2005) Cooperation on international rivers: a continuum for securing and sharing benefits. Water Int 30(4):420–427
41. Serrat-Capdevila A, Valdes JB, Gupta HV (2011) Decision support systems in water resources planning and management: stakeholder participation and the sustainable path to science-based decision making. Efficient decision support systems--practice and challenges from current to future. India, InTech
42. Smajgl A (2010) Challenging beliefs through multi-level participatory modelling in Indonesia. Environ Model Softw 25(11):1470–1476
43. Speed R, Li Y, Le Quesne T, Pegram G, Zhiwei Z (2013) Basin water allocation planning. Principles, procedures and approaches for basin allocation planning. UNESCO, Paris
44. Standa-Gunda W, Mutimukuru T, Nyirenda R, Prabhu R, Haggith M, Vanclay JK (2003) Participatory modelling to enhance social learning, collective action and mobilization among users of the Mafungautsi Forest, Zimbabwe. Small-Scale Forestry 2(2):313–326
45. Sterman JD (2001) System dynamics modeling: tools for learning in a complex world. Calif Manag Rev 43(4):8–25

46. Sukardi S, Warsito B, Kisworo H, Sukiyoto (2013) River Management in Indonesia. Directorate general of water resources, Yayasan Air Adhi Eka. International Cooperation Agency, Japan
47. United Nations (2016) Sustainable Delvelopment Goals www.un.org/sustainable development/
48. UN-Water (2006) Coping with water scarcity-a strategic issue and priority for system-wide action. UN-Water Thematic Initiatives
49. UN-Water (2013) Water security and the global water agenda: a UN-water analytical brief. UN University, Hamilton
50. Van Beek E, Arriens WL (2014) Water security: putting the concept into practice. GWP TEC Background Paper(20)
51. Van den Belt M (2004) Mediated modeling: a system dynamics approach to environmental consensus building. Island press, Washington
52. van der Krogt WNM, Boccalon A (2013) River Basin simulation model RIBASIM version 7.00. User manual. Deltares report 1000408-002
53. Vennix JA (1996) Group model building: facilitating team learning using system dynamics. Wiley, Chichester
54. Vennix JA (1999) Group model-building: tackling messy problems. Syst Dyn Rev 15(4):379–401
55. Videira N, Antunes P, Santos R (2017) Engaging stakeholders in environmental and sustainability decisions with participatory system dynamics modeling, environmental modeling with stakeholders. Springer, pp 241–265
56. Voinov A, Bousquet F (2010) Modelling with stakeholders. Environ Model Softw 25(11):1268–1281
57. Voinov A, Gaddis EJB (2008) Lessons for successful participatory watershed modeling: a perspective from modeling practitioners. Ecol Model 216(2):197–207
58. Voinov A, Kolagani N, McCall MK, Glynn PD, Kragt ME, Ostermann FO, Pierce SA, Ramu P (2016) Modelling with stakeholders–next generation. Environ Model Softw 77:196–220
59. Wahyudi SI, Ni'am MF, Le Bras G (2012) Problems, causes and handling analysis of tidal flood, erosion and sedimentation in northern coast of central Java: review and recommendation. International Journal of Civil & Environmental Enginering 12(04):65–69
60. WEF Water Initiative (2010) Water security: the water-food-energy-climate nexus. World Economic Forum (WEF). Island Press, Washington

CHAPTER 9

Challenges of Integrated Water Resources Management in Indonesia

Mohamad Ali Fulazzaky[1,2]

[1]*Institute of Environmental and Water Resources Management, Water Research Alliance, Universiti Teknologi Malaysia, Skudai, Johor Bahru 81310, Malaysia*

[2]*Directorate General of Water Resources, Indonesian Ministry of Public Works, Jalan Pattimura No. 20, Jakarta 12110, Indonesia*

ABSTRACT

The increased demands for water and land in Indonesia as a consequence of the population growth and economic development has reportedly have been accelerated from the year to year. The spatial and temporal variability of human induced hydrological changes in a river basin could affect quality and quantity of water. The challenge is that integrated water resources management (IWRM) should cope with complex issues of water in order to maximize the resultant economic and social welfare in an equitable manner, without compromising the sustainability of vital ecosystems. Even though the government of Indonesia has adopted new paradigm for water resources management by the enactment of Law No. 7/2004 on water resources, the implementation of IWRM may face the technical and managerial challenges. This paper briefly reviews the implementation of IWRM and related principles and provides an overview of potential water-related issues and progress towards implementation of IWRM in Indonesia. The availability of water and a broader range of water-related issues are identified. The recommended actions for improving the future IWRM are suggested. Challenges to improve the capacity buildings of IWRM related to enabling environment, institutional frameworks and management instruments are verified to contribute to the future directions for efficient problem-solving ability.

Keywords: capacity building; Indonesia; integrated water resources management; sustainable development; water resources management authority

1. INTRODUCTION

Population growth, urbanization, intensive agricultural development, industrial growth and environmental requirements are all increasing demands for water and land. The conversion of forest and agricultural lands to commercial and residential uses is leading to rapid transformation in agricultural production, spatial structure, social structure, land ownership and land market in the rural-urban fringe [1]. Many negative consequences of unbalanced environment such as water quality degradation, flood and drought, groundwater depletion, land subsidence, erosion and sedimentation as well as seawater intrusion are more likely to be generated from manmade objects. The negative impacts due to water pollution and its accompanied maladies are continuously jeopardizing the human health [2]. Integrated water resources management (IWRM) is now the dominant paradigm for water management in many developing countries, including Indonesia. The IWRM paradigm urges the government to involve stakeholders in each level of water management. The involvement of stakeholders to restore and maintain the natural stream flow regime of a river is a key part of building a sustainable water management initiative. The level of restored environment may also play a role in its future improvement [3].

The rationale for the IWRM approach has been accepted by many countries as the way forward for efficient, equitable and sustainable development and management of the limited water resources and for coping with conflicting demands. For example, the lessons learned from the French experience include the following: (1) IWRM at the level of river basins; (2) decentralization of water management; (3) participative management, involving all the stakeholders and the public; (4) application of the polluter-pays principle and the Water Agencies' role of economic mutualization; (5) local public responsibility for water supply and sanitation utilities; (6) experience feedback from the various methods used for managing water utilities, either managed by a public authority or by a delegated private company; and (7) transparency in the operation of services and information to the users [4]. The French experience illustrates the strengths of basin-level management and application of IWRM tools and concepts such as stakeholder consultation, economic instruments and appropriate regulation. A typical form of the participatory approaches should be effective in promoting the comprehensive plans for utilization, conservation, development and improvement of water resources. The integrated approach to water resources management advocates creating and empowering basin-level organizations (Water Agencies) to direct water management efforts in a hydrological boundary. In France, the six Water Agencies are public institutions of the Ministry for sustainable development and can have significant underlying financial requirements and challenges of coordinating among divergent water users. The private sector can play a role in providing water services and should be allowed to bring its technical expertise and efficient management practices into areas where central government has struggled to provide sustainable service. The appropriate level of decentralization depends on the nature of the specific water management problems in question, but the IWRM seeks to strike a balance between top-down and bottom-up management. Since the 1995 Mekong Agreement, the concept of IWRM to involve the participation of four lower Mekong countries (*i.e.*, Cambodia, Lao PDR, Thailand and Vietnam) has been promoted for Southeast Asia. The goal behind this agreement was to introduce a systematic process for sustainable development, allocation and monitoring of water resource use in the context of social, economic and environmental goals. Guided by this sustainable development paradigm, the Lower Mekong River Basin states attempt to balance the maintenance of water quantity with protection of water quality, and agree to cooperate and use the Mekong's water resources in a manner in which the river system's environmental conditions and ecological balance are conserved and maintained [5].

The Directive 2000/60/EC of the European Parliament and of the Council establishing a framework for the Community action in the field of water policy has been enacted since 2000 [6]. The implementations of IWRM according to such a water framework directive have been communicated for first stage towards sustainable water management in the European Union on March 2007 and then regarding programs for monitoring of water status in April 2009 [7]. Water law in the United States, which varies by region for both historical and geographic reasons, encompasses a broad array of subjects or categories designed to provide a framework to resolve disputes and policy issues relating to water [8]. In Indonesia, the implementation of IWRM needs for a participatory approach and is binding on a paradigm change, which obligates the government to reform the water resources policy (see Table 1).

Table 1. Changing paradigms in water resources management in Indonesia.

Elements	Old (1971–1998) and new (1998–present) paradigms and their implications in water resources management			
	Old Paradigms	**Implication**	**New Paradigms**	**Implication**
Governance regime	Autocracy	Strong leadership character	Democracy	Participative leadership
Administrative rule	Centralization	Concentration on national administration	Decentralization	Domination by district and provincial administrations
Decision-making process	Top down	All decisions made based on administrator interests	Bottom up	All decisions made based on stakeholders aspiration
Budgetary system	Government allocation	Financial sources from debts offering by international institutions funding and government budget	Water users sharing participation	Financial sources from water users and water polluters pay principles and government budget
Role of water in regional development	Supporting factor	Risks to environmental imbalance	Limiting factor	Priority on water balance consideration
Management responsibility	Regional authority	Regionally economic development interests	River basin and regional authority	Sustainable development interests
Role of government in construction	Provider	Risks of corruption due to power authority	Enabler	Capacity building needs for strengthening government administration
Project setting up orientation	Partial approach	Regionally (or personally) economic development interests basis	Integrated approach	Environmental sustainability interests basis
Project interest	Supremacy instruction	To increase personal popularity	People aspiration	To solve water-related problems
Project preparation	Based on short term planning	Approved based on personal/local interests	Based on master plan	Approved based on comprehensive study

The most important milestone in implementing the IWRM principles and processes occurred by the politics of administrative reform is the enactment of Law No. 7/2004 on water resources [9,10]. Efforts such as the regulations, guidelines as well as short- and long-term plans issued by local and central governments could have been attempted to implement the IWRM as prescribed by the law [11,12]. Still, the procedural integration where an agreed set of protocols is used for all the aspects of IWRM must be verified to try and make all the information accessible in a standard or known format [13]. Commitment of stakeholders, from grassroots to top decision making executives and political leaders, is important in order to ensure successful implementation of IWRM approaches [2,14]. Given such the problems of drought, flood, water pollution, lands subsidence, erosion and sedimentation are unavoidable to put a special scrutiny in the first place and to launch a national movement campaign particularly for developing a sense of crises to the entire community. Therefore, capacity building in IWRM accounting for enabling environment, institutional frameworks and management instruments should be given a profound attention in Indonesia and seems like a good reason to start strategic planning implementation to IWRM [10].

The objectives of this study are as follows: (1) to briefly overview the water availability and water-related issues in Indonesia; (2) to evaluate the progress in implementing the IWRM plans including the biases and problems being inherent inefficiency of government bureaucracy; and (3) to describe and propose the recommended actions for capacity building in IWRM related to enabling environment, institutional frameworks and management instruments.

2. WATER AVAILABILITY AND SEASONALITY OF INDONESIA'S CLIMATE

Despite the abundance of water resources, Indonesia has already experienced water shortages due to the variety of climates, geographical conditions and dry season. Total water demands to support population and economic growth for irrigation, domestic, municipal and industrial uses are currently at a flow rate of approximately 1074 $m^3 s^{-1}$ while the availability of surface water during low-flow conditions at normal climate of years is at a flow rate of approximately 790 $m^3 s^{-1}$. The pattern of variations in climate from the year to year may affect water availability to an unequal extent hence it might be predicted that the water use is already highly constrained by unbalanced conditions of demandsand the potential availability, particularly during dry season [15]. Competition for water during periods of low natural availability might be suggested that users may not have access to their full water supply requirements at all the time [16]. Growing the attention on the issues of economic and population growth development strategy is a result of the worsening environmental conditions in Indonesia today, due to the pace and pattern of growth in the past and the increasing awareness of the costs and risks of continued environmental degradation in the future [10].

3. WATER-RELATED ISSUES

3.1. Floods

In Indonesia, earlier flood management plans mainly focused on the improvement of infrastructure, to address inadequacies for flood prevention. As has been identified, Indonesia has about 5,590 main rivers; among them, 600 rivers have the worst impact on social and economic conditions [17]. To structurally mitigate flood hazard impact on these 600 rivers are required for about 30,000 km river training and 15,000 km dike. However, the capacity of the government in term of budget availability to develop the river trainings and dikes is about 300 km and 150 km, respectively; therefore, it needs to be approximately 100 years of flood hazard mitigation [10]. The remarkable fact is that the importance of these structures seems to have been addressing the only part of the problems, which makes the result ineffective, or they are too ambitious, where socioeconomic feasibility is usually not sufficient. Moreover, even though the flood risk impact factor has been introduced as an index to evaluate the impact of various sources of increased or reduced flood risk to society [18], if the plans are implemented they frequently do not deliver the results expected because of insufficient analysis of what causes the problems or because institutional capacity to support the measures is insufficient.

Flood events as shown in Figure 1 are continuing to increase from the year to year. Therefore, the government of Indonesia must take urgent actions in response to the causes of the flooding problems being encountered to recommend sustainable management solutions to them and needs to promote a more promising approach, which would be to concentrate on flood management. Floods and inundations cannot be prevented totally but preventive measures for minimizing loss and waste to be taken [19]. This implies having to live with flood risk because there is no reason how the engineers, the politicians and the public want to make sure flood never happens. Flood hazard mitigation efforts of non-structural measures are the important tools in preventing the potential damages. Whereas, structural mitigation includes physical measures and standards such as building codes, materials specifications and performance standards for construction must be provided only in the appropriate locations. Cost effective investments to reduce damages could make use of natural retention capacity [20]. By compensating for redirected flooding from the potential high damages of industry, settlement and society towards flooding of agricultural lands and natural wetlands can withstand a limited flooding without harm. The feasible investments to reduce floods should start with soil conservation in upper watershed for reducing natural or man-made bottlenecks in flood ways. In a non-structural approach, the resilience of community to live in harmony with flood is the most important component; therefore, it needs the provisions of such as early flood warning systems and flood risk mapping as well as flood-resistant building constructions [21].

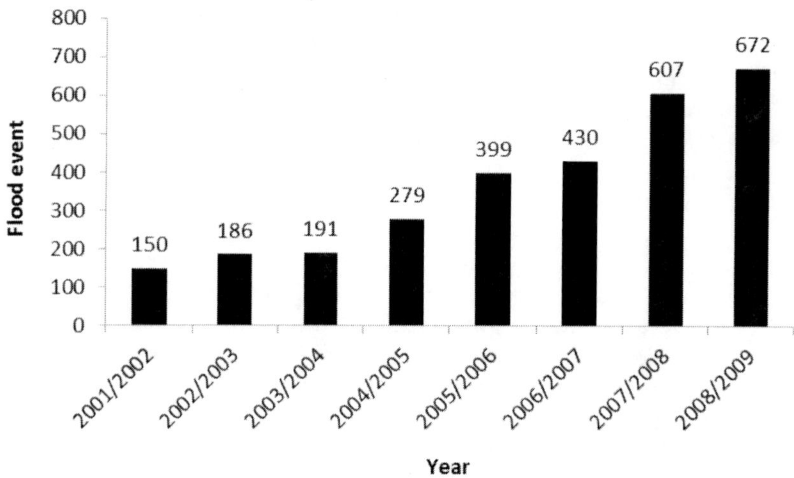

Figure 1. Flood events during the period of 2001–2009.

3.2. Droughts

Drought can change its characteristics through time [22] and mainly affects the second crop of irrigated and rain-fed rice fields, normally these rice fields have two rice cropping seasons, the first is a period from November to February and the second is that from March to June for each year [23]. Annually about a total of 0.250 million hectares, approximately 3% of the total planted rice fields, of the second crop of rice fields is affected by drought. An example of the drought impacts on society including anxiety or depression about economic losses that during the 10-

year period (1993–2003), among a total number of 2.453 million hectares of crops affected by the drought, of which 0.462 million hectares were totally lost. El-Nino phenomena are the main caused of drought in the years of 1994, 1997 and 2003, which caused a total number of 1.5 million hectares, approximately 60% of the total area affected by the 10 years of drought, of rice fields were affected. The average annual loss of rice production due to the drought was estimated to be about 300,000 tonnes or equal to IDR 550 billion or equivalent to USD 61 million [10]. Furthermore, annually these droughts have caused about 250,000 farming families that depend on rice production suffer from a decrease in rice farming.

3.3. Soil Erosion and Sludge Production

Dispose of sludge overdose into the receiving environments is an example of the negative impacts generated from misleading of such as IWRM implementation, forest land conversion and agricultural land occupation. Major sources of sludge in the country resulted from soil erosion and sedimentation are due to agricultural land occupation does not ensure that the soil conservation principles are quite well understood and implemented by farmers. Solid waste disposals and night soils as well as chemical and activated sludge produced from water and wastewater treatment plant (WWWTP) markedly contribute to an increase in quantity for sludge released into the environments [2,24,25,26]. Still the major sludge from soil erosion might eventually induce a huge accumulation of sediment in lakes and reservoirs hence the impact of decreased storage capacity may reduce water potential gradients in the stem [27,28,29]. For example, the upper Citarum river basin area in West Java with an average flow rate of 92.3 $m^3 s^{-1}$ brought about 1.05 million tonnes of suspended matter per year during the 11-year period (1981–1982). The figure has increased to about 1.47 million tonnes per year in 2004, or about 40% increase within a period of 20 years [2,30]. In another case, the garbage productions require a good practice of the solid waste management. For example, during the year 2002, three biggest cities (*i.e.*, Jakarta, Surabaya and Bandung) have produced respectively about 25.0, 9.0 and 6.5 thousands m^3garbage per day. But, the capabilities of Jakarta, Surabaya and Bandung to handle the garbage productions of about 96%, 77% and 70%, respectively, were still not quite enough [31]. Another potential source of sludge is associated with latrines or toilets referred as night-soil, having collected during the night in the past [32]. The problems with fecal sludge management is that the collected night-soil of such as for Bogor city is discharged untreated into the receiving environments due to excessive distance of disposal sites, traffic congestion and lack of suitable night-soil treatment options. Chemical and activated sludge as hyproducts of the WWWTP operation have reportedly have been increased from the year to year yet programming this matter as high priority considers, there is due to limited knowledge, technology and facility, as well as lack of the initiative to be implemented the regulation on economic instruments basis [2]. Sludge overdose may affect the discharge regime of rivers, which in turn reduces the inland water availability and worsens the quality of waters.

3.4. Lack of Adequate Water Storage

Although water resources in Indonesia are abundant, yet spatial and temporal variations in rainfall patterns and lack of adequate storage might create competition and conflicts amongst users [33]. The annual renewal water resources are estimated to be about 3085 km^3 of blue water while the estimated freshwater demands in 1990 were of the order of 50.5 km^3. In spite of the abundance in water resources estimated about 13,000 m^3 per capita per year, the water resources potential (WRP) per capita varies from the island to island. In the major parts of Java Island there are the available water less than 2000 m^3while for Papua it is greater than 282,000 m^3 per capita per year. The large variation in WRP poses challenges for the nation development and requires a sound planning and management system in water resources. The estimated surface water demand by the years of 1990, 2000, 2015, 2020 for major islands in Indonesia shows the increasing demand for all water users [10]. The figure shows that two islands *i.e.*, Java and Bali, which have a 62% country's population, could have the biggest water demands for domestic, municipality and industry. Groundwater potential in Indonesia would be a limited resource that could support only part of the urban and rural needs for water supply and provides irrigation water for very limited areas. In some regions of the country, the river basins exhibit tremendous variability in available water resources, which are inadequate to meet current and future demand more effectively. The status of surface water resources in the year 1990 showed that Java has faced water shortage during the dry season. The ongoing industrialization and urbanization have put further strain on the nation's water resources due to water quality degradation. As a consequence of the condition of scarcity, the government of Indonesia should place great emphasis on holistic water resources planning by channeling effort toward the mission driven program approaches to meet the challenges of the 21st century.

3.5. Watersheds Degradation

Many river basins have reportedly have been downgraded by illegal logging and intense land use change in these areas. Critical land areas and deforestation scattered across the country were noted to increase from approximately 13 million hectares in 1992 to 21 million hectares in 2001 and yield up to 45 million hectares in 2005 [2,10]. The population growth could affect the future changes in nature and intensity of economic activity; therefore, the issues of land and water use have become increasingly important due to the number of degraded watersheds, which have increased rapidly. The number of the critical river basin (CRB) indicating that high runoff coefficients are responsible for high flood return periods [34] were only 22 CRBs mostly in Java in 1984 and become 39 and 59 CRBs in the years of 1992 and 1998, respectively, and about 458 CRBs need to be rehabilitated since 2006 [10]. Economic and social impacts of desertification and land degradation might have caused the loss of certain spring waters [12]. The degradation of the Sengguruh Dam watershed, which covers a catchment area of 1659 km^2 in the upper Brantas River Basin, might be a good illustration of the critical land area caused by soil erosion (see Figure 2a). The Sengguruh Dam, with an initial storage of 21.5 million m^3, was built in 1988, as this construction of sediment traps on watercourse can benefit the protection of Sutami Dam in the downstream.

The Sutami Dam of the multifunction water control facility would be protected for ensuring the sustainability of electrical hydropower energy from a power generation of 29,000 kW and the important source of water supply to Surabaya city. By considering the average erosion rate of 0.58 mm per year, the design lifetime of Sengguruh Dam should be 20 years. Due to uncontrollable deforestation, land degradation and land-use change, as well as the effects of bed land-use and land-treatment for agricultural purposes, the Sengguruh Dam was completely full of sediment after 12 years (1988–2000). At a specific erosion rate of 869.8 m^3 per km^2 per year, about 1.44 million m^3 per year of the average annual volume of sediment was monitored during the 15-year period (1988–2003) [29].

3.6. Threats on Water Availability and Sustainability

Issues of water quality include surface water and groundwater contamination while issues of water quantity include the growing problem of increasing competition among water users and the declines in groundwater for urban areas of Java caused by withdrawal rates that are greater than recharge rates [35,36,37]. Even if in the aggregate, Java is well endowed with rainfall; the problem is one of seasonal and annual variations, with dry season flow in the main rivers only 20% of annual flows—and as little as 10% in a dry year. This is compounded by the fact that river basins in Java are relatively short, resulting in most of the wet season running unused water flows into the sea. A number of dams have been built in major river basins such as Citarum, Brantas, Bengawan Solo, Serayu-Bogowonto and others, and all the dams can hold a total capacity of less than approximately 5% of river flows. Several additional sites have been identified for possible future dams, but the implementation of accessibility to the built of them is likely to be constrained by high population densities and the social and economic costs of resettlement. At the current time, agricultural water use accounts for more than 90% of water use in the country, while industrial and municipal water requirements together account for less than 10%.

It is suggested that the consumption of water by households and business will grow rapidly over the next decades to support continued rapid population growth and improvements in human health and economic; however, the available water supplies will have met. Public water services currently provide at a range of 40%–60% of water supply in major cities of Java such as Jakarta, Bandung, Semarang and Surabaya while the remaining is met from groundwater extraction. The heavy reliance on groundwater to serve industrial and domestic needs for Java's large urban areas cannot continue indefinitely. This is particularly true for the northern coastal cities of Java where groundwater is being abstracted at greater than replenishment rates, leading to saltwater intrusion and land subsidence with attendant increases in floods and water logging, which in turn aggravates groundwater pollution from septic tanks and leaching pits [38].

3.7. Water Quality Problems

Unsafe water is one of the major sources of disease in Indonesia, and the lack of adequate sanitation facilities is a primary cause of fecal contamination of urban water supplies. An important indication of the threat this poses to the urban is the fact that water, with exception of bottled water, is not safe to drink. High levels of chemical contamination were detected in most well especially in the special capital region of Jakarta. Tap water samples in Jakarta in 1992 reveal a 73% rate of coliforms contamination, with a 55% rate in drinking water from wells.

The 2010 National Basic Health Research found that a high proportion of Indonesian households (32.5%) rely on low-quality drinking water. Surprisingly, approximately 83% of good-quality drinking water in urban areas, and 62% in rural areas, comes from commercial packaged water. This commercialization of drinking water adds to household spending and worsens conditions for the poor, who already spend two-thirds of their income on food [39].

(a)

(b)

Figure 2. Environmental conditions in the river watersheds; (**a**) critical land area in the upper part of the Brantas River Basin in East Java and (**b**) dirty part of the Citarum River in West Java.

A major reason for the poor quality of water supply in urban areas is that about two thirds of public water supplies are derived from increasingly polluted surface waters [39]. A 2008 water quality survey conducted by the Ministry of Environment found that the majority of main rivers in Indonesia would fall into the heavily polluted category, including iconic rivers such as the Musi River in Sumatera, Mahakam River in Kalimantan and Citarum River in Java (see Figure 2b). Since groundwater are already being over-extracted in many large coastal cities, surface water may be the only source of incremental supply for expanding public water systems. Although demand from municipal and industrial users will remain relatively small compared with the total resource base, the supply of surface water must have very high security and be used more effectively. Unfortunately, the lower reaches of most rivers are already polluted beyond the capacity of existing treatment plants.

A technically feasible option would be to upgrade the plants to handle more polluted inflows, but this would be an expensive solution and would only address the problem until pollution levels again exceed treatment capacity. The only sustainable solution, therefore, is to clean up pollution at the sources *i.e.*, to address the growing challenge of urban and industrial pollution [10].

4. A BRIEF DESCRIPTION OF THE MAJOR TASKS INVOLVED IN THE IMPLEMENTATION OF IWRM

4.1. Coordinating Perspectives in Water Management

The government of Indonesia has made special efforts on IWRM since the enactment of Law No. 7/2004 on water resources [11]. Guided by this law, the decision makers, managers and operators will always strive to do the right things to implement the strategies, programs and activities to support water management. The law and policies governing the implementation of IWRM are the legal provision for managing water resources in the river basin perspective but have not yet been synchronized effectively. Each ministry in pursuit of its mission has a mandate with responsibility for promoting social, economic, environmental and cultural well-being of communities in the present and for the future. Challenging the IWRM perspective may be the pilot projects need to be done together in order to solve a problem. This perspective seeks inter-ministerial cooperation and coordination to expedite clearances for water management activities. The goal of involvement different ministries is to help identify and implement mechanisms that encourage members to find ways to become involved in analyzing the complexity of implementation projects. Multi-criteria decision analysis is an umbrella approach that has been applied to a wide range of water management situations and helps to frame a group process that made stakeholder preferences explicit and substantive discussions about long-term river management possible [40,41]. The need for coordinated data collection has been proposed as salient features of a coordinated observation program for water resources management [42]. This paper reviews a novel purpose-oriented access control model which takes into account the purpose for which all government-related institutions involved in inter-ministerial cooperation access the IWRM perspective by operations of the pilot projects in accordance with their mandate tasks. In order to provide more detailed guidance for each water-related institution under the administration and supervision of a ministry, the sectoral ministry has developed a set of operational guidelines that are appropriate to the activities of that sector.

The operational guidelines may be reviewed and updated from time to time. The accuracy of factual information can help judge the credibility of the ministry in implementing the IWRM concepts. Accuracy of information can provide clues to possible bias in the implementation of IWRM for all institutions involved in the pilot projects.

4.2. The IWRM Pilot Projects and Reasons behind Program Delays

Strengthening inter-ministerial coordination mechanisms for the IWRM implementation needs financial sustainability, such as promotion of cost recovery, and must consider long-term management. At a ministerial meeting, the decision as to the content of the IWRM cooperation was made to select four projects to be implemented at four river basins in West Java [2,12], such that: (1) the project for capacity development of Jakarta comprehensive flood management has as its target area the upper Ciliwung River Basin in order to reduce rate of surface runoff and soil erosion. Among the initiatives and measures to be supported are the spatial planning for special region of Bogor-Puncak-Cianjur, improved knowledge of river basin environment and development of water resources management structures; (2) the pilot pollution prevention project's goal is to prevent water pollution for the Citarum River from industrial and domestic wastewater, because three large reservoirs have been built on the Citarum River over the last five decades *i.e.*, Jatiluhur (1967; 8300 Ha), Saguling (1985; 5607 Ha) and Cirata (1987; 6210 Ha). Operation principles of these three multipurpose reservoirs are to keep a stable water supply for the uses of hydropower, industry, irrigation, aquaculture, leisure, aquatic sport, domestic and municipality; (3) dynamic optimization of water flooding using optimal control of water release from reservoir has a significant potential to increase ultimate goal of water distribution service.

The maximum volume of water likely needs to be stored in the Jatigede Reservoir; therefore, reducing erosion and runoff of the upper Cimanuk River Basin is the important project to increase efficiency and extend the service life of Jatigede Dam; and (4) an overlooked effect of river withdrawal from sediment laden the Sagara Anakan Lagoon is the long-term aggradation and degradation [43] that can occur in the Citanduy River Basin. The life of the Sagara Anakan Lagoon may be extended through the project for comprehensive erosion and sediment control management practices by reducing the upstream sediment loads. The fact is that quasi-totality of the proposed projects today still has not been realized: Many program delays, cancellations and stretchouts have combined through the years to reduce effective collaborative IWRM projects. In all cases, a delay is usually a costly situation; random interviews with various types of administrative staff of water resources organizations and non-governmental organizations (NGOs) revealed preference arguments that consolidation of contingency variables, such as the institutional role, personal character of high level managers, manager's moral worst, unreliability of subordinates as well as the government bureaucracy and budgetary process are common risk factors that enable and hinder the implementation of the projects.

The economic analysis of an IWRM project helps select and design projects. The analysis should compare costs with benefits and determines which among alternative projects have an acceptable return. The costs and benefits such as for the Citarum River Basin (CitRB) projects can be identified, evidence that the Integrated Citarum Water Resources Management Investment Program funds a range of interventions across the water sector that relates to water and land management necessary to pursue the introduction of IWRM in the CitRB. A road map has been developed setting out the agreed interventions that address the IWRM key issues under the Investment Program (IP).

The estimated cost of implementing the projects of the road map could be about USD 3.5 billion over 15 years, of which the IP accounts for USD 921 million. The expected impacts of the IP are to reduce poverty and to improve health and living standards in the CitRB, and the expected outcome is to improve IWRM in the CitRB [44]. With the present condition in mind, this paper proposes one suggested implementation that integrating the issues of watershed and water resources into one management system might encourage active learning, which is widely seen as an appropriate way for capacity building of IWRM. Thus, one of the important considerations for the future IWRM in Indonesia is to improve capability of the government-related institutions, which act within the capacity building for enabling environment, institutional frameworks and management instruments.

4.3. Coordination Instruments and Water Coordinating Agency

In 2003, the government of Indonesia under coordination of Coordinating Ministry of Social Affairs has launched the National Movement for Forest and Land Rehabilitation (GERHAN). The GERHAN program focused on the efforts to decrease frequency and severity of droughts and floods and to reduce erosion and sedimentation. Even the achievement GERHAN program was not yet enough to meet the targets, about 3 million hectares of critical land have reportedly tried to rehabilitate during the 5-year period (2003–2008) [45].

The significance of this program has restored in several parts of the CRBs to reduce runoff coefficient after it was downgraded by illegal logging and intense land use. In 2005, President of the Republic of Indonesia has launched the National Movement for Water Resources Management Partnership (GNKPA). The award caps a yearlong debate program based on the value of courage and dedicated to strengthening institutional capacity building and coordination among the government agencies.

The GNKPA focuses to integrate the programs and actions related to water resources, agricultural lands and forestry in order to realize sustainable development in the country [10].

The recognition of an obligation to obey the Law No. 7/2004 must as stakeholders' representative body to establish the Water Council at national level. In 2009, a presidential decree has been issued to establish the National Water Resources Council (DSDAN), where picture of representative is the concept of equal opportunity by fifty-fifty percent chance composition members of government agencies and NGOs.

The DSDAN plays an important role in effective implementation of the IWRM programs to discuss the water-related issues during the council meeting and can influence the decision-making process on the national policy and strategy. Active participations in public policy-making processes that attempt to influence the implementation of regulatory mechanisms address to reduce the environmental impacts of implementing the IWRM at the river basin level [46].

A long-term perspective needs to be reinforced in decision making and planning to encourage research and development both for technology and organizational structure to be better able to meet the challenges of sustainability in IWRM [47].

4.4. Constraints and Weaknesses

Because the implementation of IWRM in Indonesia remains theoretical and prescriptive [48], some limitations play out in practice, such that: (1) regulatory functions and service provision functions of water resources may still intermingle; (2) law enforcement is not functional as well the frameworks for water management have not yet been legalized; (3) most human resources in the water resources sector, especially those in the fields are not professionals; (4) conflicts of interest among stakeholders, sectors, administration authority and geographical area (*i.e.*, downstream *versus* upstream); (5) strong sectoral and local ego in water resources and other sector related to water; and (6) process of formulation of framework for water management is considered by some participants as lacking adequate participation and inputs from stakeholders [10].

As a consequence, even though three large multipurpose reservoirs *i.e.*, Cirata, Saguling and Jatiluhur have been built on the Citarum River in West Java for the use of water for hydropower, recreational purpose and supply water for domestic use, industrial use, livestock watering, irrigation, fisheries and other beneficial uses, the river is suggested as one of the most polluted river in the world due to effort of the governments is still ineffective in reducing pollution loads from domesticeffluents, industrial wastewaters and solid wastes.

The Law No. 7/2004 meets the most important elements of two aspects of IWRM *i.e.*, enabling environment and institutional roles. However, management instruments, especially with respect to resource assessment, economic instruments and demand management need to be broadened. The implementing regulations may take care of some of missing IWRM elements. It is suggested that almost all stakeholders (governments, NGOs and academics) should involve in formulating the law and its related regulations [49]. As part of the democratization of public policy making, the draft bill and its regulations need to be widely discussed at public consultation meetings held at different locations throughout the major islands of Indonesia.

By eliciting opinions from the widest possible range of stakeholders in an organized and professional way, this process could significantly shorten what in the Indonesia case was a repetitive and lengthy series of discussions and a long lag time between drafting and passing the bill. In addition, there is a need to build awareness of the principles of IWRM among decision-makers and stakeholders, including NGOs to ensure that all parties are arguing from a common base.

5. CHALLENGES OF DEVELOPING THE FUTURE IWRM

5.1. Enabling Environment

It can be argued that IWRM might be considered as the means to resolve complex environmental problems caused by anthropogenic sources. The underlying causes of deforestation, critical watershed and land mismanagement are those factors which may create an environment where further imbalance between water supply and demand, water pollution, flood and drought as well as erosion and sedimentation can occur. Moreover, a number of different water-related problems caused by human activities, such as land subsidence due to groundwater withdrawal, sludge production (*i.e.*, biological and chemical sludge), solid waste and seawater intrusion, may widely affect all aspects of society, yielding some of the most systemic and most complicated feedbacks that frustrate human progress and sustainable development [50]. The fact that IWRM is a process which promotes the coordinated development and management of water, land and related resources in order to maximize the resultant economic and social welfare in an equitable manner without compromising the sustainability of vital ecosystems. Challenges of reforming enabling environment for water resources management in Indonesia can be categorized according to the eleven pillars (see Figure 3) of critical managerial elements that must be highly scrutinized by local and central governments in order to develop best practice. These pillars are grouped into two management approaches, such that: (1) an integrated approach consisting of watershed and forestry, agricultural land, flood and drought, water quality and pollution control, soil erosion and sedimentation as well as seawater intrusion should be put under authority of river basin agency supervision; and (2) a partial approach consisting of water and wastewater treatment, biological and chemical sludge, solid waste, groundwater withdrawal and land subsidence as well as night-soil should be put under responsibility of district/municipal authority. All of these pillars must be implemented by the local and central governments as well as by the private institutions and NGOs. The first step to starting IWRM is developing a river basin integrated water resources management plan (RBIWRMP) and then enacting it into official plan as the legal document. This document might guide all stakeholders corresponding program activities to the whole-of-government framework makes it possible to meet the IWRM targets.

Indonesia has undertaken a step forward the implementation of IWRM approach that is the concept of such a comprehensive approach to improve water resources management towards achieving sustainable prosperity. Since 1969, the water infrastructures have been continuously developed due to the needs to build, operate and maintain water infrastructures are usually driven by the necessity to meet the water needs for irrigations and the water supplies for domestic, municipal, industrial and other purposes from the year to year. The government of Indonesia under coordination of the appropriate regulatory agencies has issued the regulations on water supply system, irrigation, water resources management, ground water, dam, water council, water quality management and pollution control as well as water resources corporatization and financing necessary to implement the IWRM approach in accordance with the Law No. 7/2004. Though examples of attempts at government regulation are widespread, there are four stands out from the rest, *i.e.*, river and lake, swamp, watershed management as well as water use and rights.

5.2. Institutional Frameworks

Indonesia is a unitary state, which is divided into three levels of executive-administrative authority of government *i.e.*, national (central), provincial and city/district. The Law No. 7/2004 stipulates that basin water resources should be managed in an integrated and sustainable way. With all the existing problems and constraints, new institutional arrangement must consider ways of trading-off between the river basin management authority and the public administration authority. However, in many cases the programs do not synchronize with each other. The RBIWRMP should provide the acquisition planning guidance and procedures for government, river basin manager and other personnel assigned to participate in water resources management. Acquisition planning means the process by which the efforts of all personnel responsible for an acquisition are coordinated and integrated through a comprehensive plan for fulfilling the IWRM need in a timely manner and at a reasonable cost. In Indonesia, a river basin territory called "Wilayah Sungai (WS)" bordering on outer boundary of drainage basin of closed rivers might facilitate efficient water resources management. According to the ministerial regulation No. 11a/PRT/M/2006 issued by Minister of Public Works, the country is divided into 133 WS. Of these, 13 WS which are fully located in a district are termed as district WS, 51 WS which are fully located in a province are termed as provincial WS. For the remaining 69 WS, 5 WS are located crossing boundary of two countries, 27 WS in two or more provinces and 37 WS are of strategic importance are termed as national WS to be under the management authority of central government [10]. The water council at the provincial level, called "Provincial Water Council (PWC)", has been established in 13 provinces and that at river basin level called "River Basin Water Council (RBWC)" has been established for 12 WS, as required by the law. Nevertheless, there are still significant gaps in managing of water resources for rural areas of the country and analysis of matter on both water and land management necessary to pursue the introduction of IWRM. Many water councils need to be established in the future (*i.e.*, 20 PWCs at 20 other provinces and 121 RBWCs at 121 other WS). The proposed scheme to ensure the IWRM practices is zero delta policy that must maintain status of the environment before and after the development of regional land-use plans should be in the same conditions. Therefore, three basic policies must be to set up by the district, provincial and central governments, such that: (1) zero ΔQ policy to maintain run-offs; (2) zero ΔS policy to preserve erosion rate; and (3) zero ΔP policy to safeguard the pollutant loads resulting from human activities [10]. The essential elements of enforcement to implement these policies effectively are to involve the related government agencies, private institutions and other stakeholders particularly for high intensity land-use of river basin.

In implementing Law No. 7/2004, the central government has long debated the introduction of water pays for water principles in the new regulations [11]. This imposes self-financing scheme in the future IWRM perspective, specifying that water management must generate economic and financial resources to carry out its inherent tasks. The implementation of prevention, user and polluter paid principles is a recommended way of doing deep financial development, as all details for the implementation of framework legislation would otherwise need to be adopted under the lengthy ordinary legislative procedure [51]. The principle of prevention paid focuses on preventing unwanted impacts regardless of who should pay. It is closely related to precautionary principle which aims at risk reduction and pollution prevention, where investment in central wastewater treatment facility is an example. The principle of user pays focuses on overuse of water resources. For example, it can be controlled by water user fee. The principle of polluter pays emphasizes on water pollution. By adopting the environmental fee can contribute to the reduction of pollutant load discharged into the receiving waters [4,52].

5.3. Management Instruments

Water flows from the upstream to downstream areas of a river basin adhere the interests of multiparty and regions. To facilitate the implementation of IWRM needs to have a legal document of RBIWRMP, and all stakeholders should be made aware of their river basin planning framework. It is necessary to involve stakeholders at all stages in the IWRM process because the participatory approach to develop a framework will enable local authorities to implement the RBIWRMP to achieve sustainable development goals. Challenges to IWRM and good water governance should allow all stakeholders to participate directly in all aspects of policy making.

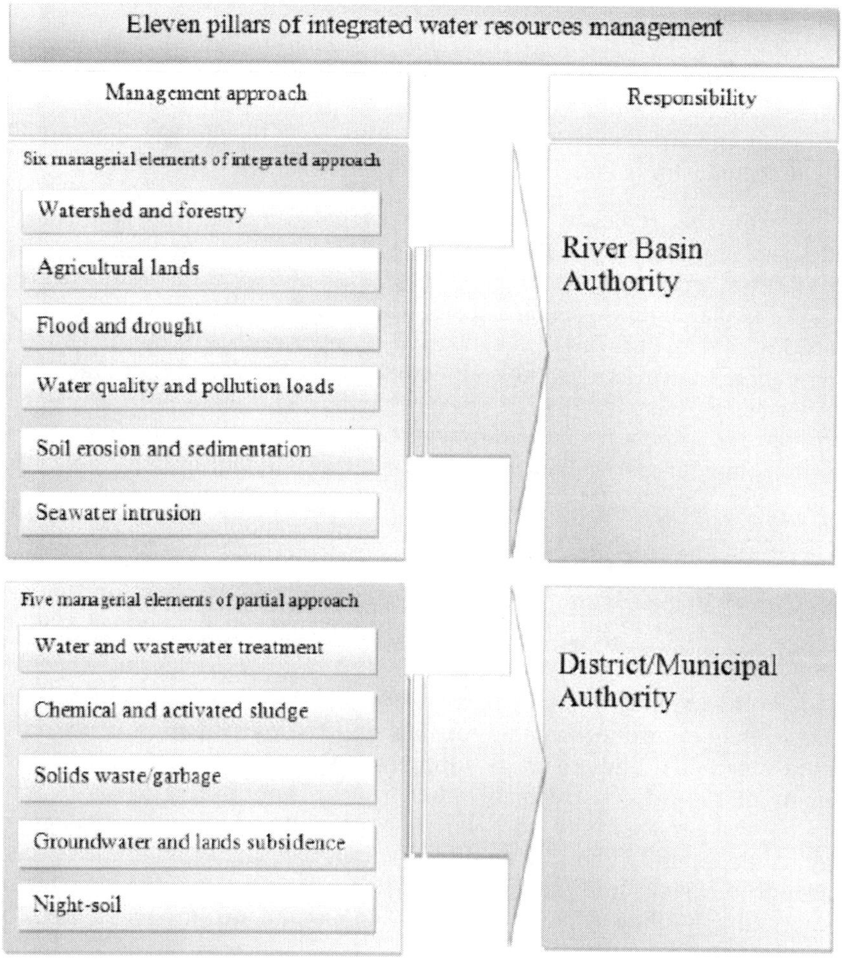

Figure 3. Elements of the future integrated water resources management.

Significant connections exist between the elements of social capital and the perceptions of stakeholders towards water allocation policies [53]. It is recognized that response indicator relates to the actions taken to mitigate loss or protect water resources. The causal links between the upstream-downstream communities in a river basin need to be established whereby the corrective actions might be implemented in response to the interests of water users and water savers. The need for a cooperative agreement included compensation to upstream communities for protecting and conserving water resources for benefit of the communities living downstream areas must be approved by the RBWC. Because the Law No. 7/2004 stipulates the representative institutions by representing the river basin diversity, it has definitely help pave the way for participatory IWRM. The need to establish a RBWC as coordinating board could provide synchronization in support of water resources program activities [11]. The specific problems to be addressed are highly diverse and vary from one basin to another.

The economic tool can be designed as priority the internalization of external impacts of pollution to make polluters use a new standard for IWRM practices. Polluter and user pay principles give a good illustration of the stakeholders' participation to the future IWRM framework. Based on the studies of farmer participation on operation and maintenance of irrigation networks system conducted in Central Java, Yogyakarta and West Nusa Tenggara [54], it seems that farmers agree with concept of volumetric pricing for irrigation water. Therefore, the implementation of water pays for water principles can be an instrument of fiscal policy for financing the IWRM activities. This is a challenge that can be recommended for avoiding the snare of debt from donor countries and/or international agencies. The recommended policy scenario has been proposed by Bryan and Kandulu [55] involved targeted information, followed by an incentive program and a proposed mandatory code of practice for IWRM.

One of the perspectives in managing of water resources is to promote the instruments and mechanisms for enacting the participation concepts to share benefits and costs resulting from land-water interactions by upstream and downstream resource users. Therefore, it needs to have much greatersensitivity to relative materiality in order to safeguard adequate quantity and safe quality of water. Inrecent years, most concern around erosion and sedimentation has focused on the impact of continuing sedimentation on the environmental quality that would be tolerable for all stakeholders. A starting set of concepts can be created by selecting the ideas based on seller-buyer approach to consolidate the interactions of multiparty entire the river basin [56]. Community group has proved to be the most appropriate in terms of manageability and positive impact with respect to conservations of forest, land and water in the upper parts of a river basin; they should be recognized as a community of sellers. Public suppliers provide water for a variety of applications such as domestic, commercial, industrial, thermoelectric power and public water use; they should be recognized as a buyers group.

It can be seen that the seller-buyer approach offers a win-win solution that must continually face severe water-related problems. Challenges to establishing markets for watershed protection services and improved livelihood opportunities for poor rural households have been studied for three river basins (*i.e.*, Brantas, Cidanau and Segara) [57]. Yet, managing water resources at a river basin level can be unrealistic, particularly when Indonesia has weak regulating institutions and limited technical and financial capacity.

The RBIWRMP lays out strategies for overcoming the water environment related problems. It sets short, medium and long term goals for IWRM, agreed by all the major stakeholders. Consequently, public consultation is a very important element in the planning process particularly in the light of future concerns about water allocation and the need for sustainable development. Guided by the RBIWRMP to assist decision makers monitor and evaluate the progress made for all proposed projects. Accurate basin-wide accounting of water use including equitable distribution based on existing legal entitlements would significantly contribute to water conservation efforts [58]. The use of water varies somewhat according to both stakeholders concerned over water rights and nature of lands. Need to establishing RBWC would be to avoid the conflicts of interest and to synchronize the access to a shared resource of water for multipurpose. Therefore, even if some manuals of standard operating procedures have been provided to support the implementation of IWRM, and web-based information systems (WBIS) have been established at national and certain WS levels, a network of the WBIS should be also established across the country to bring together different parts of the regions where water resources data and information may temporally change along the year. Consistent implementation of RBIWRMP brings success to contribute to IWRM and all key stakeholders must work toward the same compelling visions.

6. CONCLUSIONS

This paper presented a brief review on water availability and water-related issues in Indonesia reveals challenges of the future IWRM to support sustainable development. The most apparent measure toward resolving the present problems and constraints of implementing the IWRM approach is by means of the RBIWRMP, which first of all must start with massive campaigning programs to develop the public's sense of belonging, so that the communities as a whole are willing to participate in IWRM practices. Capacity buildings of enabling environment, institutional frameworks and management instruments were highlighted to contribute to pertinent policy proposals that strive for an effective implementation of the future IWRM. Picking the right policy to help stakeholders depends on accurate justification of current water conditions. The recommended actions for improving the future IWRM related to the important pillars of management approach, integrated planning, water pays for water principles, coordinating model and stakeholders' interaction on the basis of seller-buyer approach were presented to contribute to the future IWRM in Indonesia.

ACKNOWLEDGMENTS

The authors gratefully acknowledge financial support from the Universiti Teknologi Malaysia for Research University Grant: Vot. 07H01.

REFERENCES

1. Thou, A.D.M. Community and Social Responses to Land Use Transformations in the Nairobi Rural-Urban Fringe, Kenya. Available online: http://factsreports.revues.org/435 (accessed on 11 July 2014).
2. Fulazzaky, M.A.; Gany, A.H.A. Challenges of soil erosion and sludge management for sustainable development in Indonesian. *J. Environ. Manag.* **2009**, *90*, 2387–2392.
3. Ranjan, R. Environmental restoration of invaded ecosystems: How much *versus* how often? *J. Environ. Manag.* **2008**, *86*, 616–626.
4. Noël, C. *Organization of Water Management in France: Capacity Building for Better Water Management*; International Office for Water: Paris, France, 2009.
5. Bearden, B.L. The legal regime of the Mekong River: A look back and some proposals for the way ahead. *Water Policy* **2010**, *12*, 798–821.
6. Brils, J. Sediment monitoring and the European Water Framework Directive. *Ann. Ist. Super. Sanità***2008**, *44*, 218–223.
7. Özerol, G.; Newig, J. Evaluating the success of public participation in water resources management: Five key constituents. *Water Policy* **2008**, *10*, 639–655.
8. World Health Organization (WHO). *Climate Change and Human Health: Risks and Responses*; McMichael, A.J., Campbell-Lendrum, D.H., Corvalán, C.F., Ebi, K.L., Githeko, A.K., Scheraga, J.D., Woodward, A., Eds.; WHO: Geneva, Switzerland, 2003.
9. Fulazzaky, M.A.; Akil, H. Development of data and information system to improve water resources management in Indonesia. *Water Resour. Manag.* **2009**, *23*, 1055–1066.
10. Fulazzaky, M.A.; Sutardi, S. *Integrated Water Resources Management in View of Environmental Sustainability Aspects*; Penerbit UTHM: Batu Pahat, Malaysia, 2008.
11. Government of Indonesia (GOI). *Indonesian Law No. 7/2004 on Water Resources*; Indonesian Law and Regulation Document; GOI: Jakarta, Indonesia, 2004.
12. Directorate General of Water Resources (DGWR). Water Resources Conservation Problems in Indonesia. In *Ministerial Coordination Meeting amongst the Ministries of Public Works, Forestry, and Environment*; DGWR Official Document; DGWR: Jakarta, Indonesia, 2005.
13. Grover, V.I.; Krantzberg, G. *Great Lakes: Lesson in Participatory Governance*; CRC Press: Boca Raton, FL, USA, 2012.

14. Sivakumar, B. Planning and management of shared waters: Hydropolitics and hydropsychology—Two sides of the same coin. *Int. J. Water Resour. Dev.* **2014**, *30*, 200–210.
15. Bulsink, F.; Hoekstra, A.Y.; Booij, M.J. The water footprint of Indonesian provinces related to the consumption of crop products. *Hydrol. Earth Syst. Sci.* **2010**, *14*, 119–128.
16. Hughes, D.A.; Mallory, S.J.L. The importance of operating rules and assessments of beneficial use in water resource allocation policy and management. *Water Policy* **2009**, *11*, 731–741.
17. Marfai, M.A.; King, L.; Singh, L.P.; Mardiatno, D.; Sartohadi, J.; Hadmoko, D.S.; Dewi, A. Natural hazards in Central Java Province, Indonesia: An overview. *Environ. Geol.* **2008**, *56*, 335–351.
18. Morita, M. Flood risk impact factor for comparatively evaluating the main causes that contribute to flood risk in urban drainage areas. *Water* **2014**, *6*, 253–270.
19. Santato, S.; Bender, S.; Schaller, M. *The European Floods Directive and Opportunities Offered by Land Use Planning*; CSC Report 12; Climate Service Center: Hamburg, Germany, 2013.
20. Pretty, J. Agricultural sustainability: Concepts, principles and evidence. *Philos. Trans. R. Soc. B***2008**, *363*, 447–465.
21. Werner, M.; Reggiani, P.; Roo, A.D.; Bates, P.; Sprokkereef, E. Flood forecasting and warning at the river basin and at the European scale. *Nat. Hazards* **2005**, *36*, 25–42.
22. Patterson, L.A.; Lutz, B.D.; Doyle, M.W. Characterization of drought in the South Atlantic, United States. *J. Am. Water Resour. Assoc.* **2013**, *49*, 1385–1397.
23. Raman, A.; Verulkar, S.B.; Mandal, N.P.; Variar, M.; Shukla, V.D.; Dwivedi, J.L.; Singh, B.N.; Singh, O.N.; Swain, P.; Mall, A.K.; *et al.* Drought yield index to select high yielding rice lines under different drought stress severities. *Rice* **2012**, *5*.
24. Al-Malack, M.H.; Abuzaid, N.S.; Bukhari, A.A.; Essa, M.H. Characterization, utilization, and disposal of municipal sludge: The state of-the-art. *Arab. J. Sci. Eng.* **2002**, *27*, 3–23.
25. Harrison, E.Z.; Lee, T.; McBride, M.; Bossard, S.; Chase, L.; Bouldin, D.; Czymmek, K. *Considerations for Dairy Farms Regarding Use of Sewage Sludges, Sludges Products and Septage*; Cornell Waste Management Institute: New York, NY, USA, 2003.
26. Qin, M.; Jackson, R.H.; Yuan, Z.; Jackson, M.W.; Sun, B. The effects of sediment-laden waters on irrigated lands along the lower Yellow River in China. *J. Environ. Manag.* **2007**, *84*, 858–865.
27. Fulazzaky, M.A. Problems and restoration activities for eutrophicated lakes and reservoirs in Indonesia. In Proceeding of the 56th IEC-ICID Meeting and 19th ICID Congress, Beijing, China, 10–18 September 2005.
28. Maherali, H.; DeLucia, E.H. Influence of climate-driven shifts in biomass allocation on water transport and storage in ponderosa pine. *Oecologia* **2001**, *129*, 481–491.
29. Sukistijono, S. Erosion and sedimentation in Brantas upper reach and its countermeasures. In Proceeding of the First International Workshop on Water and Sediment Management on Brantas River Basin, Batu-Malang, Indonesia, 28–29 July 2005.
30. Djajadiredja, E.A.; Ibrahim, A.B. Effect of land-use change on sedimentation rate at upper Citarum river basin. In Proceeding of the First International Workshop on Water and Sediment Management on Brantas River Basin, Batu-Malang, Indonesia, 28–29 July 2005.
31. National Statistic Board (NSB). *Environmental Statistic of Indonesia Annual Statistic Publication*; NSB Publications: Jakarta, Indonesia, 2004.
32. Koottatep, T. Night soil treatment and management. In Proceeding of the International Training Course on Water Management, Water Supply and Sanitation, Hat Yai, Thailand, 19 July–6 August 1999.
33. Gleditsch, N.P. Whither the weather? Climate change and conflict. *J. Peace Res.* **2012**, *49*, 3–9.
34. Viglione, A.; Merz, R.; Bloschl, G. On the role of the runoff coefficient in the mapping of rainfall to flood return periods. *Hydrol. Earth Syst. Sci.* **2009**, *13*, 577–593.
35. Fulazzaky, M.A. Water quality evaluation system to assess the Brantas river water. *Water Resour. Manag.* **2009**, *23*, 3019–3033.

36. Fulazzaky, M.A. Water quality evaluation system to assess the status and the suitability of the Citarum river water to different uses. *Environ. Monit. Assess.* **2010**, *168*, 669–684.
37. Reba, M.L.; Daniels, M.; Chen, Y.; Sharpley, A.; Bouldin, J.; Teague, T.G.; Daniel, P.; Henry, C.G. A statewide network for monitoring agricultural water quality and water quantity in Arkansas. *J. Soil Water Conserv.* **2013**, *68*, 45A–49A.
38. Koh, K.L. Water and Health: Challenges in ASEAN Countries and Singapore's Experience. In Proceedings of the Conference on Globalising Asia: Health Law, Governance and Policy-Issues, Approaches and Gaps, Bangkok, Thailand, 27–29 October 2011.
39. Hansnata, E. Water management in Indonesia: Lessons from Australia. In *Economics, Politics and Public Policy in East Asia and the Pacific*; East Asia Forum: Jakarta, Indonesia, 2013.
40. Hermans, C.; Erickson, J.; Noordewier, T.; Sheldon, A.; Kline, M. Collaborative environmental planning in river management: An application of multi-criteria decision analysis in the White River Watershed in Vermont. *J. Environ. Manag.* **2007**, *84*, 534–546.
41. Mendoza, G.A.; Martins, H. Multi-criteria decision analysis in natural resource management: A critical review of methods and new modeling paradigms. *For. Ecol. Manag.* **2006**, *230*, 1–22.
42. Braden, J.B.; Jolejole-Foreman, M.C.; Schneider, D.W. Humans and the water environment: The need for coordinated data collection. *Water* **2014**, *6*, 1–16.
43. Kerssens, P.J.M.; van Urk, A. Experimental studies on sedimentation due to water withdrawal. *J. Hydraul. Eng.* **1986**, *112*, 641–656.
44. Asian Development Bank (ADB). *Integrated Citarum Water Resources Management Investment Program*; Project 1; Compliance Review Panel: Manila, Philippines, 2013.
45. Ministry of Forestry (MOF). *National Movement for Forests and Lands Rehabilitation: Works Plan 2003–2007*; MOF: Jakarta, Indonesia, 2003.
46. Duane, T.P.; Opperman, J.J. Comparing the conservation effectiveness of private water transactions and public policy reforms in the conserving California landscapes initiative. *Water Policy* **2010**, *12*, 913–931.
47. Palme, U. Multiple conceptions of sustainable urban water systems: Problem or asset? *Water Policy***2010**, *12*, 425–443.
48. Fischhendler, I.; Heikkila, T. Does integrated water resources management support institutional change? The case of water policy reform in Israel. *Ecol. Soc.* 2010, 15. Available online: http://www.ecologyandsociety.org/vol15/iss1/art4/ (accessed on 11 July 2014).
49. Kakabadse, N.K.; Rozuel, C.; Lee-Davies, L. Corporate social responsibility and stakeholder approach: A conceptual review. *Int. J. Bus. Gov. Ethics* **2005**, *1*, 277–302.
50. Islam, K.K.; Sato, N. Deforestation, land conversion and illegal logging in Bangladesh: The case of the Sal (*Shorea robusta*) forests. *iForest* **2012**, *5*, 171–178.
51. Sun, R.; Wang, Z.Z.; Chen, L.; Wang, W.W. Assessment of surface water quality at large watershed scale: Land-use, anthropogenic, and administrative impacts. *J. Am. Water Resour. Assoc.* **2013**, *49*, 741–752.
52. Darnsawasdi, R.; Roongtawanreongsri, S. Economic as a tool for sustainable water, wastewater and solid waste management. In Proceeding of the International Training Course on Water Management, Water Supply and Sanitation, Hat Yai, Thailand, 19 July–6 August 1999.
53. Jones, N.; Evangelinos, K.; Gaganis, P.; Polyzou, E. Citizens' perceptions on water conservation policies and the role of social capital. *Water Resour. Manag.* **2011**, *25*, 509–522.
54. Fulazzaky, M.A.; Wahyudi, S.I. Participating the farmers on operation and maintenance of the irrigation networks system in Central Java, Yogyakarta and West Nusa Tenggara Provinces. In Proceeding of the 4th Asian Regional Conference and 10th International Seminar on Participatory Irrigation Management, Tehran, Iran, 2–5 May 2007.
55. Bryan, B.A.; Kandulu, J.M. Designing a policy mix and sequence for mitigating agricultural non-point source pollution in a water supply catchment. *Water Resour. Manag.* **2011**, *25*, 875–892.

56. Bell, R. Capturing benefits from water entitlement trade in salinity affected areas: A role for trading houses? *Aust. J. Agric. Resour. Econ. Soc.* **2002**, *46*, 347–366.
57. Munawir, M.; Suhardi, S.; Salim, S.; Suyanto, A.; Vermeulen, S. *Action-Learning to Develop and Test Upstream-Downstream Transactions for Watershed Protection Services: A Diagnostic Report from Segara River Basin, Indonesia*; Centre for Water Resources and Environment—Institute for Social and Economic Research, Education & Information: Jakarta, Indonesia; International Institute for Environment and Development: London, UK, 2003.
58. Samani, Z.; Skaggs, R.K. The multiple personalities of water conservation. *Water Policy* **2008**, *10*, 285–294.

CHAPTER 10

Pushing the Governance Boundaries: Making Transparent the Role of Water Utilities in Managing Urban Waterways

Bethany Cooper[1], Lin Crase[1], Darryl Maybery[2]

[1] *University of South Australia, Adelaide, Australia*

[2] *Monash University, Clayton, Australia*

ABSTRACT

Integrated Water Resource Management (IWRM) requires simultaneous consideration of the multiple benefits that attend water. IWRM can also be more challenging in regulatory environments where the resource manager must justify choices and elements of each intervention. This is particularly challenging in the context of urban waterways that have many uses including an ecological function and a source of human amenity. To justify expenditure on maintaining and improving urban waterways for ecological and/or amenity changes regulated utilities must be able to articulate and measure these types of values with at least some degree of precision. This paper presents a generic and systematic framework for understanding the ecological and amenity values of urban waterways. We illustrate deployment of the framework in the case of Melbourne, one of Australia's fastest growing cities and a location ranked as amongst the most liveable since 2011. We also explore how the results could improve the way we measure benefits in dollar terms.

Keywords: Integrated water resources management (IWRM), Ecologicalvalue, Amenityvalue, Waterways, Delphi analysis

1 INTRODUCTION

Water pricing, regulation and management remain contentious worldwide (see, for instance, Gareau and Crow 2006). The sources of controversy include the perception that water is especially unique and requires alternative institutional arrangements to deal with hydraulic and social nuances; the notion that water is 'gifted' from nature and cannot be effectively priced or controlled to reflect scarcity; political costs associated with reassigning rights that were historically given to users without cost; recognition that water governance is at least as important as water engineering; and increased awareness that the fugitive nature of the resource makes it difficult to optimise uses across multiple demands, including environmental claims.

These issues have manifested in a number of competing approaches to water management. On the one hand, numerous analysts have urged governments to adopt a strict 'user-pay' principle to avoid excessive resource exploitation (see, for example, ICWE 1992; United Nations 1992). On the other hand, observers have noted that water provides a range of non-marketable services, such that identifying 'users' is problematic enough, let alone assigning a price to reflect the benefits they enjoy (see, for instance, Raymond et al. 2008; Lamoree and van Steenbergen 2006; Savenije and van der Zaag 2002). It is against that background that the concept of Integrated Water Resources Management (IWRM) has gained prominence.

IWRM is commonly defined as "a process that promotes the coordinated development and management of water, land and related resources, in order to maximize the resultant economic and social welfare in an equitable manner without compromising the sustainability of vital ecosystems" (Global Water Partnership – Technical Advisory Committee (GWP-TAC) 2000, p. xiv). Thomas and Durham (2003, p. 24) have defined the notion as "a sustainable approach to water management that recognizes its multidimensional character – time, space, multidiscipline and stakeholders – and the necessity to address, embrace and relate these dimensions holistically so that sustainable solutions can be brought about." IWRM has several core tenets but a key issue is the treatment of water planning at a catchment scale. The logic of this approach is that upstream choices have downstream consequences so managing water holistically offers more scope for dealing with externalities or spill-over effects. Notwithstanding its compelling logic IWRM is not without its critics. More specifically, Molle (2008) notes that IWRM can often disguise the competing nature of some objectives and the mechanisms for resolving trade-offs is not always clear. Similarly, Biswas (2004) observes that a range of interventions can easily be passed off as IWRM and Walther (1987) argues that the concept is inclined to produce 'black box' solutions.

IWRM has nonetheless become a cornerstone for water planning in many countries and in various settings. In Europe, IWRM has been embedded in proclamations related to the European Water Framework Directive (2000) but its implementation can be complicated by cross-jurisdictional waterways and competing national demands (e.g. European Commission 2000). IWRM can also occur in rural contexts, where the primary users are agricultural and environmental interests, and in urban settings. In the case of the latter pressures can come from industrial and potable demands, but increasing environmental and amenity demands also attend urban water management, especially in richer nations where amenity and environmental goods are increasingly demanded.

Whilst most of these principles are well-understood by policy makers and analysts, the contradictions that they embed are often not revealed until attempts are made to operationalise them. For example, IWRM implies that multiple uses must be considered holistically, but how is that approach reconciled with the need for clear specification of individual beneficiaries and the costs they should meet, especially when some elements of water have public good qualities?

Lamoree and van Steenbergen (2006, p. 100) highlight these challenges to answering these types of questions "Through investment and management interventions, values are created (or destroyed) in connection to these functions: productivity values, amenity values, property values, environmental conservation values, and more. The challenge in IWRM is to, at a minimum, balance these different functions and values, yet preferably to optimize them." The difficulty of striking a balance and optimising management is especially evident in the case of urban waterways used for amenity and environmental outcomes. At a broad level urban waterways have an ecological function but can also be a source of human amenity. First, urban waterways can provide habitat for indigenous plants and animals or, as a minimum, provide passage for important species between ecologically significant sites. Second, waterways can provide a sense of 'enjoyment' for those nearby, irrespective of ecological status (see, for example, Veldarde et al. 2007; Völker and Kistemann 2011). An ongoing challenge however is identifying ways of incorporating these notions into current regulations and institutional approaches to water management.

Adding to this complexity is the requirement to understand values a priori. Only by knowing values 'in advance' can the manager hope to choose optimal IWRM strategies that ultimately change some functional element of waterways. A precursor to improving our understanding of these changes is understanding the drivers of key values. In this paper, we address this by unpacking the drivers of amenity and ecological value of urban waterways to better inform how trade-offs might be made.

More specifically, this paper uses the case of urban water management in Melbourne to highlight an approach that can be used to deal with the dual tasks of managing water in an integrated fashion while retaining transparency about the benefits and costs of specific interventions.

Melbourne has been chosen specifically as it is one of Australia's fastest growing cities with the population estimated to increase by 60 % by 2050, reaching 6.5 million (OLV 2013). Growth of this magnitude substantially raises the stakes on current management of urban waterways for amenity and ecological benefit and without a clear framework for analysis the risks of irreversible and poor choices are substantial. In addition, Melbourne has been consistently rated amongst the world's most liveable cities since 2011 (Economist Intelligence Group 2013) offering a unique opportunity to explore policy choices in a setting where findings will gain immediate traction and where lessons are likely to be readily considered by other jurisdictions.

Melbourne also has other benefits as a case study for this analysis insomuch as some empirical work already exists in the context of unpacking ecological and amenity values. First, the Victorian Index of Stream Condition (ISC) provides an indication of the ecological health of Melbourne's waterways. The ISC is made up of five sub-indices: hydrology, streamside zone, physical form, water quality and aquatic life. Each river reach assessed is given an overall ISC score of between 0 and 50. Accordingly, 36 % of Melbourne's waterways are considered to have high ecological values i.e. have been given an ISC score of 30 or above. Second, in terms of the existing amenity provided by Melbourne's waterways, Melbourne Waters *Healthy Waterway Strategy* (2013) has identified amenity priority areas within management units/streams based on visitation, popularity and satisfaction of the public. Accordingly, 41 % of Melbourne's waterways have been categorized as having high amenity thus adjudged as priority areas by Melbourne Water.

In Melbourne, waterways that pass through the city are managed by the water utility that provides bulk potable water supplies to the metropolis and the same entity administers wastewater treatment. The entity is known as Melbourne Water Corporation and is owned by the state government but managed along corporate lines. Prima facie this should lead to a degree of IWRM that might not occur were the various water management responsibilities are disbursed across a range of institutions.

In order to satisfy economic regulators that all of the utility's services are provided at lowest cost, Melbourne Water Corporation is required to prepare 5-year plans that show the demand for its various activities, including those relating to amenity and ecological functioning or waterways. Finding a way to do this has proven challenging, but the lessons offered here stand to guide others in the more general search for efficient and innovative solutions to integrated water management.

In this paper we present a theoretical framework that was developed by systematically tapping expert knowledge about the ecological and amenity value of urban waterways. We use the framework and the results from a Delphi analysis to explore techniques for valuing different management actions. In addition we contemplate the implications for institutional arrangements that can accommodate IWRM and the need for transparent regulation.

The paper itself comprises five additional parts. In Section 2 a brief synopsis of the institutional arrangements that circumscribe water management in Melbourne and Victoria, Australia, is offered. This section also provides the context for the analysis of amenity and ecological water values. Section 3 presents a methodological approach for articulating hard-to-measure constructs like ecological and amenity benefits. The results from the application of this Delphi method are presented in Section 4 before the implications for efficient regulation are considered in Section 5. The final section comprises brief concluding remarks.

2 INSTITUTIONAL ARRANGEMENTS FOR URBAN WATER MANAGEMENT

IWRM has become an integral part of the water planning philosophies in all Australian jurisdictions. Australian water resources are principally controlled by state governments and each has sought to develop plans, regulations and policies consistent with the National Water Initiative (NWI) agreed in 2004. The NWI established a blueprint whereby water was recognised as an economic good; water pricing was to be introduced to reflect costs; water rights were clarified, separated from land and made tradeable; and water regulation separated from service delivery. Economic regulation has subsequently been introduced in most states, where utilities must demonstrate to regulators the value of services delivered and that demands are being met at lowest possible costs.

Australia's institutions for managing water have been praised by international observers and scholars alike (e.g. The Economist 2003) such that actions in Australia can have significant ramifications for international water management strategies. One of the major accomplishments of water management in Australia was the introduction of cost-reflective pricing as part of the National Water Initiative of 2004 and earlier reforms. This approach has now been broadly adopted in urban areas and was again ratified by the National Resource Management Ministerial Council in 2010 (Crase et al. 2015).

In Australia, regulated utilities must be able to articulate and measure different values with some degree of precision. Utilities must meet regulators' calls for lowest-cost expenditures and yet also deal with insistences at the policy level that water be managed holistically for the betterment of the community. But what is often overlooked in this setting is the need to understand the potential trade-offs that are made between different water management objectives, or to appreciate potential synergies between water management goals. For example, in the case of urban water management, there are potential trade-offs between increasing the ecologically value of waterways at the expense of amenity value (and vice versa). But what are the amenity and ecological values of urban waterways and, importantly, what are the most robust mechanisms for articulating those values? Without answering these types of questions it is not possible to both effectively employ IWRM and to also meet the needs of regulators that costs related to various services be made transparent.

As noted earlier, the responsibility of water management in Australia's states and territories lies with numerous government agencies and water corporations. For instance, in the state of Victoria the Department of Environment, Land, Water and Planning (DELWP) has responsibility for managing the state's water resources in partnership with the Victorian water corporations. In the metropolitan area of Victoria, Melbourne Water Corporation (Melbourne Water) has responsibility for bulk water supply and bulk sewerage services.

Melbourne Water is owned by the State government and since 2004 has also been the main caretaker of river health in and near Melbourne. Melbourne Water is also responsible for regional drainage, waterways and floodplain management for the region. Retail water and sewerage services in the metropolitan area are provided by three corporations; namely, South East Water, Yarra Valley Water and City West Water.

There are 13 additional regional water corporations which operate outside the metropolitan area, many controlling their own bulk water supplies (Crase et al. 2015). Melbourne Water and all of the urban water corporations are subject to economic regulation and are required to have 5-year Water Plans approved by the Essential Services Commission (ESC). In terms of the environmental performance of the urban water corporations, the Environmental Protection Agency Victoria monitors and oversees Victoria's water sector. The oversight of the sector's financial management standards lies with the Department of Treasury and Finance (DTF).

As noted earlier, a key institutional principle for water utilities is transparency and an associated requirement to be able to rationalise expenditures as legislated monopolies. This includes clarity around expenditures on urban waterways. Achieving transparency is not always straight forward, especially when some relationships between actions and values requires specialised knowledge.

3 USING DELPHI TO GAIN INSIGHTS TO MEASUREMENT DILEMMAS

In Greek mythology the 'oracle of Delphi' was able to provide the answers or the 'truth' to vexed questions. Experts in the topic of interest are used in modern day Delphi research to provide answers to vexed questions (Hasson et al. 2000). The Delphi approach consists of several rounds of open ended responses and/or brief surveys that are presented to a panel of experts who anonymously provide input. The first round initially generates ideas and information from the experts about the topic. This content is then summarised by the researchers and then re-presented to the experts for approval, modification or extension but with the aim of obtaining consensus.

The process can be repeated over a number of 'rounds' and concludes when consensus is reached (De Villiers et al. 2005). Thus, a Delphi study comprises a structured process whereby knowledge dispersed across a group is collected and condensed, usually via several rounds of interaction with experts (Maybery et al. 2009).

The Delphi technique has been employed successfully in a broad range of areas including curriculum development for health sciences education (De Villiers et al. 2005), HR research (Hatcher and Colton 2007), and designing a mental health training package (Whitman et al. 2009). The Delphi technique has also been used to assist with environmental decision-making and environmental monitoring program design (Richey et al. 1985).

In addition, the Delphi technique is often employed to enable experts in the field to more clearly define concepts that are ambiguous (see, for instance, Raine 2006). In this case, the technique has a lot to offer around clarifying concepts such as ecological value and amenity value. For instance, in the case of urban waterway management, the term 'amenity' is frequently used by water agencies as a catch-all for human values that are not directly associated with ecological characteristics but this is no easy thing to measure; the attributes that characterise high amenity in one location might do a poor job in another (e.g. Horwitz and Carter 2011). Therefore, analysis using the Delphi approach will increase our understanding of the key drivers and components of these concepts.

Accordingly, this study used a Delphi to systematically tap expertise about the ecological and amenity value of urban waterways. In this instance, experts in the ecology and amenity of waterways were subjected to three rounds of collection and synthesis using an on-line instrument.

The Delphi results aimed to deliver expert consensus on the classification of different waterway benefits along three dimensions: drivers, components and outcomes.

These three dimensions are hierarchical and enable us to distinguish between the tangible and higher-order characteristics of amenity and ecological values. For instance, in the case of amenity, the drivers are highlighted as the natural or physical qualities/characteristics of the waterways; components are the aesthetic attributes or characteristics (e.g. escape, appeal, recreation); and the outcome is personal well-being. In the case of ecological value, drivers are the measurable characteristics that have a causal relationship with the components and the final outcome is ecological well-being.

One advantage of this approach is that any descriptors developed about ecological and amenity value would be simultaneously traceable to existing standards or indices thus ensuring they are relevant and practical for management options.

Essentially, the Delphi analysis used here should enable the formation of more accurate information surrounding amenity and ecological values of Melbourne's waterways, which can then be distributed to other key stakeholders (e.g. policy makers, managers, consumers) to better inform their decision-making. For example, the linkages uncovered here could be used to clarify how values are created or destroyed by particular choices.

4 RESULTS

4.1 Recruitment

4.1.1 Selection of Participants

Critical to the Delphi methodology is the selection of experts as participants. In the initial stages a steering committee was developed to support and guide this and other components of the wider research project. The steering group included key stakeholders from the water sector and in conjunction with the research team a definition of 'experts' was developed. For the purpose of this study an expert in the amenity of Melbourne's waterways must meet one or more of the following criteria:

- Worked in urban planning, local government, tourism, recreation or for waterways for at least 3 years
- Undertaken research in the area of urban planning, local government, tourism, recreation or waterways for at least 3 years
- Worked in settings where 'liveability' has been a primary consideration in Melbourne or cities with similar levels of development
- And specific to ecology an expert must:
- hold a degree in ecology or cognate discipline or equivalent experience
- work or have worked in freshwater ecology or a related science for at least 3 years

Using these definitions recruitment for the Delphi commenced initially from within the steering committee. The panel of experts was progressively extended by approximately 20 people as a consequence of the final question in round one which asked participants to recommend other experts, even if they held opposing or different views.

The current Delphi study was conducted via a web-based survey that allowed an open response format and quantitative survey responses. Two groups of experts were employed in relation to amenity and ecology with each group responding to similarly framed questions over three separate rounds of enquiry. Participants were initially contacted by email with an outline of the study aims and approach, along with an online survey link to each round of the Delphi study. The first round of the study was open for four weeks, the second for five weeks and the final round for three weeks. To increase participation, two reminder emails were sent to participants prior to each deadline.

4.2 Application and Key Findings

4.2.1 Round One

In the initial round there were four open-ended questions. The initial question was asked in relation to the waterway system overall (question 1) and was framed as follows:

When you think of Melbourne's waterways overall, what are the things that contribute to its Ecological (Amenity) value.

The subsequent three questions were similarly framed and asked in relation to rivers (question 2), estuaries (question 3) and wetlands (question 4). Two final questions requested any additional information that may be relevant and also sought recommendations of other experts who might participate – even if they may hold opposing or different views on values. This final question sought to develop a diverse and broad panel. Participants were also given the option to 'opt out' of the study and to avoid further contact.

Specifically for the study of *amenity values*, of the 50 participants, 22 (44 %) responded to the first round of questions. Over 170 comments were generated in response to the four open-ended questions. These comments were grouped into 17 themes, using a triangulation approach. More specifically, the themes were developed independently by two of the researchers and consensus was reached by involving a third researcher. The 17 themes were then categorised into the hierarchy of 'drivers', 'components' and a single 'outcome' – in this case 'personal wellbeing'. Drivers were considered to be the tangible natural or physical characteristics of the waterways (e.g. infrastructure, plants) and components were higher order factors that were defined as the aesthetic and cognitive attributes that are beneficial (e.g. escape, appeal, recreation).

For the study of *ecological values*, 27 (34 %) participants responded to the first round of questions. A total of 91 comments were generated from the four open-ended questions, which were then grouped into five key themes. This followed a similar triangulation process described above. Again, themes were grouped into 'drivers' and 'components', however this was in the context of a single outcome referred to as 'ecological wellbeing'. In this case, drivers are antecedents that impact on components which are seen as more discrete markers of ecological well-being.

4.2.2 Round Two

Following round one, a second survey was developed and distributed to the amenity expert panel. Simultaneously, a second survey was distributed to the ecological expert panel. In this round, the study on amenity value generated a response from 20 participants, while the ecological survey generated 18 responses at this point.

In these surveys, the themes identified in round one were presented to the experts and participants were then asked to rate their importance from 1 (no importance) to 10 (extremely important). Subsequent questions gave respondents the opportunity to refine the labels of the themes or add to the existing ones. The final question asked participants if they had any further comments regarding the survey.

The 17 final themes and their general content for amenity value are shown in Table 1. The final five themes and an indicative description for ecological value are provided in Table 2 and 3.

Table 1 Amenity value – drivers, components and outcome

Drivers	Components	Outcome
•Water-related Animals (e.g. fish, birds, platypus, frogs) •Water-related Plants (e.g. riparian corridor, native vegetation) •Water-related Infrastructure (e.g. seats, paths, board walks, label or educational signage, facilities) •Cleanliness of the waterways (e.g. odour, litter, colour, turbidity, erosion) •Cleanliness of the surrounding landscape (e.g. odour, litter, colour, turbidity, erosion) •Access (i.e. both visual and physical access) to a waterway and its surrounding vegetation •Water-related Naturalness (e.g. natural components of the landscape, the geology, natural light, water level)	•Connection to Nature that Melbourne waterways provide for people •Indirect Values (e.g. the retention and protection of vegetation, knowing it is there to be enjoyed) •Escape (e.g. experience of isolation, escape, feeling of refuge) •Water-related Aesthetic Appeal (e.g. views, scent, bird noises, sound of the water, the relative impact of noise pollution, regulation of temperature) •Water-related Cultural Significance (e.g. meaning for people, heritage place, memories) •Social Interaction (e.g. communing with another, socialising) •Social Benefits (e.g. education, tourism, economy) •Ability to engage in active recreation in or on the waterway (e.g. swimming, boating) •Ability to engage in active recreation in proximity to the waterway (e.g. walk, cycle, run) •Ability to engage in passive recreation in proximity to the waterway (e.g. sit, picnic, reflect)	•Personal Wellbeing (e.g. recharge, relaxation, tranquil place, health benefits, exercise)

4.2.3 Round Three

In round 3, the amenity survey had 21 participants and the ecological survey had 36 participants. In these surveys participants rated the drivers that were identified in round 2 in relation to each of the identified components.

Table 2 Ecological values – drivers

Drivers	Description (Indicative only)
Water quality	•nutrient loads (nitrogen, phosphorus, potassium), herbicides, pesticides •sulphidic sediments •salinity levels •temperature •pollution, e.g. toxins, metals, nanoparticles, *E. coli* (faecal contamination) •concentrations of oxygen (dissolved) •pH •algae (excessive) •turbidity, clear water, suspended matter in the water column, brackish •litter load, ballast water discharge and other waste
Hydrological Regime	•flow regimes --slackwater/blackwater, pools, flooding (temporary will connect waterways), sediment loads (erosion, movement/delivery, deposition) •water availability -- volume, timing, magnitude, velocity, depth and frequency of flows •unregulated systems tend to have higher values because hydrological regime ensures maintenance of ecological processes including temperature cues for fish spawning and plant seed production and recruitment •groundwater input to stream flow (although source not always obvious) •watering regimes (esp. wetlands) -- flooding, wetting/drying cycles (magnitude, frequency, timing and duration), permanent inundation or permanent drying, flow detention times, seasonal changes and variability, depth profile (ephemeral, shallow and deep water zones), variability •flushing (esp. estuaries) -- tidal exchange, tidal fluxes, salinity gradients, fresh-water inflows •water sources •(laterally, longitudinally, vertically) between tributaries, rivers, floodplains, associated wetlands, estuaries, open to the sea, estuarine wetlands, estuarine mudflats, but also wildlife corridors, native vegetation remnants, between freshwater and marine habitats •timing, length, frequency, periodicity, seasonality of opening/closing •ensures dispersal of aquatic fauna and plant propagules, opportunities for aquatic fauna (especially fish) to access nursery and spawning habitat •barriers (weirs, dams, poorly designed culverts etc.) reduce connectivity

Accordingly, in this Delphi process both qualitative and quantitative data analysis were used (see, for instance, Hsu and Sanford 2007). The common statistics used in Delphi studies are measures of central tendency (e.g. mean) and level of dispersion (e.g. standard deviation) in order to present information concerning the collective judgements of respondents (Hasson et al. 2000).

Table 3 Ecological values – components

Components	Description (Indicative only)
Animals (presence and diversity)	•macroinvertebrates •native aquatic animal species (birds, fish, crustaceans, frogs, turtles, platypus etc) •number and diversity of species •composition and structure sympathetic to context •specific threatened animal species (e.g. Yarra pygmy perch, Macquarie perch, Australian grayling etc) •pest species, exotic biota
Plants (presence and diversity)	•number and diversity of species •cover and density •macrophytes (e.g. reeds) •vegetation for amenity, micro-climate and habitat services •composition and structure sympathetic to context •location -- fringing, in-stream, cover, catchment, intertidal, mangroves, seagrass •remnant (native plants), agricultural and exotic, pests and weeds •seed banks
Physical structure	•substrate diversity (natural mud or gravel) or human-built (channel modification e.g levees, pipes, concrete straightening), barrages/weirs •deepening, artificially opening intermittently closed (esp. estuaries) •morphologic diversity (sediments, hydraulic habitat etc.) •physical structure that provides living space, refugia and nursery environments for juveniles •structural complexity in and out of the water -- large woody debris/snags, coarse particulate organic material (CPOM) •undercut banks, stable banks, degraded river banks •pool size and depth variability, channelization, incision

The mean scores and the standard deviations for the relationship between the amenity drivers and components are presented in Table 4. More specifically, respondents were asked to rate the impact that each of the drivers had on each of the indicators of amenity. As noted, these relationships were rated on their importance from (1) no importance to (10) extremely important. The questions were framed as follows:

"Now, we would like you to rate the impact that each of the DRIVERS has on each of the COMPONENTS of amenity. We have listed each DRIVER separately and ask that you rate the impact of each DRIVER on each of the COMPONENTS.

For example, if the DRIVER Animals has an extremely important impact on the COMPONENT Connection to Nature you might select '10' in the following question. Equally you might indicate that a particular DRIVER has no impact at all (i.e. tick a 0) on a particular COMPONENT."

In this case, the driver with the highest average score is 'Access' with a mean of 7.9. Therefore, the collective judgement of the amenity experts advocates that access (i.e. both visual and physical access) to a waterway and its surrounding vegetation is the most important driver relative to the impact that it has on the components.

Table 4 Amenity value – Mean score of the importance of the relationship between drivers and components (with standard deviation in parentheses)

COMPONENTS	DRIVERS Mean (Standard Deviation)						
	Water-related **Animals**	Water-related **Plants**	Water-related **Infrastructure**	**Cleanliness of the waterways**	**Cleanliness of the surrounding landscape**	**Access to a** waterway and surrounding vegetation	Water-related **Naturalness**
Connection to Nature that Melbourne waterways provide for people	8.33 (1.91)	9.19 (0.74)	6.86 (2.85)	8.52 (1.16)	8.71 (1.27)	8.43 (1.28)	8.76 (1.17)
Indirect Values (e.g. the retention and protection of vegetation, knowing it is there to be enjoyed)	7.19 (2.25)	8.71 (1.45)	4.81 (3.07)	7.14 (2.71)	6.76 (3.06)	6.19 (2.44)	8.33 (1.98)
Escape (e.g. experience of isolation, escape, feeling of refuge)	6.05 (1.93)	8.62 (1.46)	5.52 (2.42)	7.90 (1.70)	8.28 (1.42)	7.57 (1.59)	8.28 (1.67)
Water-related **Aesthetic Appeal** (e.g. views, scent, bird noises, sound of the water, the relative impact of noise pollution, regulation of temperature)	7.57 (2.06)	8.67 (1.19)	5.71 (2.68)	8.76 (1.04)	8.52 (1.28)	7.57 (1.66)	8.52 (1.24)

Water-related **Cultural Significance** (e.g. meaning for people, heritage place, memories)	6.67 (2.52)	7.14 (1.85)	5.33 (2.26)	6.76 (1.86)	6.57 (1.83)	6.09 (2.44)	7.00 (2.75)
Social Interaction (e.g. communing with another, socialising)	5.00 (2.14)	6.90 (1.81)	8.33 (1.49)	7.38 (1.43)	8.00 (1.09)	8.19 (1.28)	5.90 (1.89)
Social Benefits (e.g. education, tourism, economy)	7.24 (1.61)	7.10 (1.92)	8.38 (1.49)	7.66 (1.62)	7.76 (1.78)	7.95 (1.53)	6.90 (1.48)
Ability to engage in **active recreation in or on the waterway** (e.g. swimming, boating)	4.14 (2.88)	5.14 (2.66)	8.29 (1.90)	9.09 (1.14)	7.14 (2.3)	8.85 (1.59)	6.04 (2.26)
Ability to engage in **active recreation in proximity to the waterway** (e.g. walk, cycle, run)	3.95 (2.45)	5.81 (2.71)	9.14 (1.11)	6.85 (1.98)	7.66 (1.85)	9.04 (0.97)	5.9 (2.41)
Ability to engage in **passive recreation in proximity to the waterway** (e.g. sit, picnic, reflect)	5.29 (2.88)	6.95 (2.80)	8.71 (1.42)	7.57 (1.83)	8.38 (1.63)	9.19 (0.87)	6.85 (2.08)
Total Average	**6.142857143**	**7.42381**	**7.109524**	**7.766667**	**7.780952**	**7.909524**	**7.257143**

In addition, the driver and component relationships that have a mean score greater than 9 are highlighted in Table 4.

This indicates that the experts collectively regard these particular drivers as having a strong and important impact on the corresponding component. Obviously, having some understanding of the relative importance of drivers and components enables us to better predict the consequences of specific management interventions. Put differently, these data indicate the critical role of access in driving amenity outcomes, in the context of urban waterways.

The mean scores and standard deviations for the relationship between the ecological drivers and components are presented in Table 5. As was the case with the amenity survey, respondents were asked to rate the impact that each of the drivers had on each of the indicators of ecological value. These relationships were also rated on their importance from (1) no importance to (10) extremely important. The questions were framed as follows:

"Now, we would like you to rate the impact that each of the DRIVERS has on each of the COMPONENTS of the ecological value of waterways. We have listed each DRIVER separately and ask that you rate the impact of each DRIVER on each of the COMPONENTS.

For example, if the DRIVER Water Quality has an extremely important impact on the COMPONENT Plants you might select '10' in the following question. Equally you might indicate that a particular DRIVER has no impact at all (i.e. tick a 0) on a particular INDICATOR."

Table 5 Ecological Value – Mean score of the importance of the relationship between Drivers and Components (with standard deviation in parentheses)

	DRIVERS Mean (Standard Deviation)	
COMPONENTS	**Water Quality**	**Hydrological Regime**
Water-related **Plants**	7.50 (1.99)	8.63 (1.55)
Water-related **Animals**	9.22 (0.89)	8.88 (1.32)
Physical Structure	3.75 (2.78)	8.97 (1.00)
Total Average	**6.82**	**8.83**

In this case, the driver with the highest average score is 'Hydrological Regime', where the mean score is 8.83. Here, the collective judgement of the ecological experts supports hydrological regime as the most important driver relative to the impact that it has on the components (see Table 2 for a description of hydrological regime). Moreover, the driver and component relationships that have a mean score greater than 9 are highlighted in Table 5. This indicates that the experts collectively regard these particular drivers as having a strong and important impact on the corresponding component. Thus, the critical importance of the hydrological regime in supporting higher ecological values is highlighted by these data.

The drivers of amenity and ecological value identified in the Delphi analysis act as a novel and sound foundation for making transparent actions that relate to amenity and ecological value of waterways. More specifically, the study has revealed drivers that can be mapped to existing standards and guidelines to ensure the amenity and ecological characteristics of waterways can be meaningfully described. For example, the 'hydrological regime' and 'water quality' drivers of ecological value can be described in terms of the Victorian Index of Stream Condition (ISC), which was briefly outlined in section 1.0. Accordingly, waterways can be given a hydrological regime and water quality score that provides a meaningful way to measure these particular drivers. Similarly, the drivers of amenity value can also be mapped onto existing standards and guidelines. For instance, the 'access' driver can be described in terms of whether a waterway meets the strategies and guidelines developed by Melbourne Water Corporation as best management practices for Melbourne's waterways. For example, do particular waterways meet: (a) Melbourne Water's Shared Pathways Guidelines (includes bridges and crossings) and (b) Melbourne Water's Signage Guidelines. Accordingly, the extent to which 'access' to waterways exists can be meaningfully measured. Table 6 lists each of the drivers identified for amenity and ecological values and outlines how these are linked to the standards and guidelines relevant to the driver.

Ultimately, the drivers identified in the Delphi study provide a means to accurately describe the components of waterways and can also form the basis for additional measurement tools such as non-market valuation. For instance, utilities in Australia have increasingly used non-market valuation studies to gauge customer's willingness to pay for specific interventions. Clearly articulating the link between actions and outcomes can substantially improve the accuracy of these measures. This issue is explained in more detail in the following section.

Table 6 Drivers mapped onto standards or guidelines

Drivers	Standards and guidelines
Water-related Animals	1. The extent to which there are water-related animals using the waterways as their habitat or passing through them without harm. 2. The extent to which the water-related animals are native species of significance as listed under the Flora and Fauna Guarantee Act and Environmental Protection and Biodiversity Conservation Act. 3. The extent the waterway does not obstruct water-related animals (e.g. fish birds, platypus, frogs) and meets the strategies and guidelines developed as best management practices in Victoria's catchments. For example, do the waterways meet: •The Department of Sustainability and Environment's Guidelines for Waterway Management for 'Fish Passage Through Culverts' section 3.3.19. •The design requirements for the safe passage of platypus (see Serena and Williams 2008). •Melbourne Water's guidelines for birds and bird roosting sites.
Water-related Plants	1. The extent to which the water-related plants are native species of significance as listed on the Ecological Vegetation Classes (EVC). 2. The extent to which the waterway harms water-related plants, both native and introduced (e.g. shrubs, grasses and trees) and is consistent with the strategic direction of Victoria's Native Vegetation Management: A Framework for Action.
Water-related Infrastructure	1. The extent to which the infrastructure (e.g. paths, board walks, label or educational signage) surrounding these waterways does not obstruct the native environment – is the primary purpose of this infrastructure is to create access to nature? 2. The extent to which the infrastructure surrounding these waterways meets the strategies and guidelines developed as best management practices for Melbourne's waterways. For example, do the waterways meet: •Melbourne Water's Shared Pathways Guidelines (includes bridges and crossings) •Melbourne Water's Signage Guidelines
Cleanliness of waterways	1. The extent to which the waterways are clean in terms of odour, litter, colour, turbidity, and erosion. 2. The extent to which the cleanliness of the water meets the Australian and New Zealand guidelines for fresh marine water quality 3. The extent to which the maintenance of the waterways meets the maintenance checklist of Melbourne Water's Constructed Wetlands Guidelines, where inspection is carried out every 3 months. For example, the waterways are inspected and maintained in terms of: •Evidence of odours •litter accumulation •evidence of damage or vandalism •evidence of dumping •condition of the aquatic vegetation (disease, pest infection, dead plants)
Cleanliness of surrounding landscape	1. The extent to which the surrounding landscape of these waterways is clean in terms of odour, litter, colour, and erosion. 2. The extent to which the maintenance of these waterways meets the maintenance checklist of Melbourne Water's Constructed Wetlands Guidelines, where inspection is carried out every 3 months. For example, the waterways are inspected and maintained in terms of: •Evidence of odours •litter accumulation •evidence of damage or vandalism •evidence of dumping •condition of the aquatic vegetation (disease, pest infection, dead plants)

Drivers	Standards and guidelines
Access	1. The extent to which there is easy physical and visual access to these waterways and the surrounding vegetation. 2. The extent to which the means of access meets the guidelines developed as best management practice of Melbourne's waterways. For example, these waterways meet: •Melbourne Water's Shared Pathways Guidelines (includes bridges and crossings) •Melbourne Water's Access for Maintenance guidelines
Naturalness	1. The extent to which these waterways take 'natural-form' and resemble natural components of the landscape. 2. The extent to which the form, roughness, and pools and riffles of the waterway meet Melbourne Water's Constructed Waterways in Urban Developments Guidelines. 3. The extent to which the form of these waterways meet the Australian Rainfall and Runoff guidelines, section 1.7 and 1.8 and Melbourne Water's Land Development Manual.
Water quality	1. The extent to which these waterways meet the Australian and New Zealand guidelines for fresh marine water quality 2. The Victorian Index of Stream Condition (ISC) is a tool used to categorise the ecological health of our waterways. The ISC is made up of five sub-indices: hydrology, streamside zone, physical form, water quality and aquatic life. Each river reach assessed is given an overall ISC score of between 0 and 50. This score is then categorised into one of five broad condition bands – excellent (>70 %), good (51–70 %), moderate (31–50 %), poor (11–30 %) or very poor (<10 %). What is the water quality score of the waterway.
Hydrological Regime	1. The Victorian Index of Stream Condition (ISC) is a tool used to categorise the ecological health of our waterways. The ISC is made up of five sub-indices: hydrology, streamside zone, physical form, water quality and aquatic life. Each river reach assessed is given an overall ISC score of between 0 and 50. This score is then categorised into one of five broad condition bands – excellent (>70 %), good (51–70 %), moderate (31–50 %), poor (11–30 %) or very poor (<10 %). What is the hydrological regime score of the waterway.

5 IMPLICATIONS FOR EFFICIENT REGULATION AND SERVICE PROVISION

In the previous section we illustrated how expert knowledge could be systematically employed to provide transparent links between activities by waterway managers and outcomes. However, to make use of this information to improve economic efficiency, mechanisms are needed that translate management choices into costs and benefits.

The basic tenet of cost-reflective pricing is that water and the services that relate to it should be used only up to the point where benefits equate costs. The theoretical elegance of this approach is that water is then deployed efficiently, in the economic sense at least.

However, there remains a major gap in this method as not all benefits that attend water are clearly visible in markets; for example, there is no market for the environmental or amenity benefits from improved management of urban waterways.

Some benefits can be easily inferred from risk assessment and market data, like flood mitigation, but others are far more complex and often overlooked or aggregated. Against that background there is a risk that environmental and amenity benefits from these streams will be under-supplied because agencies responsible for waterway management, like Melbourne Water, are constrained to deliver only those services for which there is a measurable demand. Set against this is the possibility of over-supply of such services where potential beneficiaries, frustrated by the absence of a clear measure of environmental and amenity benefits, mobilise political forces that result in over-provision.

Economists have spent the last 50 years developing and refining techniques that can reasonably measure demand where there is no market. When carefully used, such techniques have the potential to limit under-provision and/or over-provision of a range of non-market goods (Rolfe et al. 2000). These techniques fall into two broad categories: market-related measures and stated preference approaches. The former uses existing markets to act as a proxy for demand for the underlying environmental good (e.g. Tapsuwan et al. 2012), but such approaches are notoriously data intensive and, in any case, do not capture all benefits. In contrast, the stated preference approaches 'create' a hypothetical market and then use survey data to enumerate a generalizable value for environmental change. There are several stated preference approaches but a technique known as Choice Modelling has been widely recommended for many water-related applications (e.g. Cooper et al. 2011).

Choice Modelling 'creates' a market for goods with each comprising different attributes. Respondents are asked to compare goods and choose one over the other. The attributes used to describe each good must include a price but other relevant features, such as indicators of ecological health, accessibility for recreation or features that might add to other elements of amenity could be included. By repeating the choice experiment with different levels and across a sample of the population it is then possible to statistically identify how different segments of the community trade off price against other attributes. This becomes the basis on which willingness to pay, or demand, can be estimated.

Whilst major advances have been made in many areas of Choice Modelling, an ongoing challenge relates to the description of complex attributes in a form that is accessible for those participating in choice experiments. As a general rule, practitioners of Choice Modelling seek to generate attributes that: (1) can be traced back to policy or management options, (2) are grounded in at least some scientific information, and (3) are also sufficiently simple to encourage a high response rate from the general population. In the context of water quality, attributes are thus commonly described by suitability for use - potable, swimmable, suitable for boating (see, Rolfe and Windle 2011) - albeit with limited direct reference to water chemistry or related disciplines. In the case of something as complex as the ecological health of waterways, attributes have been described in terms of the number of species that are protected or the length of waterway that is of a particular standard (e.g. Kragt and Bennett 2011). Arguably, these measures act as proxies that illustrate community preferences, albeit within the limited information sets provided to respondents.

Perhaps not surprisingly, Choice Modelling is not without its critics, in part because of the necessity to simplify important scientific information in order to generate responses from the general population (Hanley et al. 2001). In the context of urban waterways in particular, the measurement challenge for practitioners of Choice Modelling is exacerbated by the suite of 'other' psychological relationships between well-functioning urban waterways and human well-being, or amenity.

In this regard, we use the term 'amenity' to describe the genre of so-called 'non-ecological' benefits, although the literature is somewhat divided on this point. For example, in the lexicon of the ecosystem services approach amenity forms a subset of cultural values (Costanza et al. 1997) but amenity is also seen as separate and yet overlapping with tourism, recreation and non-use values (MacDonald et al. 2011). We noted earlier that in the case of urban waterway management, the term 'amenity' is frequently used by water agencies as a catch-all for human values that are not directly associated with ecological characteristics.

Völker and Kistemann (2011, p. 458) conclude that "urban environments, although the everyday environment for most of us and therefore important for any goals towards health promotion are underrepresented in current literature on blue therapeutic landscapes" and "there is little respect for water and health in urban planning issues".

It is our contention that the method and results presented in Section 4 specifically address these challenges. Having systematically identified the relationship between the drivers and components of ecological and human wellbeing we are able to plan and compare alternative actions that relate to waterway management.

Importantly, this also provides much greater confidence around the application of non-market techniques which are commonly used to assign values to management options.

6 CONCLUDING REMARKS

Waterway management is complicated by the requirement to integrate multiple values, including ecological and amenity values. This is becoming increasingly challenging in urban contexts and where increasing incomes require greater attention to environmental and amenity values. Set against the challenges of IWRM is the need to simultaneously align benefits and costs of various waterway users in order to fulfil economic efficiency criteria.

Without decomposing the factors that underpin ecological and amenity value it is not possible to systematically analyse the actions of waterway managers. This stands to undermine efforts to apply IWRM, since the links between management choices and multiple benefits are not clear. In addition, the absence of this information makes it difficult to describe the benefits that accrue and thus compare these against costs.

Efficient strategies that aim to maximise the benefits of waterways within a given budget require an understanding of the costs and benefits of investment decisions. Accordingly, it is imperative that we are able to trace how amenity and ecological benefits of waterways are generated. The results presented in this paper provide that information, at least in the context of Melbourne, Australia.

The results of this Delphi study also provide a sound foundation for informing the initial stages required for non-market valuation studies in this context. From a regulatory perspective, quantifying the benefits of urban waterways is of particular value. More specifically, the regulatory environment that shapes the operating environments of Melbourne Water Corporation requires that benefits be measured, where practicable, to guard against over or under investment. From a technical perspective, this study has improved the methodological developments in the measurement of non-market goods generally, and more specifically for the concepts ecological value and amenity value associated with Melbourne's waterways.

The Delphi results also provide additional insight into somewhat vague concepts such as liveability. The Victorian Competition and Efficiency Commission sought to offer a global definition of liveability in 2008 and argued that the concept reflected "the wellbeing of a community and compris[ed] the many characteristics that make a location a place where people want to live now and in the future". Melbourne Water subsequently announced that integrated management of waterways was a central component of the city's liveability. The Delphi has improved our understanding of the linkages between the management of urban waterways, human well-being and liveability.

ACKNOWLEDGMENTS

This research was undertaken with financial assistance provided by the Australian Research Council. We also acknowledge the helpful insights provided by expert waterway managers across Australia, especially Dr. Terry Hillman.

REFERENCES

1. Biswas A (2004) Integrated water resources management: A reassessment. Water Int 29:248–256
2. Cooper B, Burton M, Crase L (2011) Urban water restrictions: Avoidance and attitudes. Water Resour Res 47:W12527. doi:10.1029/2010WR010226
3. Costanza R, d'Arge R, de Groot R, Farberk S, Grasso M, Hannon B, Limburg K, Naeem S, O'Neill RV, Paruelo J, Raskin RG, Suttonkk P, van den Belt M (1997) The value of the world's ecosystem services and natural capital. Nature 387:253–260
4. Crase L, Pawsey N, Cooper B (2015) Water pricing in Australia: Unbundled politics, accounting and water pricing. In: Dinar A, Pochat V, Albiac J (eds) Water Pricing Experiences and Innovations. Springer, New York, pp. 15–39
5. De Villiers M, De Villiers P, Kent A (2005) The delphi technique in health sciences education research. Med Teach 27:639–645
6. European Commission (2000) Directive 2000/60/EC Establishing a framework for community action in the field of water policy. EC, Brussels
7. Gareau BJ, Crow C (2006) Ken Conca, governing water: contentious transnational politics and global institution building. Int Environ Agreements 6:317–320
8. Global Water Partnership – Technical Advisory Committee (GWP-TAC) (2000) Integrated water resources management. TAC Background Paper, No.4. GWP, Stockholm, Sweden
9. Hanley N, Maurato S, Wright RE (2001) Choice modelling approaches: a superior alternative for environmental valuation? J Econ Surv 15(15):435–462
10. Hasson F, Keeney S, McKenna H (2000) Research guidelines for the Delphi survey technique. J Adv Nurs 32(4):1008–1015
11. Hatcher T, Colton S (2007) Using the internet to improve HRD research. The case of web-based Delphi research technique to achieve content validity of an HRD-orientated measurement. J Eur Ind Train 31:571–587
12. Horwitz, P. and M Carter (2011) Access to inland water for tourism: Ecosystem services. Water Policy, Tourism and Recreation: Lessons from Australia, L. Crase and S. O'Keefe (eds). Washington, Resources for the Future 3352–3366
13. Hsu C, Sanford BA (2007) The Delphi technique: making sense of consensus. Practical Assess Res Eval 12(10):1–8
14. ICWE (1992) The Dublin Statement and Report of the Conference. International Conference on Water and the Environment: Development Issues for the twenty-first century. 26–31 January. Dublin
15. Kragt M, Bennett J (2011) Using choice experiments to value catchment and estuary health in Tasmania with individual preference heterogeneity. Aust J Agric Resour Econ 55(2):159–179
16. Lamorce B, van Steenbergen F (2006) From value to finance: making IWRM work, journal of contemporary water research and. Education 135:100–106
17. MacDonald, D. H., S. Tapsuwan, S. Albou, and A. Rimbaud (2011) Challenges of Estimating the Value of Tourism and Recreation. Water Policy, Tourism and Recreation: Lessons from Australia S. O'Keefe and L. Crase. Washington, Resources for the Future 33–52.

18. Maybery D, Reupert A, Goodyear M, Patrick K, Crase L (2009) Prevalence of children whose parents have a mental illness. Psychiatr Bull 33:22–26
19. Melbourne Water (2013) Healthy Waterways Strategy: A Melbourne Water strategy for managing rivers, estuaries and wetlands, http://www.melbournewater.com.au/aboutus/reportsandpublications/key-strategies/Documents/MW-HWS-Chapter1_Executive%20%20summary%20Introduction.pdf
20. Molle F (2008) Nirvana concepts, narratives and policy models: insights from the water sectors. Water Alternatives 1:131–156
21. Office of Living Victoria (2013) Melbourne's Water Future. Melbourne, OLV Raine S (2006) Defining the Bobath concept using the Delphi technique. Physiother Res Int 11(1):4–13
22. Raymond CM, Bryan BA, MacDonald DH, Cast A, Strathearn S, Grandgirard A, Kalivas T (2008) Mapping community values for natural capital and ecosystem services. Ecol Econ 68(5):1301–1315
23. Richey JS, Mar BW, Horner RR (1985) Delphi technique in environmental assessment. I. Implementation and effectiveness. J Environ Manag 21:135–146
24. Rolfe J, Windle J (2011) Assessing community values for reducing agricultural emissions to improve water quality and protect coral health in the great barrier reef. Water Resour Res 47. doi:10.1029/2010WR010355
25. Rolfe J, Bennett J, Louviere J (2000) Choice modelling and its potential application to tropical rainforest preservation. Ecol Econ 35:289–302
26. Savenije HG, van der Zaag P (2002) Water as an economic good and demand management paradigms with pitfalls. Water Int 27(1):98–104
27. Serena M, Williams B (2008) Distribution and management of platypus in the Greater Melbourne Region, July 2008, (Report to Melbourne Water), Australian Platypus Conservancy, Melbourne
28. Tapsuwan S, Hatton-MacDonald D, King D, Poudyal N (2012) A Combined Site Proximity and Recreation Index Approach to value Natural Amenities: An Example from a Natural Resource Management Region of the Murray-Darling Basin. J Environ Manag 94:69–77
29. The Economist (2003) Survey: liquid assets. 368: 13
30. The Economist Intelligence Group (2013) Liveability Rankings. Retrieved 20 May, 2014, from http://www.eiu.com/Handlers/WhitepaperHandler.ashx?fi=WEB_ Liveability_rankings_Promotional_August_2013.pdf&mode=wp&campaignid=Liveability2013
31. Thomas J, Durham B (2003) Integrated water resource management: looking at the whole picture. Desalination 156(1–3):21–28
32. United Nations (1992) Chapter 18. Protection of the Quality and Supply of Freshwater Resources: Application of Integrated Approaches to the Development, Management and Use of Water Resources, Agenda 21
33. Veldarde M, Fry G, Tveit M (2007) Health effects of viewing landscapes – landscape types in environmental psychology. Urban For Urban Green 6:199–212
34. Völker S, Kistemann T (2011) The impact of blue space on human health and well-being – Salutogenetic health effects of inland surface waters: a review. Int J Hyg Environ Health 214(6):449–460
35. Walther P (1987) Against idealistic beliefs in the problem-solving capacities of integrated resource management. Environ Manag 11:439–446
36. Whitman E, Maybery D, Reupert A, Fudge E (2009) Use of a web-based Delphi study in the development of a training resource for workers supporting families where parents experience mental illness. Int J Ment Health Promot 11:42–54

CHAPTER 11

Water Supply Network Losses in Jordan

Nadhir Al-Ansari[1*], N. Alibrahiem[2], M. Alsaman[2], Sven Knutsson[1]

[1]*Lulea University, Lulea, Sweden*

[2]*Al al-Bayt University, Mafraq, Jordan*

ABSTRACT

Water supply network losses are an international problem especially in countries suffering from water scarcity like Jordan. Jordan is one of the poorest countries in its water resources and it is estimated to be below the water poverty line. Jordan is located in the Middle East and has a surface area of approximately 90,000 km^2. Its population is around 6.3 million and it is estimated that the population will be 7.8 million in 2022. The gap between water supply and demand is widening due to development and a relatively high population growth rate. In addition, global climate change is expected to intensify the water shortage problem in Jordan. Thirteen years of complete records obtained from the Ministry of Water and Irrigation were analyzed. According to these records, water losses in Jordan reach about 50%. In view of the evaluation of the data and the case study conducted in this research, it is believed that Jordan can overcome the water shortage problem by adopting a water demand management strategy. In this context, efforts should be focused on reducing water losses. If this is achieved, it will save huge quantities of water and revenue.

Keywords: Jordan; Water Supply Network; Water Losses; Neamie

1. INTRODUCTION

Water supply network losses are an international problem in developed countries [1-4] as well as developing countries [5-7]. The situation becomes more critical in countries suffering from water shortages like Jordan. Jordan has a surface area of approximately 90,000 km^2 (**Figure 1**) and its population is around 6.3 million. It is estimated that the population will be 7.8 million in 2022 [8]. Jordan is located in the Middle East, at the eastern end of the Mediterranean Sea. The most distinctive topographic features of Jordan are the Rift Valley, a narrow strip of highlands (with a maximum elevation of 1600 m above sea level), the steppe, the desert zone and the Dead Sea (426 m below sea level in 2010).

The climate is of a Mediterranean type in western Jordan which is characterized by dry, hot summers, mild wet winters and extreme variability in rainfall during the year as well as from year to year. The climate in the highlands is characterized by cold winters and mild summers. The Jordan Rift Valley and Aqaba have a subtropical climate—hot in summer and warm in winter. The desert region has a continental climate with large variations in temperature. Precipitation is very low and ranges from 30 to 600 mm annually. About 93.5% of the country has less than 200 mm of rainfall, and only 0.7% of the country has annual precipitation of more than 500 mm [9]. The rainy season starts in November and ends in April. Most of the country (83%) is composed of desert and desert steppe.

The total precipitation amounts to 8.2 km^3 and about 92% of it is lost by evaporation. The country is estimated to be below the poverty line by 682 million m^3/year, The developed surface water potential was approximately 295 million m^3 in 2007, and is projected to reach 365 million m^3 by 2022 [9].

Jordan relies mainly on its groundwater resources which accounts for about 54% of water supply in from 12 groundwater basins (**Figure 2**). Renewable groundwater resources are 450 million m^3/year with a safe yield of 275.5 to million m^3/year [10]. Abstraction of groundwater is beyond sustainable limits causing a deficit of 151 million m^3/year in 2007 as an example [9]. Surface water resources form about 37% of the total water supply from 16 basins (**Figure 3**). Yarmouk accounts for almost 50% of the country's surface water resources. Other water sources are fossil water, treated wastewater (110 million m^3 in 2009) and brackish water. The contribution of treated wastewater is increasing with time where it was only 7 million m^3 in 1984 and increased to about 67 million m^3 in 1996 [11]. For these reasons, Jordan is considered to be one of the poorest countries in terms of its water resources.

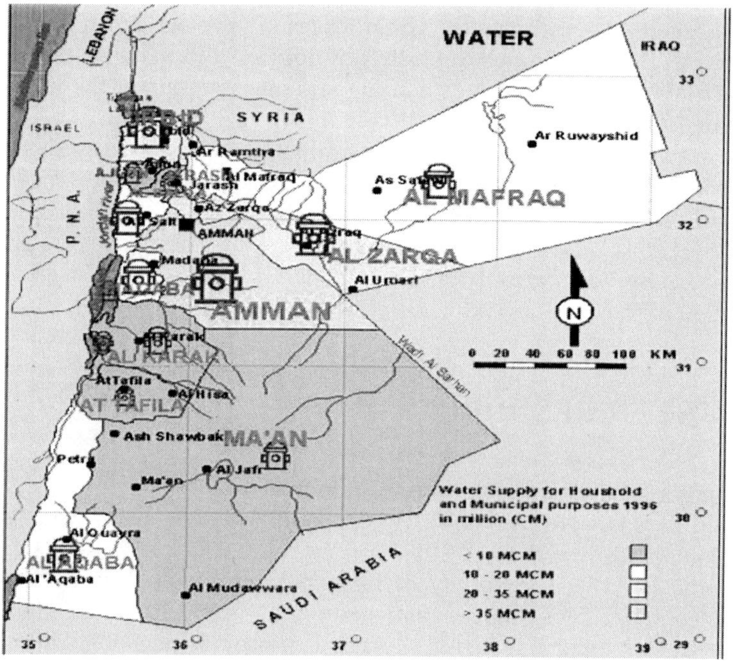

Figure 1. Location map of Jordan.

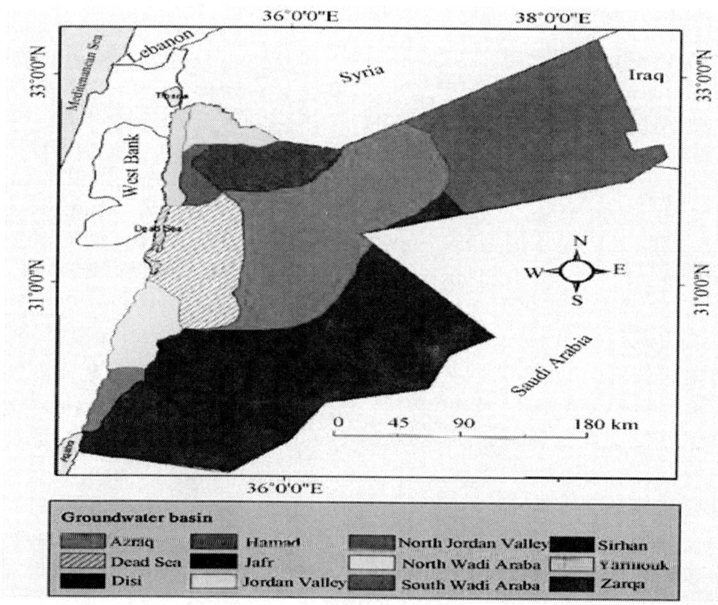

Figure 2. Ground water basins in Jordan.

The difficulties to estimate the water losses in developing countries are the access to the data and the distribution of the data within several departments. This research is a study and investigation of the information obtained from the Ministry of Water and Irrigation about the extracted, supplied and sold water quantities. Complete records were studied for the period 1986 to 1998 in particular. Water losses in Jordan reached about 50% of the supplied water quantities [5] where the accepted rate should be less than 25% [12]. The benefits of reducing these losses are discussed and a case study was carried out to help decision makers to overcome the problem of shortages in water supply.

2. WATER RESOURCES OF JORDAN

A number of authors have reported shortages of water resources in the Middle East [13-16]. At least 12 Arab countries have acute water scarcity problems with less than 500 m^3 of renewable water resources per capita available [17,18]. The supply of fresh potable water is essential to life, socioeconomic development, and political stability in the region. It was reported that one cubic meter of water can provide drinking water for one person for one year or the same quantity can produce only one kg of food grain when used for irrigation in a dry climate [19]. The need for a rationalized holistic management of this most vital natural resource is paramount in order to attain a sustainable society.

Jordan is among the most arid countries in the world and the allocation per capita reaches less than 150 l/capita/day [17,18,20]. Aquastat [21] stated that the water withdrawal by source in 2005 was, Groundwater 59%, surface water 31% and wastewater contributions and others 10%. Due to the sharp increase in population the per-capita water availability, dropped from 3600 m^3 in 1946 to 145 m^3 in 2008 [9]. By 2022, the population may exceed 7.8 million, and total water demand may reach 1673 million m^3. Water deficit of 659 million m^3 (2009) could be reduced to 457 million $m^3$by 2022 [9]. The Jordanian government constructed 28 dams between 1950 and 2008 with a total storage capacity of 368 million m^3 and at the same time, locations were identified for a number of reservoirs that would give the potential to add 444 million m^3 to Jordan's water storage capacity.

Municipal water consumption increased by an average of 8% per year while it was 6.6 and 3% within the industrial and agricultural sectors respectively. The increase of water consumption in these sectors is due to the increase in investments in these sectors of the economy. According to the records the average consumption per capita in the municipal sector reaches 138 l/capita/day while it reached 147 and 145 l/capita/day in 2010 and 2011 respectively [22]. When water losses are taken into consideration, this amount decreases to 70 - 75 l/capita/day. Within the industrial and agricultural sectors the allocated water per capita (as an average) reaches about 126 and 493 l/capita/day respectively. In 2005, the allocation was 165 m^3/capita/day [21], and expected allocations in 2015, 2025 and 2050 are 130, 121, 100 m^3/capita/day respectively [23]. El-Fadel and Maroun [24] stated that Jordan has a shortfall in supply by 53%.

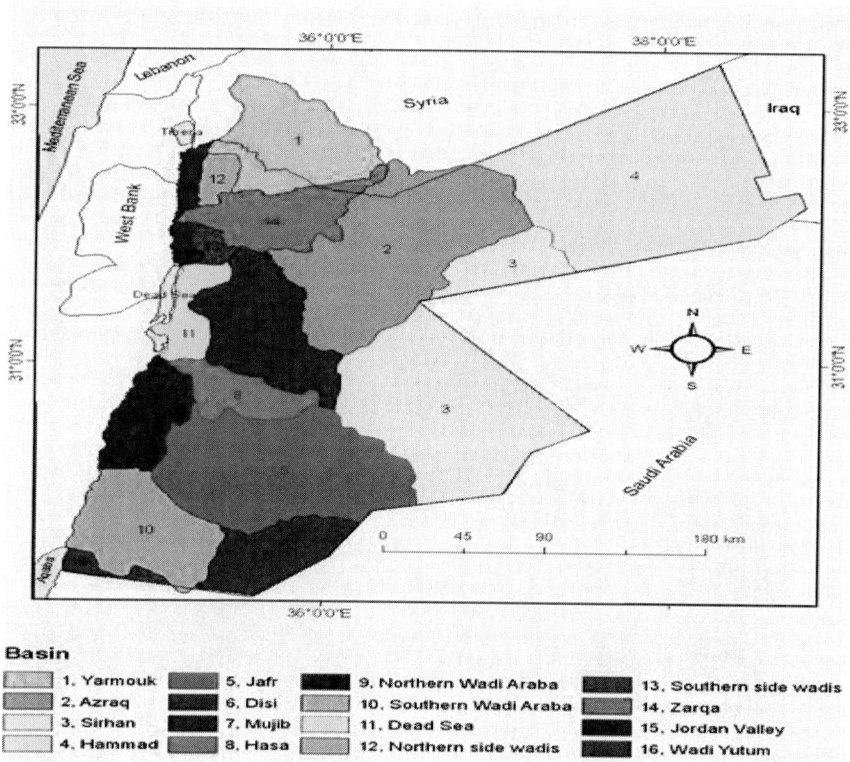

Figure 3. Surface water basins of Jordan.

The Ministry of Water and Irrigation [22] stated that municipal water supplied in 2009, 2010 and 2011 reached 313.4, 330 and 330.1 million m^3 respectively. This gives 144,147 and 150 l/capita/day for these years respectively. Furthermore, the total amount of water released for drinking and irrigation purposes in 2010 and 2011 were 229.675 and 222.272 million m^3 [22].

The highest percentage of water consumption is in the field of Agriculture followed by municipal and then industrial sectors [20]. Nimah [23] indicated that by 2050 the available water in Jordan will be 100 m^3/capita/year. The Ministry of Water and Irrigation [22] published the projected water demand for the year 2025 and the type of water use (Figures 4 and 5). In this report the municipal, industrial and agricultural demands were 613,117, and 700 million m^3 respectively. It was also noticed that they added 100 and 29 million m^3 as nuclear power plant and touristic water demands. In addition, the same report showed that the total demand will reach 1695 million m^3 in 2030 (710, 150, 700, 100, 35 million m^3 for municipal, industrial, agricultural, nuclear and touristic water demands respectively). Jordan's Water Strategy Plan [8] indicated that the available water in 2022 is 1632 million m^3 (**Figure 6**).

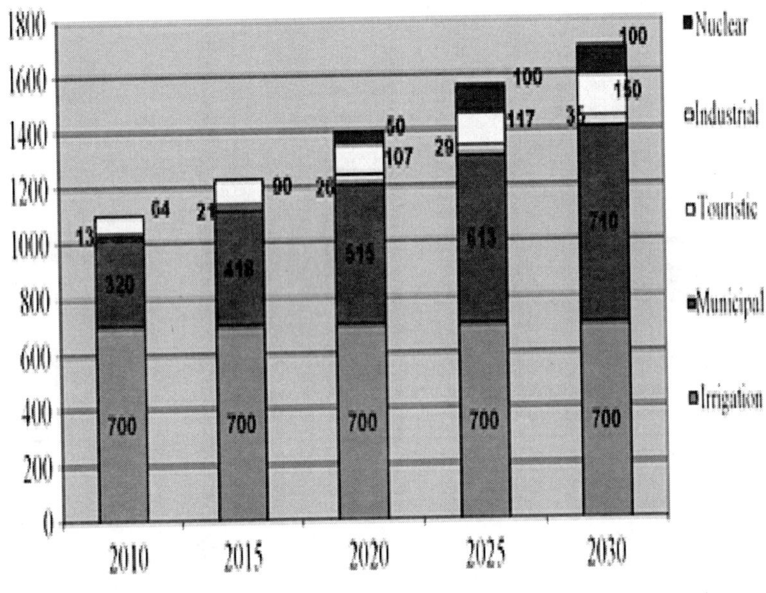

Figure 4. Projected water demand in Jordan [22].

The largest consumer of water across the region is agriculture which accounts for 66% of demand, [25] and therefore the water shortage problem cannot be objectively analyzed nor adequately addressed without a tho rough consideration of agriculture in the region [26]. For example an approximate 10 percent transfer of water away from agriculture would produce a 40 percent increase in domestic water supply for Jordan [26]. Postel [27], argues that rather than diverting precious water to agriculture this water could be saved by importing the food/grain. However, this is not the case in many Middle Eastern countries which have unrealistic aspirations of food self-sufficiency and in it would require a most fundamental change in national outlook [28]. Sadik and Barghouti [26] emphasized that the extent of the problem is so severe that "the future challenges in meeting the growing demands for water are beyond the capabilities of individual countries". Mitigating supply shortfalls could be achieved by re-allocation of current agricultural supplies [13].

The Ministry of Water and irrigation usually satisfy the demand through supply management practices and with the diminishing opportunities to increase water supply and the escalating costs of the projects to augment supply, the ministry should give greater emphasis to demand management [11,26,29,30].

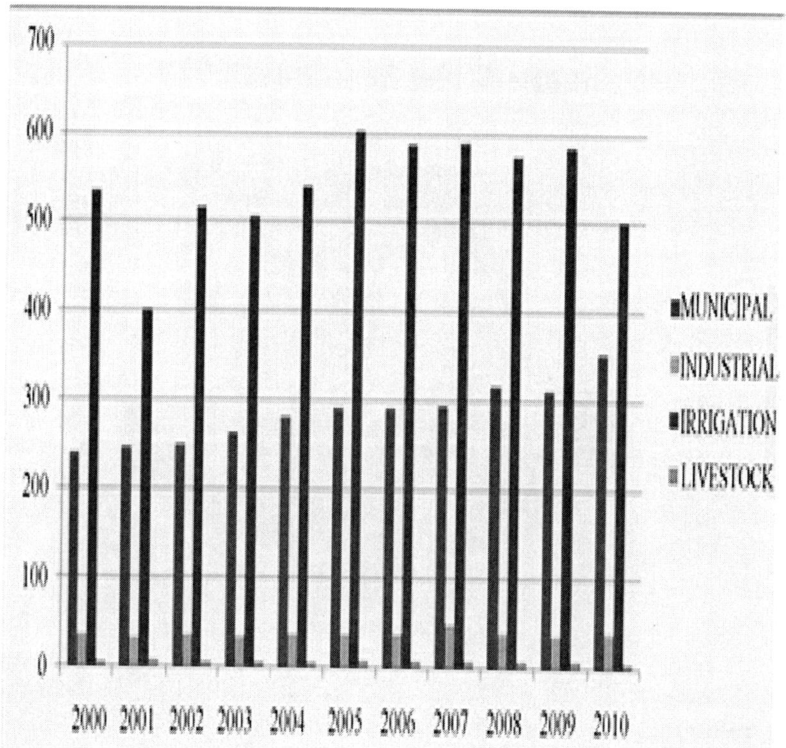

Figure 5. Type of water use in Jordan in million cubic meters (MCM) [22].

In addition to the above, the effect of climate change is expected to intensify the water shortage problem in Jordan. The Middle East and North Africa region (MENA) contains hyper-arid, arid and semi-arid zones [31]. Several research projects have concluded that arid and semi-arid regions are highly vulnerable to climate change (e.g. [32]). It is expected that the region will suffer from higher temperatures and intense heat waves affecting inhabitants and crop yields, and will also affect marine ecosystems and fisheries. Less but more intense rainfall, coupled with higher temperatures, will likely cause more droughts and greater flooding, sea level rise, more intense cyclones and new areas exposed to dengue, malaria, and other vector and waterborne diseases.

Al-Ansari and Baban [33] and Al-Ansari et al. [34,35] indicated that future rainfall forecast is decreasing with time in Jordan. The drought will affect the agricultural life and water supply [36]. This is due to the fact that most of the agricultural areas of the MENA region are rain-fed [37] and decreases groundwater recharge which is already depleting [38].

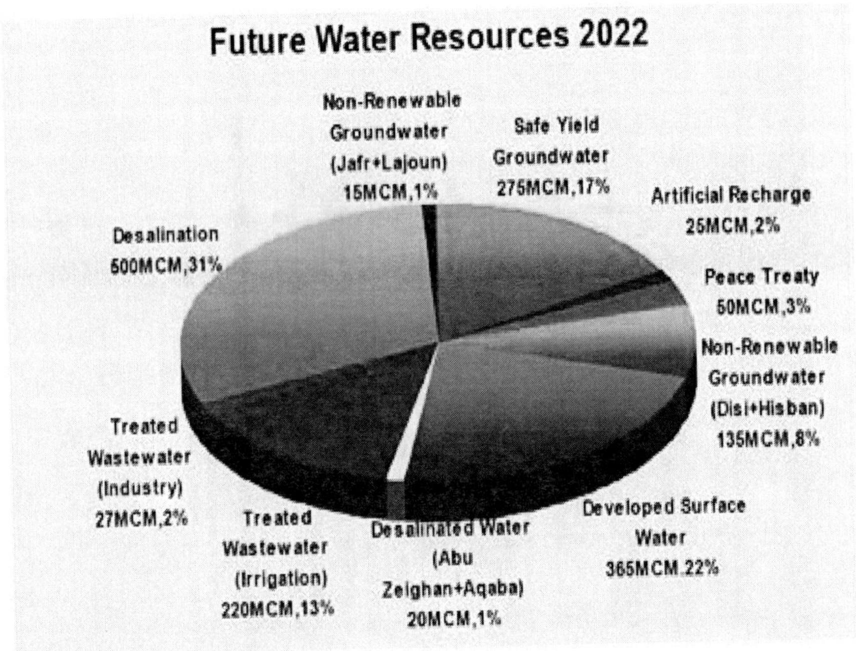

Figure 6. Future available water resources in Jordan 2022 [12].

3. WATER LOSSES

Water losses are defined here as the water which has no revenue [5]. These can be classified as:

Technical losses: they result from breakdown or seepage within the distribution pipes systems (**Figure 7**). This type of loss can be divided into three groups:

Apparent technical loss: this is due to apparent breakdown of the distribution net systems.

Figure 7. Examples of technical losses: left-broken pipe; middle-leaking pipe; right–breaking pipe to supply water to livestock.

Unseen loss: such loss is due to the breakdown of the pipes in the distribution network under the ground.

Administrative losses: these are due to the existence of broken water quantity meters, human errors (e.g. issuing wrong bills), not installing meters correctly, using the water illegally, centralization, and weak governance (**Figure 8**). Charalambous [39] has summarized what happens typically to the distributed water (**Table 1**).

To calculate the loss the following equation is used:

Quantity of water loss (L) = Net supplied water (Ss) – Net sold water (So) + Water consumed outside water meters (C).

Records extending from 1986 to 1998 were investigated to see the percentage of water loss in different governorates in Jordan in addition to the data given by the World Bank for the period 1999 to 2004 (**Table 2**). Accordingly, it was evident that almost all the governorates lose about 50% of the supplied water (**Table 2**).

Table 1. Water distribution [39].

Distributed Water	Consumed Water	Water with revenue	Waters with bills
			Valued water
		Water without revenue	Free water
			Desert station water
	Water loss	Administrative loss	Illegal Connections
			Broken counters
		Technical loss	Seepage fom distribution net
			Seepage from pipes between towns and villages
			Seepage and excess water from tanks

The Mafraq governorate had the highest water losses (69.13%). This is believed to be due to the very large area of the governorate (26,541 km^2) and this is in part due to the length of the water supply pipelines which amounts to about 10,000 km. Consequently, long supply pipelines make it hard for the inspection and maintenance teams to report or repair seepages and leakages [40]. In addition, this governorate's population livelihood is based on livestock. In many cases, livestock owners deliberately make seepages in the supply pipe lines using guns, or they break the lines in valleys, in order to secure water needs for their livestock [41]. Furthermore, it seems that as the water supply increases with time, so do the losses where it is increasing with an annual average of about 5%. The records indicate that the administrative losses were between 27% - 32.8% and the illegal water use had the highest percentage of 11.8% followed by faults in water meters (10.6%). This is followed by faulty readings (5.3%) and faulty estimation of the volume of water consumed (3%). On the other hand, technical losses reached an average of 25.6% from the total volume of water supplied. These figures are considered very high compared to developed countries where it is about 10% [43,44].

Figure 8. Examples of administrative loss in Mafraq: leftthe meter is removed and replaced by a T-joint; middle-pipe connection passing the meter; right-connection to supply the housing unit before the meter.

Table 2 . Percentage water losses in governorates of Jordan (Sources of data from 1999 to 2004 is [42]).

Governorate	Amman	Irbid	Zarqa	Madaba	Balqa	Karak	Tafela	Maan	Mafraq	Ajloun	Jerash
1986	54.9	55.2	52.5	-	54.2	49.2	56.5	49	54.2	-	-
1987	55.1	56.2	51.8	-	52.6	51.4	54.3	51.6	55.1	-	-
1988	57.4	55.3	52.1	-	53.8	53.3	54.6	52.8	57.8	-	-
1989	58.1	55.6	53.2	-	54.5	55.1	55.8	51.6	60.3	-	-
1990	59	56.6	54.4	-	56.1	55.9	55.5	53.6	68.2	-	-
1991	58.3	50.9	52.2	-	55.8	53.8	50.9	51.1	68.7	-	-
1992	60.5	57.1	55.8	-	60.3	59.8	65.8	54.6	72.3	-	-
1993	57.8	52.3	52.3	-	67.3	54.1	53	50.6	75.3	-	-
1994	52.6	52.5	50.5	-	66.5	52.7	46.2	52.2	70.8	-	-
1995	54.4	55	55.1	-	62.2	53.6	53.8	53.8	76.2	-	-
1996	50.25	49.6	54.3	67.2	59.6	50.2	52.7	56.7	68.66	57.8	50.2
1997	48.54	49.5	54.2	77.34	61.2	57.1	44.8	65.8	78.9	60	47.4
1998	49.5	48.9	56.4	86.1	62.1	59.4	43.9	67.3	78.5	60.3	60.1
1999	50.03	45.71	55.34	75.03	56.13	56.77	47.51	62.27	76.39	47.96	42.94
2000	50.29	44.34	54.75	58.21	54.73	56.35	41.34	59.86	74.57	44.67	44.71
2001	50.03	41.92	55.03	58.33	46.76	56.59	38.85	53.51	74.34	37.19	33.56
2002	47.29	42.16	55.78	50	53.94	51.91	44.58	52.31	70.61	40.56	37.18
2003	48.53	39.57	51.48	46.46	51.59	47.47	43.31	46.78	68.21	37.41	23.8
2004	44.89	35.08	51.2	46.5	53.86	47.51	41.25	45.19	64.5	28.19	29.38
Average	53.02	49.66	53.6	62.8	57.01	53.8	49.72	54.24	69.13	41.4	36.93

The high percentage of water losses affect the quantity of water available and this usually forces the water authorities to search for new water resources to compensate for the losses rather than manage demand and leakage. In 1986, the allocation per capita per day was 92 l/capita/ day, while the allocation drops to 42 l/capita/day after counting the losses. In 1998 however, the allocation capita/day was 144 liter and drops to 62 liter after counting the losses. The decrease in allocation per capita is ranging from 42% to 45% (see **Table 3**).

It should be noted that the administrative loss does not affect the total water supply, simply because these quantities of water do exist but were consumed in an illegal way while the technical losses affect the total volume of water supplied. The records indicate that the volume of supplied water decreased to 97.6 and 162.3 MCM in 1984 and 1998 respectively while the technical losses were 26.8 and 23.5% respectively (**Table 4**). This leads to an estimated loss 35.9 and 56.8 million Jordanian.

Dinars (JD) (51.3 - 81.14 US Dollars) in 1986 and 1988 respectively. This was estimated using the using the following equations:

$$\mathbf{WP} = f\{\mathbf{WS}, \mathbf{Tr}, \mathbf{Ar}\} \quad (1)$$

$$\mathbf{WP} = \mathbf{a} + \beta\mathbf{WS} - \mathbf{Tr}(\mathbf{WS}) + \mathbf{Ar}(\mathbf{WS}) \quad (2)$$

where:

WP: Volume of Consumed water

WS: Volume of supplied water

Tr: Percentage technical water loss

Ar: Percentage administrative water loss

Using the least squares method, then we get:

$$\mathbf{WP} = -10.25 + 0.95(\mathbf{WS}) - 0.69(\mathbf{WS}) + 0.04(\mathbf{WS}) \quad (3)$$

This indicates that technical water loss will cause a decrease of 0.69 m^3 of the offered water and administrative loss will cause an increase of 0.04 m^3.

If the technical water loss is decreased by 15% then the volume of consumed water will increase (see **Table 5**). For 1998 records the water consumed will increase from 205.3 to 217.4 million m^3.

If we consider the water allocation is 200 l/capita/day then the required water in 1986 will be 204.1 million m^3 and increases to 335.9 million m^3 for the year 1998. For the purpose of finding the effect of the water losses on the required quantities of water, the following equation was used:

$$Wd = Wn + Tr(WP) - Ar(WP) \tag{4}$$

where:

Wd: Needed water

Wn: Total water required

Wp: Water supply

Tr: Percentage technical loss

Ar: Percentage administrative loss

This shows that the administrative loss will decrease the quantity of the required water (**Table 6**). If we take 1986 as an example then the quantity of needed water is 204.1 million m^3 and if we add the total losses then this quantity increases to 276.5 million m^3 and we already know the quantity of water which is referred to as administrative loss (36.4 million m^3) already exist within the quantity required (**Table 6**).

As for the year 1998, the total quantity of water required, including the total loss, is 471.9 million m^3 and when we disregard the administrative loss the required water quantity is 370 million m^3 (**Table 6**).

For future predictions, it was assumed that the water losses will be dealt with and will decrease to 20, 15 and 10% in 2015, 2020 and 2025 respectively (**Table 7**). In this case the required water quantities will be 476.9, 610.7 and 747.2 million m^3 in 2015, 2020, and 2025 respectively (**Table 7**). From an economic perspective, 70% of the cost was considered for water supply while the remainder, 30%, was considered as sewerage services costs [42,45].

The cost of one cubic meter of extracted water was calculated through the actual expenditure involved and it was found to be 0.170 and 0.390 JD for the years 1986 and 1988 respectively indicating an increase of 125%. Furthermore, the selling price of one cubic meter of water was 0.37 and 0.88 JD for the years 1986 and 1998 respectively. Using the financial balance sheets of the water authorities during the study period, it is found that revenue from the sale of one cubic meter of extracted water will be 0.1 to 0.166 JD for the years 1986 and 1998 respectively indicating an increase of 66%.

Table 3 . Water supplied and lost for the period 1984-1998.

Year	Population (1000)	Water (1000 m^3)			Lost water %	Supplied Water	Sold Water
		Supplied	Sold	Lost		l/capita/day	
1984	2596	87	39.8	47.2	54.3	92	42
1985	2700	93	42.5	50.5	54.3	94	43
1986	2796	134	61.6	72.4	54	131	60
1987	2896	150.3	68.5	81.8	54.4	142	65
1988	3001	164.5	72.7	91.8	55.8	150	66
1989	3111	170.1	74.4	95.7	56.3	150	66
1990	3468	174.6	74.1	100.5	57.6	138	59
1991	3701	178.5	78.1	100.4	56.2	132	58
1992	3844	208.1	81.5	126.6	60.8	148	58
1993	3993	218.6	92.3	126.3	57.8	150	63
1994	4138	219.9	97.9	122	55.5	146	65
1995	4152	246.2	103.3	142.9	58	162	68
1996	4261	242.3	108.5	132.8	55.2	155	69
1997	4441	239.8	108.7	131.1	54.7	148	67
1998	4601	241.5	105.6	135.9	56.3	144	63

Table 4. Volume of supplied water and percentage of lost water.

Year	Supplied water MCM	Lost water %	Technical water loss %
1984	87	54.3	27.5
1985	93	54.3	27
1986	134	54	26.8
1987	150.3	54.4	26.6
1988	164.5	55.8	26.9
1989	170.1	56.3	27.8
1990	178.6	58.5	26.8
1991	178.5	56.2	26.7
1992	208.1	60.8	25.1
1993	218.6	57.8	26.7
1994	219.9	55.5	24.4
1995	246.2	58	26
1996	242.3	55.2	22.9
1997	239.8	54.7	22.9
1998	241.5	56.3	23.5

Table 5. Effect of decreasing technical water loss on actual consumed water.

Year	Extracted water MCM	25% Technical loss	Consumed Water MCM	15% Technical loss	Consumed Water MCM	10% Technical loss	Consumed Water MCM
1986	134.7	33.7	101	20.2	114.5	13.5	121.2
1987	150.3	37.6	112.7	22.5	127.8	15	135.3
1988	164.5	41.1	123.4	24.7	139.8	16.5	148.1
1989	170.1	42.5	127.6	25.5	144.6	17	153.1
1990	174.6	43.7	131	26.2	148.4	17.5	157.1
1991	178.5	44.6	133.9	26.8	151.7	17.9	160.7
1992	208.1	52	156.1	31.2	176.9	20.8	187.3
1993	218.6	54.7	164	32.8	185.8	21.9	196.7
1994	219.9	55	164.9	33	186.9	22	197.9
1995	246.2	61.6	184.7	36.9	209.3	24.6	221.6
1996	242.3	60.6	181.7	36.3	206	24.2	218.1
1997	239.8	60	179.9	36	203.8	24	215.8
1998	241.5	60.4	181.1	36.2	205.3	24.2	217.4

When the revenue of one cubic meter of sold water is calculated it will be 262 and 381 JD for the years 1986 and 1998 respectively. This implies that the cost of lost water was 8.74 million JD (1986) and 22.6 million JD (1998) showing an increase of 158% (**Table 8**).The actual price or cost of the lost water is its cost plus the revenue if it was sold as follow:

$$TC = WL * Acp + WL * Ars \tag{5}$$

where:

TC: Total cost

WL: Quantity of lost water

Acp: Average cost of 1 $m^3$of extracted water

Ars: Average revenue 1 m^3 of sold water

Using the above equation, the cost of lost water reached 21.2 million JD (1986) and 75 million JD (1998) (**Table 8**) while the financial budget of the water authority was showing a negative balance of 3 million JD (1986) and 32.8 million JD (1998).

The total value of lost water reached 131%, and 186% of the income of the water authority in 1986, and 1998 respectively (**Table 8**). There is a decrease noticed in 1997 which is due to the introduction of a new water tariff (**Table 8**). If we consider the total studied record (1986-1998) then the accumulated total cost of the lost water reaches 552 JD. In case we consider that water losses were reduced by 30% of the total supplied water, then we get a surplus of 9.44 million JD in 1986 and 40 million JD in 1998. The cost of reducing water losses is about 2.5 million JD plus 10% for contingencies [46]. Accordingly, if each water authority in Jordanian Governorates is given one million JD to overcome the water losses then the overall surplus in the income of the water authority will be about 75 million JD.

4. CASE STUDY

Neamie city was selected as a case study for the water losses. Neamie city is located within the Irbid Governorate in the northern part of Jordan (**Figure 9**). Its area is approximately 16 km^2 and its population is about 13,000. The area is mountainous and for this reason water is distributed through the supply water network using electrical pumps. Complete records of the year 1998 were used in these analyses of the water losses.

Table 6 . Effect of water losses on the quantity of required water.

Year	Water needed MCM	Water loss (MCM)			Quantity required including total loss MCM	Quantity required-Administrative Loss MCM
		Total	Administrative	Technical		
1986	204.1	72.4	36.4	36	276.5	240.1
1987	211.4	81.8	41.8	40	293.2	251.4
1988	219.1	91.8	47.5	44.3	310.9	263.4
1989	227.1	95.8	48.5	47.3	322.9	374.4
1990	253.1	104.5	56.6	47.9	357.6	267
1991	270.2	100.3	52.7	47.6	370.5	274.7
1992	280.2	126.5	71.6	54.9	406.7	308
1993	291.5	126.4	68	58.4	417.9	328.6
1994	302.1	122	67.7	54.3	424.1	334.5
1995	303.1	142.8	78.8	64	445.9	355.5
1996	312.2	133.7	78.3	55.4	446.9	357.5
1997	324.2	131.2	76.3	54.9	455.4	358
1998	335.9	136	79.2	56.8	471.9	370

Table 7. Effect of decreasing water losses on expected water requirements.

Year	Water loss %	Expected required water million m^3	Administrative water loss		Technical water loss		Volume of required water million m^3
			%	million m^3	%	million m^3	
2015	20	561.1	5	28.1	15	84.2	476.9
2020	15	678.5	5	33.9	10	67.9	610.7

Table 8. The monetary value of water losses and the impact of reduced water losses return.

Item/Year		1986	1987	1988	1989	1990	1991	1992	1993	19941	1995	1996	1997	1998
Extracted water MCM		134	150.3	164.5	170.1	178.6	178.5	208.1	218.6	219.9	246.2	242.3	239.9	241.5
Sold water MCM		61.8	68.5	72.7	74.3	75	78.5	83.5	92.3	97.8	104	108	109	105.6
% lost water		53.9	54.4	55.8	56.3	58	56	59.9	57.8	55.5	57.7	55.4	54.7	56.3
Water revenue Million JD		16.2	15	16.6	19.5	18.4	20.7	20.3	25.8	26.9	30	31.9	34.7	40.2
Average Revenue (JD/ m^3)	Extracted water	0.121	0.1	0.101	0.115	0.103	0.116	0.098	0.118	0.122	0.122	0.132	0.143	0.166
	Sold water	0.262	0.219	0.228	0.262	0.245	0.264	0.243	0.280	0.275	0.288	0.295	0.319	0.381
70% of total cost		23.1	22.2	25.9	35.9	40.5	44.4	49.1	51.4	62.6	69.6	72.4	74.3	93
Average cost of m^3 JD	Extracted water	0.17	0.15	0.16	0.21	0.23	0.25	0.24	0.24	0.28	0.28	0.3	0.31	0.39
	Sold water	0.37	0.32	0.36	0.48	0.54	0.57	0.59	0.56	0.64	0.67	0.67	0.68	0.88
Lost water MCM		72.2	81.8	91.8	95.8	103.6	100	124.6	126.3	122.1	142.1	134.3	131.1	135.9
Actual cost of lost water		21.2	20.2	23.7	31.2	34.2	36.5	41.6	44.6	49.7	57.5	57.8	59.6	75
% lost/income		13	135	143	160	186	176	205	173	185	192	181	172	186
30% water loss	Percentage	40.2	45.1	49.4	51	53.6	53.6	62.4	65.6	66	73.9	72.7	72	72.5
	Cost JD	11.8	11.2	12.8	16.6	17.7	19.5	20.8	23.2	26.9	29.9	31.3	32.7	40
Surplus JD		9.38	9.09	11	14.6	16.5	16.9	20.7	21.4	22.8	27.6	26.5	26.9	35
20% water loss	Percentage	26.8	30.1	32.9	34	35.7	35.7	41.6	43.7	44	49.2	48.5	48	48.3
	Cost JD	7.86	7.44	8.5	11.1	11.8	13	13.9	15.4	17.9	19.9	20.9	21.8	26.6
Surplus JD/m^3		13.2	2.8	15.2	20.1	22.4	23.5	27.7	29.2	31.8	37.6	37	37.8	48.3

The records of the Water Authority of Irbid indicate that water losses reached 54.2% during the studied period. Analyses of the records revealed that:

About 92% of the families of the beneficiaries were composed of 8 persons.

Water services reach 90.4% of the inhabitants.

The length of the water supply network pipelines is 27 km. According to operation and maintenance records, about 20% of the supply network pipes are older than 30 years and are in a medium to good condition. The remainder, 80%, is almost new.

Water is supplied through three wells about 500 m away from the city.

From an economic perspective, the relationship between high prices of a "Normal Good" and its consumption rate is inversely related while the relationship is directly related between income and consumption rate. For an "Inferior Good" the relationship is inverse between income and consumption rate. Consumed water can be evaluated using the following equation:

$$\mathrm{Wc} = f(\mathrm{P}, \mathrm{Y}, \mathrm{Ed}, \mathrm{F}, \mathrm{G}) \tag{6}$$

where:

Wc: Consumed water

P: price of water

Y: Income Ed: Educational level

F: Number of persons in the family

G: Garden

Using the above equation, and taking into consideration that the most effective variables on water consumption in Neamie city are: the number of persons within the family and their educational level, the existence of a garden, the price of water and the family income.

Three months of records for July, August and September, 1998 were analyzed (Tables 9-11). During the summer the quantity of supplied water increased to 140,000 m^3 i.e. an increase of 64,000 m^3 over winter figures (46%). This indicates a strong relationship between the season and the quantity of consumed water. It was also noticed that, according to these tables, the allocation per capita per day was 96 l. Using the records indicates that the percentage of losses reaches 48.9% in Irbid Governorate and 55% in Neamie city. Using water meter readings, water losses reached

58.1% where the consumed water was 58,590 m^3 with a difference of 6451 m^3.

It was also noticed that the type of consumed water were 93.6% household use, 2.3% commercial, 2% governmental, 1.3% agricultural and 0.8% industrial uses. In addition 59% of the consumers had gardens with an area of 1000 to 1500 m^2. Most of the gardens were planted with olive, fig, grapevines and some ornamental trees. Rural consumers formed 0.9% and the total area of their farms was 472 acres. About 77% of the water counters were not properly protected, 17.2% of them were not working and 6% of them were not properly working. Furthermore, 5.6% of the water counters were installed in the opposite direction, or there was a by-pass before the counter or they were tampered with.

To calculate the technical losses, all the malfunctions during the investigated period were studied through the operation and maintenance reports. Since there was no accurate measure or standard to determine the losses, the following equations were used:

$$WTL = wo/t \times T \tag{7}$$

where:

WTl: Water lost in cubic meters

Wo/t: Average loss per unit time (minute)

T: Time

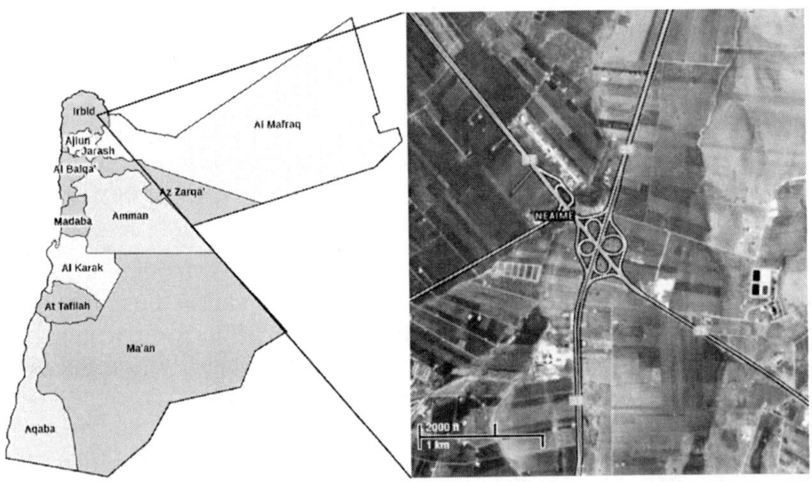

Figure 9. Location map of Neamie city.

Table 9. Water losses in Neamie city in 1998 assuming water year starts October 1997 till September 1998.

Period	Water supplied m^3	Water Sold m^3	Water losses m^3	% losses
Oct.-December, 1997	87,498	40,208	47,290	54
Jan-March, 1998	75,483	32,062	43,421	57.2
April-June, 1998	120,384	52,335	68,049	56.5
July-September, 1998	139,779	65,041	74,738	53.5
Total	423,144	189,646	233,498	55.2

Table 10. Three months analyses for actual water readings of consumed water for the period July-September, 1998.

Period	Supply m^3	Counter Reading	Sold m^3	Estimated Difference m^3	Quantity lost m^3	% lost
July. 1998	45,249	18,985	21,097	2112	26,264	58
Aug.,1998	48,692	19,822	23,372	2550	28,870	59.3
Sep. 1998	45,838	19,783	20,572	789	2605	65.8
Total	139,779	58,590	65,041	6451	811,089	58.1
Average	46,593	19,530	21,680	2150	27,063	58.1

Table 11. Type of water losses during the investigated period.

Type of loss	Number of cases	Average loss (m^3/hr)
From 0.5 inch diameter pipe	53	1
From 0.75 inch diameter pipe	11	5
From 1 inch diameter pipe	7	10
From 2 inches diameter pipe	10	54
From 3 inches diameter pipe	2	148
From 4 inches diameter pipe	5	300

$$WTL = \sum NLi \times wo/h \times T \tag{8}$$

where:

WL: Quantity of water lost

$\sum$NL: Number of malfunctions

i: Diameter of pipe

T: Time

The results showed that the quantity of water lost was 15,243 m^3 which is equal to 10.9% of the total supplied water for the period July-September and it was 18% of the total water lost that year.

The quantities of lost water due to administrative errors were calculated using the records and it is the difference between the quantities actually consumed, and the amount of water sold. This quantity was 6,451 m^3 representing 4.6% of the total supplied water. Water quantities lost due to errors of officers reading the water counters were about 1893 m^3 which represents 1.4% of the total water supply. As far as the water counters that are not functioning are concerned, this caused the loss of 6634 m^3 representing 4.7% of the total water supplied. Finally, the losses due to illegal use of water were calculated by adding all the losses and subtracting it from the total loss. This indicated that it was 50968 m^3 representing 36.5% of the total supplied water. This quantity was the highest compared to other types of losses.

5. DISCUSSION

There is no international unified recipe to reduce water losses that can be applied or used. Several different methods can be used to reduce the losses. This depends on several factors; the most important is the technical conditions, and economic potential of the Water Authority, in addition to the desirable size of the reduction in losses. This problem of overcoming the water losses is the most important problem facing all countries without exception.

The strategy for reducing water losses depends directly on the management techniques used in the water sector. Before thinking about finding new water sources or establishing new pumping stations or drilling of new wells to meet the growing water needs, we must think how we can draw on all the capabilities and potential to reduce water losses. This will provide significant quantities of water that already exist within the network to the public. It does not cost a lot of either effort or funds or require extra extraction, processing, distribution and transport. To reduce the losses we have to secure the economic base, technical competence administrative structures. Therefore there are many of the administrative, technical solutions and economic measures that can be adapted to solve the problem.

These measures can be summarized as follows:

1) Administrative measures

These measures are as important as technical measures to reduce water losses. A secure economic, well qualified and efficient administrative and technical staff can greatly improve the administration of the water authority. This can be achieved by:

Setting a special law and regulations for the water authority. This will improve the action of the authority.

Restructuring the water authority. Although the water authority enjoys the independence on financial and administrative levels, but its budget is subject to the procedures and provisions that apply to any ministry or government department. The water authority should be restructured on the basis of actual financial and administrative independence, without the intervention of the Ministry of Water and Irrigation, or any other governmental entity, and therefore will allocate the money necessary to overcome the water losses and can have a good selection process to get highly qualified staff.

Granting other water departments of the Water Authority the required autonomy in the implementation of water projects and make decisions that work to reduce water losses.

Improving the internal administrative staff by selecting highly qualified and trained personnel Financial incentives should be used to encourage staff to increase productivity. Intensive training programs should be implemented for the rehabilitation of the staff as well as to train them using international experts and try to learn from experiences of developed countries.

Reconsidering consumers contracts so that they hold the responsibility of excessive unjustified water wasting and be subject to the issue of accountability and trial for illegal water use. Each consumer should have a file. In this file full information about the consumer (e.g. monthly consumption rate, number of family members, annual income, unserved area, and type of water use) and the documentation of the public network on the ground. Under charts and maps mounted with locations and type and size of pipes, fittings and knowing conduct of the pipeline and that part buried in the ground to avoid long time search through the site for waterproofing, in order to speed in finding water losses in the cases of faults for maintenance and repair and that will reduce water losses.

2) Technical Measures

New technologies should be adopted in finding covered (under surface of ground) seepages or leakages.

The life span of the water supply network should be considered.

Human capacity building program is to be adopted for workers in the maintenance of public networks on scientific grounds, through the provision of qualified technical personnel.

Providing suitable methods and materials for maintenance.

Maintenance groups should work round the clock to repair leaks and monitor the distribution network.

Periodic maintenance of the water counters at home should be carried out and protecting these counters from adverse weather (e.g. major waves of snow).

Adopting the idea of pumping water to water central tanks, and then to the public network, rather than direct pumping to the network which requires high pressure. High pressure reduces the accuracy of the counters, and potential damage to the network. In addition, it will avoid the destruction of counters and leakage of weak sites in the public network.

Discontinuous pumping of water to the public network should be avoided because it helps to form a rusty layer in the metallic parts of the water counters and the pipes of the network.

Water counters should be installed in accessible locations so that water authority personnel can monitor them and pin point any illegal acts to bypass these counters.

3) Economic Measures

The water authority should adopt water demand management, which includes direct and indirect actions to monitor water use that affects the behavior of the consumer. This includes financial incentives, and public awareness. Such programs are considered as of low-cost projects but they are effective in reducing water losses.

Public awareness programs should be adopted and should address all the public sectors.

Internal and external monitoring teams should be established to monitor performance. Water quality and economic distribution methods should be adopted. These teams should monitor all systems of accounting, financial conditions, legal and the economic situation of the authority and check monthly and annual reports. They should set highly effective economic and financial indicators to measure the performance level of the water authority.

Water prices should be set on the basis of the overall cost. It is assumed that the revenues from the water should cover the cost of production. Restructuring of water rates on the basis of total cost of production will reduce the losses, in addition to the reduction of unjustified water consumption.

Privatization of some of the duties of the water authority will raise the performance in the water sector. As an example the monitoring and maintenance operations could be given to the private sector.

6. CONCLUSION

Thirteen years of complete records of the Jordanian Ministry of Water and Irrigation were investigated. Furthermore, water losses in Neamie city were also analyzed as a case study. The data analyzed from these records showed that Jordan is severely suffering from a water shortage problem. This problem is expected to intensify in future due to the effect of global climatic change. It was also noticed that water losses through the distribution network reach about 50%. To overcome this problem, a water demand strategy should be adopted. Water allocation to the agricultural sector should be decreased due to the fact that most of the water is consumed by this sector. In addition, minimizing the water losses will save huge quantities of water and revenue. Technical, administrative and economic measures to be taken by the Ministry of Water and Irrigation were also identified in this research project.

ACKNOWLEDGMENTS

Nadhir Al-Ansari, N. Alibrahiem, M. Alsaman, Sven Knutsson Most of the work presented in this paper forms part of the second Author's Master Dissertation submitted to Al al-Bayt University in Jordan. Thanks to Dr. Ali El-Naqa of the Hashemite University, Jordan for his assistance. Professor Ian Foster of Northampton University gratefully revised the paper and gave fruitful suggestions. The research presented has been financially supported by Luleå University of Technology, Sweden and by "Swedish Hydropower Centre SVC" established by the Swedish Energy Agency, Elforsk and Svenska Kraftnät together with Luleå University of Technology, The Royal Institute of Technology, Chalmers University of Technology and Uppsala University. Their support is highly appreciated.

REFERENCES

1. R. Liemberger, "The New German Water Loss Regulations in Context with Other International Applications of the IWA Water Balance and Real Loss Performance Indicators," IWA Conference, Limassol, 2002.
2. O. Lambert and M. Fantozzi, "Recent Advanced in Calculating Economic Intervention Frequency for Active Leakage Control, and Implication for Calculation of Economic Leakage Levels," IWA International Conference on Water Economics, Statistics and Finance, Greece, 8-10 July 2005.
3. H. Sewilam, "Capacity Development for Drinking Water Loss Reduction: Challenges and Experiences," UN-Water Decade Programme on Capacity Development, United Nations University, 2011.
4. J. Koelbl and R. Gschleiner, "Austria's new guideline for water losses", Water Supply Networks and Pipelines, 2009.
5. N. A. Al-Ansari, S. Al-Oun, W. Hadad, and S. Knutsson, "Water Loss in Mafraq Governorate, Jordan", Journal of Natural Science, 2013, Vol. 5, No. 3, pp. 333-340. http://dx.doi.org/10.4236/ns.2013.53046
6. M. Fantozzi, A. Pretner and D. Marshallsay, "Management Optimization of Water Distribution Systems: The TILDE Project (Tool for Integrated Leakage Detection)," In: A. Hamdy and R. Monti, Ed., Food Security under Water Scarcity in the Middle East: Problems and Solutions, CIHEAM-IAMB, Bari, 2005, 444 p.
7. M. Tabesh, A. H. Asadiyani Yekta and R. Burrows, "An Integrated Model to Evaluate Losses in Water Distribution Systems," Water Resources Manage, Vol. 23, No. 3, 2009, pp. 477-492. http://dx.doi.org/10.1007/s11269-008-9284-2
8. Jordan's Water strategy, "Water for Life," Jordan's Water Strategy 2008-2022, 2008. http://web.idrc.ca/uploads/user-S/12431464431JO_Water-Strategy09.pdf
9. UNESCO, "Managing Water under Uncertainty and Risk," The United Nations World Water Development, Report 4, Chapter 39, 2012.
10. FAO (Food and Agriculture Organization of the United Nations), "Irrigation in the Middle East Region in Figures: AQUASTAT Survey 2008," FAO, Rome, 2008.www.fao.org/nr/water/aquastat/countries_regions/jordan/jordan_cp.pdf
11. N. A. Al-Ansari. N. Alibrahiem, M. Alsaman and S. Knutsson, "Water Demand Management in Jordan," Journal of Engineering, Vol. 6, No.1, 2014, pp. 19-26.
12. S. Haffner, D. Bley, G. Klein and K. Rudolph, "Water Supply and Sanitation Systems—the First Point of Action," In: H. Sewilam, Ed., Capacity Development for Drinking Water Loss Reduction: Challenges and Experiences, UN-Water Decade Programme on Capacity Development, UN University, 2011.
13. P. Rogers and P. Lydon, "Water in the Arab World," Massachusetts, Harvard University Press, Cambridge, 1994.
14. K. Biswas, "International Waters of the Middle East —From Euphrates, Tigris to Nile," Oxford University Press, Bombay, 1994.
15. N. A. Al-Ansari, "Water Resources in the Arab Countries: Problems and Possible Solutions," UNESCO International Conference on World Water Resources at the Beginning of the 21st Century, Paris, 3-6 June 1998.
16. N. A. Al-Ansari, "Management of Water Resources in Iraq: Perspectives and Prognoses," Journal of Engineering, Vol. 5, No. 8, 2013, pp. 667-68.
17. C. Cherfane and S. E. Kim, "Arab Region and Western Asia, UNESCWA, in Managing Water under Uncertainty and Risk," UN World Water Development Report 4, Chapter 33, 2012.
18. J. Barr, S. Grego, E. Hassan, M. Niasse, W. Rast and J. Talafré, "Regional Challenges, Global Impacts, in Managing Water under Uncertainty and Risk," UN World Water Development Report 4, Chapter 7, 2012.
19. J. C. Perry and J. Bucknall, "Water Resource Assessment in the Arab World: New Analytical Tools for New Challenges," In: N. J Jagannathan, A. S. Mohamed and A. Kremer,

Eds., Water in the Arab World: Management perspectives and Innovations, World Bank, Middle East and North Africa Region, Washington, 2009.
20. R. Alkhaddar, W. Sheely and N. A. Al-Ansari, "Jordan's Water Resources, Supply and Future Demand," Water International, Vol. 30, No. 3, 2005, pp. 1-11. http://dx.doi.org/10.1080/02508060508691870
21. Aquastat, FAO, "Countries, Regions, Transboundary River Basins," Jordan, 2013. http://www.fao.org/nr/water/aquastat/countries_regions/JOR/index.stm
22. Ministry of Water and Irrigation, Annual Report, 2011.
23. M. Nimah, "Water Resources," In: Tolba and N. Saab. Eds., Arab Environment and Future Challenges, Chapter 5, the Arab Forum for Environment and Development, 2008.
24. M. El-Fadel and R. Maroun, "Virtual Water Trade as an Adaptation Demand Management of Climate Change Impact on Water Resources in the Middle East," In: F. Zereini and H. Hotzl, eds., Climatic Changes and Water Resources in the Middle East and North Africa, Springer, Berlin, 2008, pp. 93-108. http://dx.doi.org/10.1007/978-3-540-85047-2_9
25. M. Hiniker, "Sustainable Solutions to Water Conflicts in the Jordan Valley," Green Cross International, Geneva, 1999.
26. K. Sadik and S. Barghouti, "The Water Problems of the Arab World: Management of Scarce Water Resources," In: P. Rogers and P. Lydon, eds., Water in the Arab World, Harvard University Press, Cambridge, 1994, pp. 4-37.
27. S. Postel, "Last Oasis—Facing Water Scarcity," World Watch Institute WW Norton & Co, New York, 1992.
28. Charrier and F. Curtin, "A Vital Paradigm Shift to Maintain Habitability in the Middle East: The Integrated Management of International Watercourses," Water for Peace in the Middle East and Southern Africa, Geneva: Green Cross International, 2007.
29. Salameh and P. Udluft, "Towards a Water Strategy for Jordan," Hydrogeologie und Umwelt, Wurzburg, 2001.
30. W. Owens and K. Brunsdale, "Solving the Problem of Fresh Water Scarcity in Israel, Jordan, Gaza and the West Bank," Water for peace in the Middle East and Southern Africa, Green Cross International, Geneva, 2000, pp. 67-74.
31. WRI, World Resources Institute, "Drylands, People, and Ecosystem Goods and Services: A Web-Based Geospatial Analysis," 2002. http://www.wri.org
32. IPCC (Intergovernmental Panel on Climate Change), "Climate Change Impacts, Adaptation and Vulnerability," Cambridge University Press, Geneva, 2007.
33. N. A. Al-Ansari and S. Baban, "Rainfall Trends in the Badia Region of Jordan," Surveying and Land Information Science, Vol. 65, No. 4, 2005, pp. 233-243.
34. N. A. Al-Ansari, E. Salameh and I. Al-Omari, "Analysis of Rainfall in the Badia Region, Jordan," Al-Bayt University Research Paper No.1, 1999, 66 p.
35. N. A. Al-Ansari, B. Al-Shamali and A. Shatnawi, "Statistical Analysis at Three Major Meteorological Stations in Jordan," Al Manara Journal for Scientific Studies and Research, Vol. 12, 2006, pp. 93-120.
36. M. Medany, "Impact of Climate Change on Arab Countries," Chapter 9, 2008. www.afedonline.org/afedreport/english/book9.pdf
37. T. Oweis and A. Hachum, "Water Harvesting and Supplemental Irrigation for Improved Water Productivity of Dry Farming Systems in West Asia and North Africa," New Directions for a Diverse Planet, Proceedings of the 4th International Crop Science Congress, Brisbane, 26 September-1 October 2004, Published on CDROM, 2004.www.cropscience.org.au.
38. K. A. Voss, J. S. Famiglietti, M. H. Lo, C. Linage, M. Rodell and S. Swenson, "Groundwater Depletion in the Middle East from GRACE with Implications for Transboundary Water Management in the Tigris-EuphratesWestern Iran Region," Water Resources Research, Vol. 49, No. 2, 2013, pp. 904-914. http://dx.doi.org/10.1002/wrcr.20078
39. B. Charalambous, "Leakage Management: A Practical Approach," HHBS, 2002.
40. Water Authority of Jordan (WAJ), "Analysis of Water Systems in the Mafraq Governorate, Final Report," Main Report, Prepared with SAFEGE and SIGMA and Financed by French Government, Jordanian Ministry of Water and Irrigation, Amman, 1998.

41. N. Akel, "Water Crises and Industrial Production," 1st Edition, Moasasat al Murshed for Advertisement and Publications, Amman, 2005.
42. World Bank, "Water Sector Review," Report No. 17095-JO. World Bank, Washington, DC, 1997.
43. L. C. Stephen and B. D. Brooks, "Watershed, the Role of Fresh Water in the Israeli-Palestinian Conflict," International Development Research Center, Ottawa, 1994.
44. S. Sharma, "Water Losses in the Distribution System: Performance Indicators of Water Losses in Distribution System," 2008. http://www.switchurbanwater.eu/outputs/pdfs/GEN_PRS_PI_of_Water_Losses_AC_Apr08.pdf
45. N. A. Fawah, "Economics of Drinking Water in Jordan," MSc. Thesis, University of Jordan, Jordan, 1996. (Unpublished)
46. Ministry of Water and Irrigation (MWI), "Study of Water Losses in Jordan," Ministry of Water and Irrigation, Amman, 1998.

CHAPTER 12

Water Resources of Uganda: An Assessment and Review

Francis N. W. Nsubuga[1,2*], Edith N. Namutebi[3], Masoud Nsubuga-Ssenfuma[2]

[1]*Department of Geography, Geoinformatics and Meteorology, University of Pretoria, Pretoria, South Africa*

[2]*National Environmental Consult Ltd., Kampala, Uganda*

[3]*Ministry of Foreign Affairs, Kampala, Uganda*

ABSTRACT

Water resources of a country constitute one of its vital assets that significantly contribute to the socio-economic development and poverty eradication. However, this resource is unevenly distri- buted in both time and space. The major source of water for these resources is direct rainfall, which is recently experiencing variability that threatens the distribution of resources and water availability in Uganda. The annual rainfall received in Uganda varies from 500 mm to 2800 mm, with an average of 1180 mm received in two main seasons. The spatial distribution of rainfall has resulted into a network of great rivers and lakes that possess big potential for development. These resources are being developed and depleted at a fast rate, a situation that requires assessment to establish present status of water resources in the country. The paper reviews the characteristics, availability, demand and importance of present day water resources in Uganda as well as describing the various issues, challenges and management of water resources of the country.

Keywords: Water Resources Management, Water Resources Utilisation, Climate Change, Water Resources Development, Uganda

1. INTRODUCTION

Of all the renewable resources, water has a unique place. It is essential in sustaining all form of life, food production, promoting economic development and for general wellbeing. Water is impossible to substitute for most of its uses, difficult to de-pollute, expensive to transport but a manageable natural resource capable of diversion, transport, storage and recycling [1] There are two distinct water resources categories (surface and ground water) which are part of the earth's hydrologic cycle. Surface and groundwater resources play a major role in domestic water supply, watering livestock, industrial operations, hydropower generation, agriculture, marine transport, fisheries, waste discharge, tourism, and environmental conservation. The water resources of a region, conceived as a dynamic phase of the hydrologic cycle are influenced by climatic, physiographic and geological factors [2] .Orography on the other hand, plays a significant role in influencing rainfall and other climatic elements (temperature, humidity and wind) that affect evapo-transpiration, which determines the totality of water resources [2]

Water resources in Uganda comprise of large lakes like; Lake Victoria, Kyoga, Albert, George and Edward; wetlands and rivers, such as the Nile River, Katonga, Semliki, Malaba; rainfall, surface water runoff and ground water [3] . Two or more of these water resources are shared by other riparian states that are part of the Nile Basin, shown in the (Table 1). In a regional context, the whole of Uganda's water resources is part of the Nile. Uganda is a downstream riparian to Burundi, Democratic Republic of Congo (DRC), Kenya, Tanzania and Rwanda and an upstream riparian to South Sudan, Sudan and Egypt [4] .

Much of the literature on water resources in Uganda is scattered and embedded in policy documents, strategic plans and reports produced by consultants and international organisations. There are also several issues and challenges surrounding water resources, which are aggravated by climate change and variability and population growth among others. The current water management practices in Uganda may not be robust to cope with these challenges which impact on water resources and increase water use requirements. With rapid growing population and improving living standards, the pressure on Uganda's water resources is increasing and per capita availability of water resources is reducing day by day. Spatial and temporal variability in precipitation is another challenge the country faces, which often result into floods, landslides and droughts. The quality of surface and ground water resources is also deteriorating because of increasing pollutant loads and non-point sources. Climate change is expected to impact on rainfall and water availability. Currently, the data collection, processing, storage and dissemination have not received the deserved attention. All these aspects require an assessment of the current water resources status. The review therefore, focuses on the character, potential, governance, challenges, availability and demands of water resources of the country using secondary data sources.

2. HYDROLOGY OF UGANDA

2.1. Surface Water Resources

Uganda is a landlocked country that occupies 241550.7 km^2 of land. Open water and swamps constitute 41743.2 km^2 of area [4] with about 16% of total land area of wetlands and open water, plus the annual water supply of 66 km^3 in form of rain and inflows. One would therefore conclude that, Uganda is fairly well endowed with wa- ter resources. Since direct rainfall is the most important source of water resources in Uganda, understanding the spatial and temporal variability of rainfall is therefore paramount when assessing water availability [5] . Water availability has determined the local water resources, land-use potential and population distribution.

Table 1. The major shared water bodies and river courses of Uganda.

Water resource	Countries sharing	Basin wide
River Nile	Uganda, South Sudan, Sudan, Egypt	Burundi, DRC, Egypt, Ethiopia, Kenya, Rwanda, South Sudan, Sudan, Tanzania, Uganda, Eritrea
River Achwa	Uganda, South Sudan	Uganda, South Sudan
River Kagera	Tanzania, Burundi Rwanda, Uganda	Nile Basin Riparian States-Burundi, Tanzania, Rwanda, Uganda
River Semliki	DRC, Uganda	Nile Basin Riparian States
River Malaba	Kenya, Uganda	Kenya, Uganda
River Sio	Kenya, Uganda	Kenya, Uganda
Lake Victoria	Uganda, Kenya, Tanzania	Nile Basin Riparian States
Lake Albert	DRC, Uganda	Nile Basin Riparian States
Lake Edward	DRC, Uganda	DRC, Uganda
Sango Bay Swamp	Tanzania, Uganda.	Kenya, Burundi, Rwanda

The Department of Water Resources Management of Uganda has categorized the surface water resources into eight main drainage sub-basins [4] . These include; Lake Victoria, Lake Kyoga, River Kafu, Lake Edward, Lake Albert, River Aswa, Albert Nile and Kidepo Valley (Figure 1). The yield from these sub-basins, though small compared with the total Nile flow, dominates the water resources potential with in Uganda. Major water bodies include lakes Victoria, Kyoga, Albert, George, Edward (Table 2) and another 149 smaller lakes spread across the country covering an area of 38,500 km^2 [3] . The lakes are interconnected by a river system, which developed as a result of river reversal and ponding that formed the two major lakes [6] (and references there in). In the north-eastern part of the country, many of the water courses are seasonal. The main river basins of Uganda are shown in the Figure 1.

The most prominent hydrological feature in Uganda is Lake Victoria, which is the second largest fresh water lake in the world [7] , with an area of 69,000 km^2. River Nile, which is the only outflow from the lake, has its source at the point where Lake Victoria spills over Ripon Falls. The 130 km stretch of the Nile from Lake Victoria to Lake Kyoga is termed the Victoria Nile. Lake Kyoga is drained through the Kyoga Nile which, after a relatively flat reach downstream from the lake, enters a series of rapids and falls before it flows into Lake Albert at a level 410 m lower than Lake Kyoga. The plateau immediately to the north of the Rwenzori drains to Lake Albert via the Muzizi River.

In Lake Albert, the Nile is joined by river Semliki which drains Lakes George & Edward found in the rift valley and the high rainfall area of the Rwenzori Mountains. Lakes George & Edward are connected through the Kazinga Channel. The Nile flows from Lake Albert with a gentle slope to the Sudanese boarder. This reach of the river is called the Albert Nile.

Major rivers include the Nile, Ruizi, Katonga, Kafu, Mpologoma and Aswa. The rivers and lakes of Uganda are within the upper part of the White Nile Basin [4] , with the exception of those in the tiny North-eastern catchment of Kidepo, which drains into the Lake Turkana basin in Kenya (Figure 1). The North-western slopes of Rwenzori watershed in Uganda drain to Lake Edward via the Ishasha, Chiruruma, Nchwera and Nyamweru rivers, and also by several streams, which enter the western flowing part of the Katonga River. The North-east- ern part of the Virunga range watershed however, drains directly to Lake Victoria via a series of swampy lakes and streams culminating in the Kibaler River, which enters Lake Victoria through the swamps at Sango Bay [8] . The Ugandan slopes of Mt. Elgon and the central highlands along the Kenyan border drain via rivers with swampy valleys or seasonal floodplains to Lake Kyoga, while the north eastern highlands and most of the northern plateau drain directly to the Bahr el Jebel in Sudan via the Achwa River that also joins the Nile.

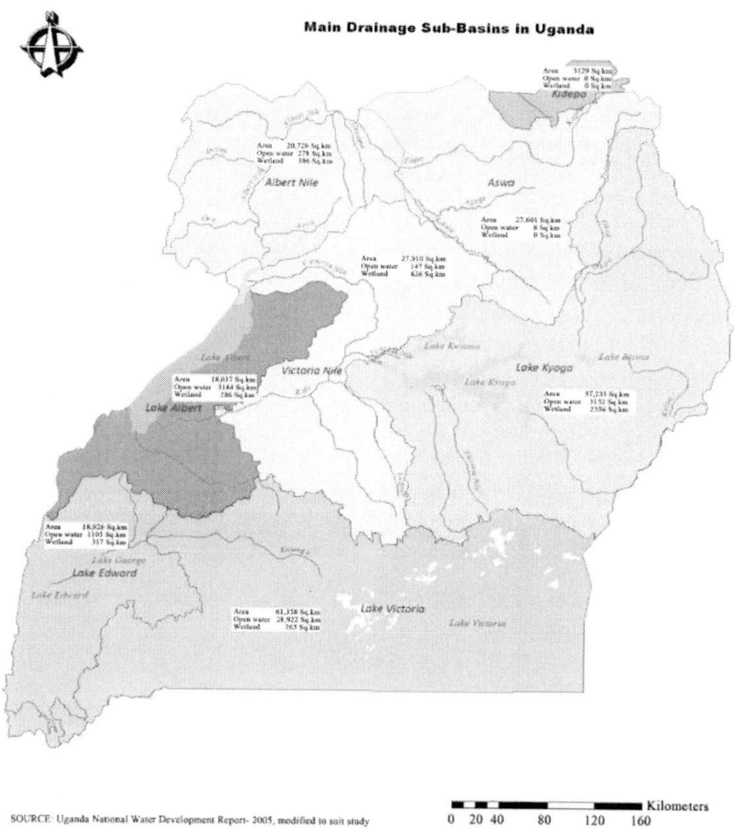

Figure 1. Main drainage sub-basins in Uganda including rivers and lakes.

Because of warping of the landscape, many of the perennial streams of the plateau are clogged with swamp [6] . About 10 Percent of the country is covered by swamps (wetlands), of which one third is permanently inundated. In the south and west of the country, swamps form an extensive low gradient drainage system in steep V-shaped valley bottoms with a permanent wetland core and relatively narrow seasonal wetland edges. In the north, they mainly consist of broad flood plains. In the east they exist as a network of small, vegetated valley bottoms in a slightly undulating landscape [6] .

2.2. Wetland Resources

There are basically two broad distributions of wetland ecosystems in Uganda: 1) the natural lakes and lacustrine swamps found around major lakes; 2) the riverine and flood plain wetlands which are associated with the major river systems in Uganda. Wetlands, cover about, 29,000 km^2, or 13% of the total area of the country [9] . The wetlands comprise swamp (8832 km^2), swamp forest (365 km^2) and sites with impeded drainage (20,392 km^2). Wetlands also include areas of seasonally flooded grassland, swamp forest, permanently flooded papyrus, grass swamp and upland bog. By 2008, the coverage had declined to 10.9% (Table 3). This decline has been generally observed around Lake Victoria and Kyoga drainage basins. The decline according to the Uganda water and environment sector performance report of 2011 is largely attributed to encroachment for expansion of urban centres, settlement, industrial developments and extension of agricultural land. Except for Sango Bay, the bulk of Uganda's wetlands lie outside protected areas [3].

Table 2. Major Lakes of Uganda and their associated characteristics.

	Surface Area (km^2)	Area in Uganda (km^2)	Mean Elevation above Sea Level (m)	Maximum Depth (m)	Volume in Uganda (km^3)	Mean Depth (m)
Lake Victoria	68,800	28.655	1134	84	1237	40
Lake Albert	5659	2850	618	56	80	25
Lake Kyoga	2636	2636	1034	10.7	7.9	3
Lake Edward	2324	638	912	117	16.8	34
Lake Kwania	540	540	1033	5.4	2	4
Lake Wamala	250	250	1290	9	1.2	4
Lake Bisina	150	150	1030	N/A	0.5	N/A
Lake George	228	228	914	7	0.8	2.4
Lake Bunyonyi	61	61	1974	39.3	0.2	
Lake Kachira	39.6	39.6	1235	4.8	0.2	
Other 149 Minor Lakes		2453			7.2	
Total		38500.6			1353.8	

Source: UN-Water. 2006 in [6].

Wetlands perform a number of functions, such as; mitigating the effects of both floods and droughts, providing fish resources and support cropping and grazing along their margins. They are centres of high biodiversity and productivity as well as valuable refuges and sources of food like fish. Furthermore they are active biological filters in the treatment of effluents, but due to this function they are also sensitive to the accumulation of pollution.

2.3. Yield of Rivers

All rivers in Uganda ultimately reach the Nile through various interconnections. Thus an understanding of the Nile flows, gives us an insight into the flow regime of major rivers in Uganda. The flows of the Nile are highly variable from year to year. This annual variability has been monitored and reported in previous studies [3] [4] evidenced in Figure 2. The plot shows the long-term variations in flows monitored along Victoria Nile at Jinja, Kyoga Nile at Masindi Port and Albert Nile at Panyango. It shows the flow increased from 1960s, after which there is a steady decline in the flow at all monitoring points. The long-term average outflow from Lake Victoria has been 840 m^3/s and the range of outflows is between a minimum of 345 m^3/s and a maximum of 1720 m^3/s. At the 95% monthly reliability level, the flow of the Victoria Nile is of the order of 495 m^3/s.

Using annual rainfall data series for the period 1940 to 1999, we computed a Drought Severity Index (DSI) for the drainage sub-basins in Uganda. Results presented in Table 4 have been compared with plot in Figure 2, which was produced by DWRM. There is a close correlation between DSI and the river Nile outflows monitored for 1899-1997 period.

Table 3. Changes in wetland coverage by drainage basin between 1994 and 2008.

Drainage Basin	1994 (Area km^2)	%	2008 (Area km^2)	%	% Change
Albert Nile	1736.3	6.21	1255.2	4.71	27.7
Aswa	3028.0	10.83	2168.9	8.24	28.4
Kidepo	168.1	0.60	197.2	0.74	−17.3
Lake Albert	2838.6	10.15	2421.7	9.20	14.6
Lake Edward	1671.1	5.97	1096.3	4.16	34.4
Lake Kyoga	15008.3	53.67	11028.5	41.92	26.5
Lake Victoria	7167.6	25.63	3310.2	12.58	53.8
Victoria Nile	5786.3	20.69	4829.4	18.35	16.5

Source: Uganda water and environment sector performance report, 2011—modified to suit study.

Following steps in [10] , one can discover the years and seasons when the sub-basins were experiencing droughts. It is evident from the plot (Figure 2) and Table 4 that, the early 1940s and 1950s in the first climatic normal were years of below normal rainfall. The situation is similar in the second normal when the early 1980s were drought years, consequently affecting the outflow of river Nile. This re-affirms the fact that the main source of water for the water resources in Uganda is rainfall and the whole of Uganda's water resource is part of the Nile.

Table 4. Drought years for selected drainage sub-basins in Uganda.

Series Variables	Climatic Normal	Albert Nile	Lake Kyoga	Lake Victoria
Annual	1940-1969	1943, 1949, 1952, 1953, 1954, 1965	1943, 1949, 1953, 1957, 1958, 1965	1943, 1946, 1952, 1953, 1959, 1965
	1970-1999	1971, 1983, 1984, 1989, 1993	1973, 1974, 1979, 1980, 1984, 1986, 1993	1970, 1971, 1980, 1982, 1983, 1985

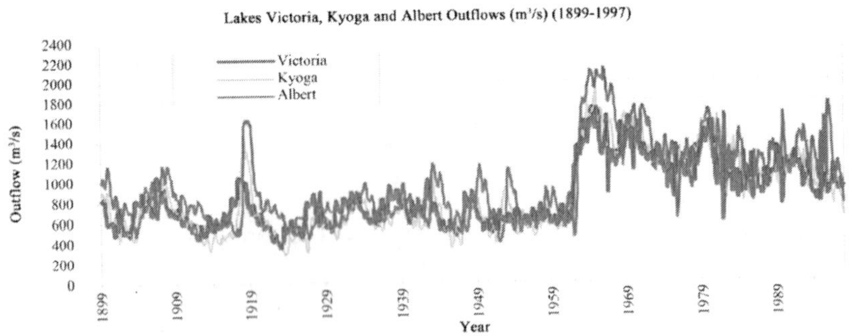

Figure 2. Long-term variations in river Nile flows (figure adopted from WRM subsector reform study report 2004).

Annual rainfall data for the period 1940 to 1999 were exposed to Sen's T test to establish the behavior of rainfall trends in Uganda. The results of Sen's T test applications to annual rainfall of the main drainage sub-basins are shown in Table 5. The Sen's T test results indicate a negative trend of annual rainfall in six of the sub-basins, namely; Albert Nile, River Aswa, Lake Albert, Lake Edward & George, Lake Kyoga and Lake Victoria. Significant decreasing trends were observed in River Aswa and Lake Albert sub-basins. Results for Victoria Nile sub-basin show a positive trend, which is not statistically significant at 0.05. Consequently the decreasing rainfall affects water resource distribution in the country.

Despite Uganda's significant water resources, their spatial and temporal variability often renders many parts of the country water stressed over long periods of the year. Water resource management sub-sector reform study of 2004 showed that, districts in the north-eastern and south-western parts of the country have the least per capita water availability. The study also reveals that by 2015 more than 75% of the country will be water stressed (Figure 3).

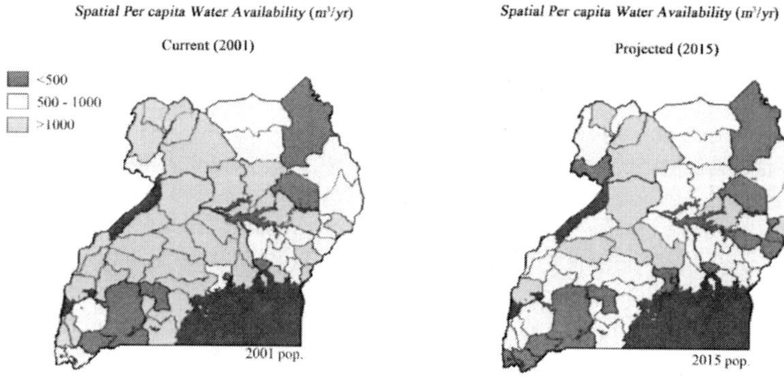

Figure 3. Spatial per capita surface water distributions (m^3/yr). Source: WRM sub-sector Reform study report, 2004.

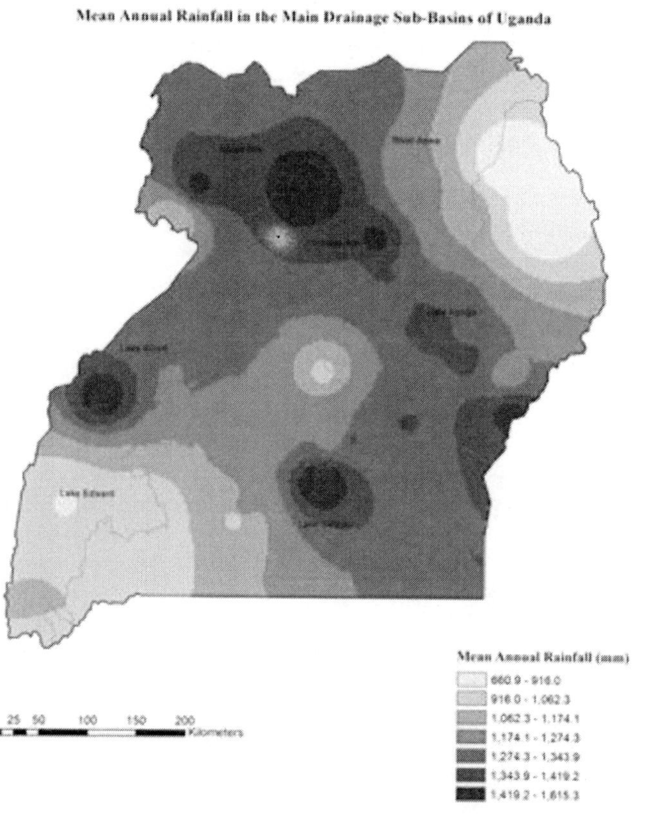

Figure 4. Distribution of Mean annual rainfall in main drainage sub-basins in Uganda 1940-2009.

The methodology used in computing the water availability only considers the runoff generated within each district. It is used here to give a picture of the average yield of major catchments in Uganda.

For example the south-western and north-eastern parts of the country have the lowest annual run-off (<1 litre/s/km^2). These are semi-arid areas, which receive very low rainfall. The Lake Victoria basin has the highest run-off (>10 litres/s/km^2) [11] . Mean annual rainfall calculated for the different drainage sub-basins during 1940 to 2009 is depicted in Figure 4. Analysis supports the correlation between rainfall and run-off generated from each sub-basin. The average yield of major catchments in Uganda is also represented in Figure 4.

A recent study by DWRM in Rwizi, Wamala and Victoria catchments, in the Victoria water management zone concluded that, reduced water levels occur annually in the dry season with in the three catchments [12] . The lowest levels are witnessed at the end of every dry season, until the rains set in, but not reaching the original high levels. This however paints a bleak picture especially when the rainy seasons become unpredictable [12] .

3. STATUS OF WATER AVAILABILITY

Two organizations, Directorate of Water Development (DWD) and National Water and Sewerage Corporation (NWSC) have been tasked to ensure that water is availed to Ugandans. Available information from NWSC shows that there has been a steady increase in total water production e.g. from 46.7 million·m^3/yr, in 2000/01 to 72.14 million·m^3/yr in 2009/10 financial years. Comparing the water requirements with the available renewable freshwater, Uganda had a capacity to utilize only 1% of the current renewable freshwater for consumptive purposes. Out of the total water withdrawal, domestic water supply accounted for about 51%, agriculture 41% and industry 8% [8] . Withdraw per capita rose from 12 m^3/yr in 2002 to 21 m^3/yr in 2008 (Figure 5).

Table 5. Trend results for the main drainage sub-basins. Sen's T-test was performed on area average rainfall data.

Sub-Basins	Rain Series	Albert Nile	River Aswa	Lake Albert	Lake Edward	Lake Kyoga	Lake Victoria	Victoria Nile
Year of Rainfall	Maximum	1961	1983	1961	1996	1961	1961	1961
	Minimum	1943	1943	1943	1979	1957	1957	1943
SEN's T-Test	Area mean	−0.31	−3.91	−3.13	−0.91	−1.09	−1.36	0.19

Percent change is equivalent to 112% in a seven-year period. Notable changes were realized in domestic withdrawals from 45% to 51%, while industrial withdrawals fell by 7%. The use of water all over the world is a function of population change, food consumption, economic policy, technology, lifestyle and societies' views of the value of freshwater. Since Uganda has one of the fast growing populations (3.2%), its demands will lead to increased water withdrawal. For instance, total volume of water sold according to NWSC annual reports had increased from 31,151,380 million·m^3/yr in 2002/03 to 47,027,817 million·m^3/yr by June 2010. Much of the total water produced was sold to domestic consumers (48.9%), industrial and commercial enterprises (25.4%), institutions and governments (20.8%), and (4.8%) was sold through public standpipes. This positive trend shows how effective the organizations have been in achieving national targets.

The major water sources exploited are protected springs, deep boreholes and shallow wells. These are also sources through which natives access water as individuals or licensed independent water providers. Records from water permit database of the DWRM, for example put water abstractions of registered users for the year 2003 in Lake Albert basin at 1500 m^3/yr abstracted from surface water while ground water abstraction accounts for 63 m^3/yr. Water abstraction particularly from the Albert basin is expected to increase following the recent discovery of oil reserves. Water resources management actions within the Albert basin will require vigilance and strict monitoring during the process of drilling and processing oil.

3.1. Groundwater Occurrence

The primary source of freshwater for drinking and irrigation in the world is groundwater. Ground water supplies 75% of all safe sources of drinking water in Africa [13] . In Uganda for example, 61% of the country's water is from a ground water source, accessed from springs and boreholes around Lake Victoria and south-western Uganda. Several studies, like [13] -[17] have assessed ground water occurrence in Uganda from different perspectives mostly at a catchment scale. These assessment studies have been a basis for water resources planning in the country.

According to the Monitoring and Assessment Division report of 2011, ground water resources of Uganda were estimated in 2010 during the National water resources study to be as shown in Table 6 for the different basins. The report produced a map of estimated exploitable groundwater resource per district. This exploitable ground water represents the proportion of renewable resources that can be exploited on a sustainable basis without seriously affecting the environment.

The Victoria Nile and Kyoga basin have sustainable ground water, which is more than 36 mm/yr, while Kidepo has the least amount of sustainable ground water equivalent to 6.3 mm/yr (Table 6). This is a true reflection of the catchment size from which these basins draw water (Figure 1).

3.2. Groundwater Development

Groundwater is the major source of water supply in the rural, semi-arid and arid areas in Uganda. Groundwater development has been going on since the 1930s through construction of deep boreholes, shallow wells and protected springs. There are approximately 20,000 deep boreholes, 3000 shallow wells and 12,000 protected springs in the country constructed mainly for rural domestic water supply. In the 1990s ground water was intensified to provide for town water supply. For example, 782 small towns were identified for the provision of piped water by June 2006, and 70% of this water was to be provided by ground water sources like the deep boreholes [13] . Boreholes and shallow wells are normally installed with hand-pumps with capacity of 1 m^3/hour and their yields are usually low. Ground water has advantages like, good quality water that requires little or no treatment costs can easily be operated and managed by users, which makes investment and operational costs to be lower than those of surface based systems [13] . Boreholes with yields greater than 3 m^3/hour are thus normally considered for installation with motorized pumps for piped water supply.

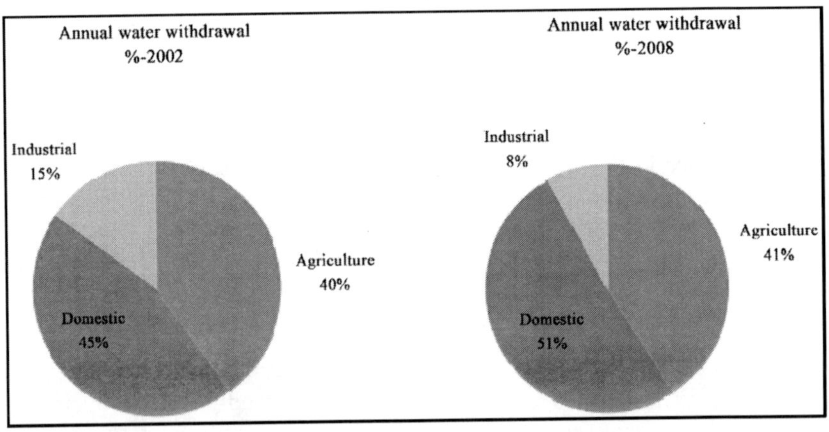

Figure 5. Water withdrawals in Uganda. (Data was sourced from WRMD quarterly reports, joint water sanitation sector program support, DWRM).

Table 6. Average sustainable groundwater for major river and lake basins.

Drainage Basin	Area km²	Land Area km²	Sustainable Ground Water (mm/yr)
Lake Edward	18.946	17,855	20.3
Lake Victoria	61,886	32.924	24.7
Lake Albert	18,079	14,882	23.7
Victoria Nile	27.961	27.807	39.9
Lake Kyoga	57.236	53.899	36.1
Albert Nile	20,727	20,484	24.4
Aswa	27.637	27.635	17.3
Kidepo	3229	3228	6.3
Miscellaneous	5716	5679	15.0

Source: Monitoring and Assessment Department, DWRM.

In 1996 GoU initiated groundwater assessment studies, to fully understand the nature, extent and reliability of the country's groundwater resources. The study has provided Information on the distribution and behaviour of aquifers, groundwater recharge, aquifer vulnerability to pollution, impact of motorized abstraction on groundwater resources and a conceptual model of groundwater dynamics has been developed.

For example, groundwater assessments in three catchments namely Ruizi, Wamala and Victoria concluded that, groundwater resources were declining [11] . The causes of decline according to the monitoring and assessment department of the DWRM report, includes among others; climate change/global warming, changes in land-use, especially deforestation, unsustainable water withdrawals, poor catchment management, prolonged droughts, reduced rainfall in the catchment (see Table 5). This information, though still scanty, forms the basis for the current groundwater resources planning and management in the country.

3.3. Groundwater Recharge

Recharge assessments have recently been carried out in Apac in northern Uganda, Mbarara in western Uganda, Wobulenzi in central Uganda, Nkokonjeru in eastern Uganda and Hoima in mid-western Uganda. Results obtained using the various methods on groundwater recharge reveal a range between 90 and 220 mm per annum and accounts for 7% and 20% of the average annual precipitation in Uganda [18] . Dynamically downscaled climatological data have been applied on a soil moisture balanced model to estimate ground water recharge and runoff in south-western Uganda along river Mitano [13] . Their results under the A2 emissions scenario, for the period 2070-2100, relative to 1960-1990 indicated an increase in precipitation (14%) and a modelled potential evaporation (53%) with a substantial increase in rain intensity [13] . Hydrological projections also gave a rise to increases in recharge (53%) and runoff (137%).

From the above figures it can be stated that groundwater recharge in Uganda is quite high compared to current abstraction volumes and will not be a limiting factor in groundwater development for a few years to come. However, there is a need to carry out more detailed recharge and water balance studies in the country to ensure that groundwater development is carried out in a sustainable manner.

3.4. Ground Water Potential

The potential of groundwater in various areas of the country is exhibited by presence of deep boreholes, shallow wells and springs. These are assessed in the paragraphs that follow.

3.4.1. Deep Boreholes

Deep borehole potential can be assessed by a means of a number of borehole parameters such as regolith thickness, aquifer yields and rest water levels. Uganda is characterised with a clayey regolith especially in the upper layers where relatively low permeability dominates. The regolith thickness across the country can be described as low to medium varying between 20 - 45 m. This leads to medium to high ground water potential through provisional storage. Aquifer yields on the other hand vary from one part of the country to another according to the formation in which they are drilled and their degree of fracturing and weathering. For example, potential yield from deep aquifers were estimated during the AQUASTAT survey, to be above 3 m^3/hr in the southwest, southeast, northwest and along the eastern border of the country [8] . In large areas of central parts of the country potential yields are between 2 and 3 m^3/hr, while in some areas it is below 1 m^3/hr [8] . Rest water levels also give an indication of the groundwater potential of an area. Shallow water levels (<20 m) indicate that the aquifer has high potential for yielding groundwater while deeper water levels indicate the reverse. Rest water levels (static water levels) in the country vary between 1 and 45 m below ground level.

3.4.2. Springs

Springs occur either where the flow of unconfined groundwater is interrupted by an impermeable formation or where the head of confined groundwater is released by flow to the surface [8] . There are 2 major types of springs in Uganda namely; contact and fracture springs. Fracture springs are usually very susceptible to contamination and drying up while contact springs are more reliable [14] [16].

3.4.3. Shallow Wells

The potential of shallow wells is quite high, especially in the valleys. Their potential is favoured by the thick regolith that is fairly coarse grained. From Uganda's experience, shallow wells are a very reliable source of water supply to the communities although precautions need to be taken to ensure that they are not contaminated. Springs and wells are a major source of water for the country [4] . Through DWD, shallow wells have been constructed and improved to provide good quality water.

4. WATER RESOURCES SECTOR DEVELOPMENT TRENDS

Much of the national documentation in Uganda refers to a series of development plans, strategies and action plans which were planned and executed to varying degrees of success, without an explicit sector policy framework before the 1980s. In the 1990s a number of guidelines were prepared and studies undertaken by government which paved the way for the review of water legislation and sector policy development process. Since then, GoU has taken major steps to rationalize water resources management, development and the delivery of Water and Sanitation Services (WSS). The development is aligned to achieve the Millennium Development Goal (MDG) of 62% by 2015 on water.

To achieve the set targets, the GoU has reformed the water and sanitation sector in order to ensure that services are provided and managed with improved performance and cost effectiveness. In this regard reform studies were carried out that resulted in development of Sector Strategic Investment Plans (SIPs) with appropriate policies and strategies like;

- The rural water and sanitation SIPs which aim at increasing the safe water supply coverage to 77% or 95% by 2015.
- The urban water and sanitation reform SIPs that aim at providing 100% urban population with safe water supply and sanitation by 2015.
- The water for production SIPs, which stipulates the participation of the private sector and farming community in financing this regard.
- The water resources management SIPs, which aims at promoting integrated water resources management and development at both national and local levels.

The strategies above required a Sector Wide Approach to Planning (SWAP). SWAP framework for the WSS sector was adopted at the second GoU/Donor review for the water sector in September 2002. The move towards sector wide approach called for GoU/donors to promote; uniform fund disbursement rules, uniform and stronger fund accountability rules, common indicators, joint appraisals and reviews. Since then reviews have been carried out every year to monitor progress in the water sector.

The National Water Policy was adopted in 1999 and sets the stage for water resources management and guides development efforts aimed at improving water supply and sanitation in Uganda. It promotes a new integrated approach to water management that guides the allocation of water and the associated investments.

This new approach is based on the continuing recognition of the social value of water, while at the same time giving much more attention to the economic value of water. Key aspects include; application of a participatory demand driven approach to planning which promotes user ownership and management of services, backed by measures to strengthen local authorities and private institutions in implementing and sustaining water and sanitation programmes.

The National Water Policy embraces international and regional resolutions, declarations and guidelines [19] because more than 90% of Uganda's waters are trans-boundary or river courses. This therefore means that the use and development of Uganda's water resources is controlled by international laws and principles and colonial agreements.

Because much of Uganda's water ends up with the Nile, therefore, Uganda's approach to the water resources development and management is a critical element within the Nile Basin and hence requires clear trans-boundary policy and institutional frameworks and also legal and harmonized laws with riparian states.

The current strategic framework for the development and management of water resources in Uganda recognizes two important principles.

First, water is fundamental, to achieving the national objective of poverty eradication through the promotion of rapid economic growth, good health, food security, and social equity. Secondly, within the regional context, according to the ministerial policy statement, cooperative development of the trans- boundary water resources e.g. the Nile, the Kagera, Sio-Malaba River, Lake Victoria, Lake Albert, Lake Edward and others can serve as a catalyst for a broader range of cooperation and economic integration.

5. WATER RESOURCES UTILIZATION IN UGANDA

Water for Agriculture/irrigation: Rain fed agriculture is the most practiced land use method in the Uganda. Currently this practice is threatened with climatic variability and a fast growing population, which are impacting on food security levels. As a result national policy on agriculture aims at increasing agricultural production per unit area e.g through a more efficient use of land and water resources which will improve food security. The increase in population has increased the need to raise crops in areas that do not get enough rainfall, hence requiring irrigation. A Food and Agricultural report study in 1987, estimated the irrigation potential in Uganda to be more than 280,000 ha. If fully developed a water demand of 2.7 billion·m^3/year will be required from Uganda's water resources hence, placing a considerable demand on the national and trans-boundary water resource like the river Nile. In order to manage water fairly, the Shared Vision Program (SVP) on efficient water use for agricultural production was set up to establish a forum to assist stakeholders at regional, national and community levels to address issues related to efficient use of water for agricultural production in the Nile basin and link it with the trans-boundary water policy. This program provides the appropriate linkage under the Trans boundary Water Resources Program (TBWRP) being formulated with NBI-SVP for improved agricultural Production.

Water for power: Most waterpower is used to generate Hydro-Electric Power (HEP), which is used to light homes and to run factories. A study by Nile Basin Capacity Building Network (NBCBN) in 2005 estimated the hydrological resources to have a power production potential of over 2500 MW, of which over 2000 MW is mainly concentrated on river Nile and the rest is scattered in parts of the country.

By 2005, less than 10% of this potential was exploited [20] . Of the 2000 MW potential along the river Nile, 630 MW (31.5%) have been tapped, and the unexploited potential is well over 1300 MW [21] . Installed capacity stands at 872 MW including thermal and hydropower plants, of which HEP contributes 621.5 MW. This shows that power supply of Uganda is almost totally dependent on hydropower. Unfortunately, installed capacity is grossly inadequate to meet the current and rapidly increasing demands. For example in 2008, the number of consumers by type was recorded at 347,433 for domestic small general service; 28,810 for commercial small general service; 324 for large industrial; 1194 for general-medium industrial and 195 for street Lighting in 2008 [22] . Given the current economic growth of 7%, domestic power demand is expected to increase to 1130 MW in 2023 [23] . One of the national development plan objectives is to increase power generation to between 780 MW and 820 MW, as well as increase rural electrification by 10% and promote efficiency programmes and renewable sources countrywide.

The main sites that have been identified for the development of major schemes are all located downstream along the river Nile. The other small-scale hydropower potentials exist along the rivers draining Mt. Elgon, the extreme southwest of Uganda, rivers draining west Nile near Arua and the rivers draining the Rwenzori Mountains. The country has a big potential for medium, small and micro hydropower stations. Development of large schemes on the Nile should be the right way to go by the government of Uganda, because the markets are available in the region.

The national policy provides for regional co-operation for optimum hydropower development. For example, the Nile Basin Power Trade of NBI-SVP has the objectives of establishing the institutional means to coordinate the development of regional power markets and improve access to reliable and low cost power among the Nile Basin countries. In order for Uganda to fully exploit her hydropower potential on the Nile, this program will be the main avenue for the linkage for cooperation under the trans-boundary water policy and to also promote the benefit-sharing concept.

Water for industry: The role of water for example in industry contributed 24.2% in 2008/2009 to the total Gross Domestic Product (GDP) and increased to 26.9% in 2010/2011 [24] . Water is highly consumed in the construction industry [25] especially during this period when the country is undergoing infrastructural development. These industrial activities (Construction, manufacturing, mining and quarrying) that consume a lot of water contributed 12.3%, 7.5%, and 0.3% of the total GDP in 2008/09 respectively. Water, which is vital in transforming industry and eradicating poverty, will be in high demand in future, given the trend of economic growth.

Water for fish: Inland water bodies support a thriving fishing industry in Uganda. Fish catches vary from place to place, with the largest yields associated with major lakes and rivers (Table 3). According to [24] fishing activity was worth 2.9% share of the GDP by 2008/09 fiscal year and many people are indirectly dependent on this industry. Fish is often a large part of diet of people living in the country and abroad. Tonnes of fish are also harvested from wetlands and ponds on fish farms. Fishing activities have become so popular that are fast increasing in importance and numbers that exerts pressure on water resources especially on quality of water resources.

Water for transport: People depend on water transport to carry goods for use and trade from one region to another, using boats, ferries and ship. Water transport links Ugandans living on islands with the mainland and also those involved in trade across borders towards DRC, Tanzania and

Kenya. Transport as an activity though not fully developed, has impact on the quality of water resources, which requires further study.

Water for environment and tourism: Development of tourism in Uganda is a high priority area. Tourism development is dependent on the natural beauty and quality of the environment, including wetlands, range lands, lakes, rivers and the associated sceneries like the waterfalls, beaches and rare fauna species which provide an array of tourism activities. The splash and flow of water in the streams, hot springs, geysers and fountains also soothes and inspires many people who love being near water. Most of the recreational places in Uganda provide water sports, such as swimming, fishing, Water rafting and sailing. People also enjoy the beauty of a quiet lake, a thundering waterfall (like at Karuma, Kabalega and Budhagali) or even roaring surf. Business community in Uganda has built recreational areas along the lakes, such as Entebbe, Lutembe, Sesse-Kalangala, Kigo, Jinja where local tourism is flourishing. This study discovered that climate variability was impacting on the activities at the shores of the Lake Victoria, and the subsequent income of the entrepreneurs.

These touristic facilities generate huge financial and socio-economic benefits for the country. For example in the year 2008, a total of 844,000 tourists to the country were recorded, of which 144,000 declared their purpose of visit as leisure and holiday. And about 138,000 persons (foreign and national) visited the national parks in 2008, 190,000 in 2010 and 208,000 in 2011. Between 2010 and 2011 the industry grew by 9.4% and contributed 21% to GDP in 2011, earning the country US $805 million in 2011/12 financial year according to the Daily Monitor Newspaper, 26.09.2012.

Water for Sanitation: This is a field of public health necessary for controlling and preventing disease. It includes personal cleanliness, sewerage treatments, waste disposal systems, water treatment and so forth. Sanitation deteriorates where water conditions are poor and uncertain which directly affects the wellbeing of communities. Likewise sewerage and sanitation service requirements increase in step with improvements in water supplies and have important health implications. Because water, health and sanitation are set targets for 2015 MDGs, African Development Bank (ADB) financed water and sanitation projects to small towns and rural areas in Uganda to the tune of USD 28 million and USD 60 million respectively [26] . GoU has also signed agreements in 2012 to finance projects to improve health services delivery and access to water and sanitation in both urban and rural areas worth USD 155.8 million of investment [27] . This development is expected to put more pressure on the water resources of the country as discussed in sections above.

Water supply: Water supply is a key factor of production in e.g. manufacturing industry, power generation, mining, construction and agriculture. It sustains the natural environment; therefore its quantity and quality are important indicators of what the country has achieved. Access to water has levelled off at 64% for rural and 77% for urban areas but set targets are 100% and 77% for urban and rural areas respectively by 2015.

The challenges faced in Uganda with regard to water supply issues are: to ensure that water for both human and livestock consumption is available in sufficient quantities and quality all the time; that the demand is quantified; water is used efficiently and sanitation is promoted; and shared waters and watercourses are used and developed to meet this demand in particular with regard to the new strategy of bulk water supply.

6. VULNERABILITY TO CLIMATE CHANGE AND ADAPTATION STRATEGIES

The changing climate in Uganda presents very serious national challenges and risks across various sectors such as agriculture, water resources and energy, which support the economy and the wellbeing of its people. Climate change has significant impacts, which manifest itself in form of droughts, variations in groundwater and surface water levels, melting of ice caps and incidences of diseases, high temperatures and wild fires. For example, [10] identified 8 seasonal droughts within the Lake Victoria basin in the period between 1990 and 1999. Our analysis has also established drought years and seasons in the various drainage sub-basins in Uganda (Table 4). Prominent was the 2004/5 drought period when the water level in Lake Victoria dropped by a meter below the 10 year average [28] . Droughts significantly affect the country's water resources, hydropower production and agriculture.

Cycles are observed in rainfall patterns in Uganda for the period 1940-2009. Analysis shows decades of normal and beyond normal rainfall (Figure 6) during the period of analysis. Three long spans of below normal rain are evident in the 1940s, 1960/70 and 1980s. Other particularly wet periods were experienced in early 1950s, 1960s, late 1970s and the 1990s (Figure 6). The contributing factors to the below average annual total rainfall for Uganda have not been established in this study despite the 1980s being drought years with consequences on food security for the continent [29] .

From a hydrological perspective, Lake Victoria, the largest water resource exerts a big influence in Uganda's climate. The main source of water for the lake is rain, but due to rainfall anomalies, the lake has consequently displayed large and rapid changes. This has been demonstrated by [30] [31] who find a significant correlation between lake rainfall series and lake levels. A number of studies highlighted by [15] and [32] have concluded that the lake is a sensitive indicator of climate change. Models on Sensitivity of lake levels, predict a fall in levels during the 2021-2050, and levels pick up by 2070-2099, time horizons in both emission scenario. This circumstance is explained by the predicted decrease in direct rainfall with a corresponding increase in lake potential evaporation by 2021-2050 [15] .

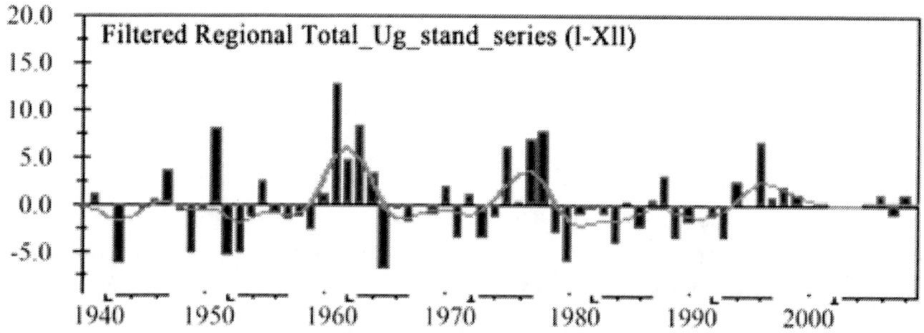

Figure 6. Regional total rainfall (mm) for the period 1940-2009.

Melting glaciers have been a centre of attraction of late as a source of water. The interaction between glaciers and climate represents a particular sensitive process on the water resources of Uganda. A number of authors have been cited by [11] to have comprehensively studied the effect of this interaction in the Rwenzori catchment area. Recent research in the 1950s, 1990s and 2000s indicate that the area covered by alpine glacier has reduced from 7.5 km^2 in 1906 to 1 km^2 in 2003 on Rwenzori Mountains [33] . Between 1955 and 1990, Rwenzori ice caps according to the Sector Performance Report of 2011 have retreated by 40%. This drastic trend calls for further studies and urgent response from policy makers.

The effect of deglaciation on water resources has been investigated to determine the impact of glacial melt on river discharge during the dry season and the risk of flooding during the rainy season [34] . Their findings, also held by [35] showed that current glacial recession has a negligible impact on alpine river flow since melt water discharges contribute less than 2% of the total discharge of the principal river receiving melt water (River Mubuku) at the base of Rwenzori Mountains during both dry and wet seasons. Consequently the effect of melt water has insignificant impact on the water resources, but may have adverse effect on tourism of Uganda. Melting of ice caps of the Rwenzori is expected to have an effect on water catchments downstream and eco-tourism as well as the overall economy [36].

The effects of climate change on ground water recharge and base flow in upper Ssezibwa catchment in Uganda has been studied using global climate models [17] . Projections from their study indicate that rainfall will generally increase, from 20% to 100% by the 2080s. Simulations from the same study showed an increase in recharge, base flow and total flow in the future. The mean annual daily discharge is expected to increase by 40% - 100% from the current 1.47 m^3/s in the coming 20 to 80 years.

7. WATER GOVERNANCE IN UGANDA.

Water governance is seen as a range of political, social, economic and administrative systems that are in place to develop and manage water resources and the delivery of water services at different levels of society [37] . The manner of allocation and regulatory process and interaction in the management of water resources that embraces formal and informal institutions [38] including poverty eradication [39] are a focus of governance. Some of the necessary conditions for good governance are inclusiveness, accountability, participation, transparency, predictability and responsiveness. A major challenge is to understand how all these different processes in concert, determine certain policy outcomes and how change in governance regimes occurs.

For instance policies tend to focus on standard policy solution like liberalization and pricing of water services, which have had counter-productive results. Currently Climate change and the related increase in extreme weather events have exposed the vulnerability and lack of resilience of water resource management regimes. At the same time, policies to accommodate the prospects of climate and global change are absent. Consequently the conditions under which water resources managers have to perform have become increasingly unpredictable especially in a developing country like Uganda.

The broad strategies in Uganda for implementing activities and programs in the Water sector are articulated in the Poverty Eradication Action Plan (PEAP) of 2004. PEAP which formed the basic planning framework for the formulation and implementation of all sector policies, strategies and programs, identifies water and sanitation as part of the priority areas for poverty eradication. The strategies are embedded in the three pillars, namely; to increase production and competitiveness of Uganda's products to attain increased household incomes including water for production and water resources management; Strengthening security, conflict resolution and disaster management (includes water for security in north-eastern Uganda and provision of water and sanitation services; and human development among others.

Implementation of the PEAP is guided by a comprehensive sector policy and legal framework for management of the water and sanitation sector. These include: the 1995 Constitution of the Republic of Uganda, Local Governments Act Cap 243, Water Act Cap 152, and accompanying regulations, the 1998 Water Resources Regulations, the 1998 Waste Discharge Regulations, Water Supply Regulations of 1999, Sewerage Regulations of 1999, National Environment Management Authority Act Cap. 153, National Water & Sewerage Corporation Statute Cap 317, Uganda Water Action Plan of 1995, National Water Policy of 1999, National Environment Management Policy of 1994, National Health Policy and Health Sector Strategic Plan of 1999, the 1997 National Gender Policy, and the Community Mobilisation and Empowerment Strategy of 2006.

Since Uganda ascribes to international water conventions and declarations, it has prepared a Water Action Plan (WAP). The WAP is a flexible and dynamic framework for the protection and development of Uganda's water resources in conformity with the Water Act.

The government of Uganda further developed a water policy that provides for the overall policy framework for water resources management and development. The overall objective of the water policy is "To manage and develop the water resources of Uganda in an integrated and sustainable manner, so as to secure and provide water of adequate quantity and quality for all social and economic needs of the present and future generations with the full participation of all stakeholders" [19] .

The National Water Policy for Uganda identifies a need to enhance co-ordination and collaboration between the water and sanitation sub-sectors, so that they both adequately address environmental health and sanitation issues. This is done through the Water Policy Committee (WPC), Water Sector Working group (WSWG), Donor Forum, Non-Government Organisations (NGOs) forum under the umbrella of Uganda Water and Sanitation Network (UWASNET), Sector reviews by GoU and Donors and District Water and Sanitation Committee (DWSC).The policy also acknowledges the need for cooperation on trans boundary water resources management issues and promotes decentralization of water management functions. Water supply and sanitation, water for agriculture, hydropower and industrial use are priorities areas for allocation of water through planning and operation of systems.

The key player in water governance is the Ministry of Water and Environment (MWE), which has the overall responsibility of setting national policies and standards, and priorities for water development and management. It also monitors and evaluates sector development programmes to keep track of their performance, efficiency and effectiveness in service delivery. Under the Ministry there are two agencies for water sector namely; DWD as the lead agency responsible for managing water resources, coordinating and regulating all water and sanitation activities and providing support services to local Governments and other service providers. The National Water and Sewerage Corporation (NWSC) an autonomous parastatal entity established in 1972, and recognised by NWSC statute of 1995, is responsible for the delivery of water supply and sewerage services.

The other players in water governance are Ministry of Finance, Planning and Economic Development (MoFPED) mobilises funds, the Ministry of Health (MoH) is responsible for hygiene and sanitation promotion for households, while the Ministry of Education and Sports (MoES) does similar work in primary schools, the Ministry of Gender, Labour and Social Development (MoGLSD) is responsible for gender responsiveness and community mobilisation, the Ministry of Agriculture, Animal Industry and Fisheries (MAAIF)spearheads agricultural development. This includes the use and management water for production including Irrigation, animal production and aquaculture at farm level. Local Governments (Districts, towns, Sub-Counties) are empowered by the Local Governments Act of 1997 to provide water services. Communities are responsible for demanding, planning, contributing a cash contribution to operating and maintaining rural WSS facilities. While (NGOs) and Community Based Organisations(CBOs) are active in the provision of water and sanitation services, training of communities and local governments, hygiene promotion as well as advocacy and lobbying.

Since 1993, Uganda has been undergoing a decentralization of responsibilities from central government level to district local government level. This shift of responsibility has had a profound impact on the players in the Water Supply and Sanitation (WSS) sector. Districts became the main implementation agents and the human dimension is being emphasised in governance issues.

8. CATCHMENT BASED WATER RESOURCES MANAGEMENT

There is a lot of diversity presented in literature relating to the implementation of Integrated Water Resources Management (IWRM), especially in developing countries. This diversity is attributed to differences in institutional set up, governance structures, legal framework and level of development. The framework of IWRM stipulates that water resources are better managed at a catchment unit. This is why the WRM sub-sector reform study in Uganda recommended a paradigm shift from a centralized to a catchment based water resource management. Uganda is divided into four main management zones namely; Victoria, Kyoga, Albert and Upper Nile water management zones.

A pilot study on river Rwizi catchment area located in Victoria water management zone has revealed some aspects in the initiation of catchment management. Such as identification of threatened wetlands systems, integration of catchment management issues in district development plans, promotion of community participation, forestation, and dissemination of weather outlook/forecasts among others [40].

One of the key lessons learnt during piloting was that IWRM is better embraced in an area that is experiencing serious water resources problems. Secondly, because the process requires involvement of multiple stakeholders for its success, it involves a great deal of consultations and consequently substantial amounts of funds are required. Involving political, administrative and technical representatives in catchment management structures was found to be a viable aspect, thus it is highly encouraged.

9. WATER RESOURCES MANAGEMENT CHALLENGES AND ISSUES

Water resources in Uganda are estimated at 66 km^3/year corresponding to about 2800 m^3/person/year. The spatial and temporal distribution of water resource is uneven, which pauses a big challenge to their management. Some areas like those in north-eastern Uganda have less water resources while those in the central have plenty of water resources. That is why there are increasing incidences of water use conflicts in the water scarce parts of the country, especially in the cattle corridor. Pastoralists migrate from place to place in search of water and vegetation. This movement not only creates a security threat, it is also a major health hazard as diseases are transferred from one part of the country to another in this process.

There is also increasing pressure on water resources due to rapid population growth, increased urbanization and industrialization, uncontrolled environmental degradation and pollution. A combination of some of these factors is cited to be responsible for the recent landslides in eastern Uganda. This pressure still remains a big challenge to the sustainable management and development of country's water resources.

The frequent recurrence of extreme weather events (floods and droughts) and increasingly erratic rainfall are a big challenge to the management of Uganda's water resources. The variability in seasonal rainfall, has significantly affected the different socio-economic activities that are heavily dependent on rainfall.

Ground water is presently the major source of rural domestic water supply and is planned as the main source of water for small towns (40 completed and 80 new ones) and rural growth centres (800) by 2015. Some of the aquifers however, are limited in yield, extent, hydraulic characteristics, and recharge is low in certain parts of the country. The availability and quality of groundwater for larger rural water supply projects is a significant future challenge. Given the limitations of ground water supplies, it would appear that groundwater abstractions for many large projects such as irrigation and municipal water supplies might not be sustainable. There is still little knowledge on how climate change will impact on this important resource.

Given the increasing demand for water and the high rates of population growth of 3.2%, it is of utmost importance that an analysis of the hydrological impacts of climate change for the country are intensified.

Despite all the above planned developments, there is still very limited knowledge of the country's groundwater resources, making it difficult to guarantee sustainable groundwater development for the current and future needs.

Groundwater recharge varies considerably across the country and is extremely sensitive to land use and the amount and intensity of precipitation falling in a given area. However, due to inadequate data and resources, very few groundwater recharge assessments have been carried out in Uganda and thus recharge estimates for most areas remain unknown.

Government should be aware that, the development of hydropower schemes, though a non-consumptive use, creates reservoirs, which have adverse effects and environmental consequences on water resources. These include; Siltation/sedimentation and aquatic weed encroachment in the reservoir, diminished downstream river discharge, eutrophication in reservoirs due to domestic and industrial waste, increased loss of water due to evaporation and others.

The unique geographical location of Uganda in the Nile Basin, i.e. being a lower and an upper riparian state imposes constraints and responsibilities to the country [41] . As a country it would be affected by the use of the water both in quantity and in quality by the riparian countries.

Another setback in Uganda's effort especially to exploit HEP potential is the current colonial agreement of the agreed curve policy [41] . This colonial agreement has been rejected by Uganda even though the main beneficiaries of the agreement are still upholding it. It is therefore expected that regulation of Lake Victoria with a more scientific friendly policy supported by Decision Support Systems (DSS) may overcome this problem.

The challenges that Uganda faces in this respect include: failure to regulate Lake Victoria with a better regulatory policy, fluctuation of levels of lake

Victoria and other equatorial lakes in the Nile Basin due to climatic change and variability, maintenance of water quality to be free of aversive weeds, sedimentation etc.

10. CONCLUSIONS

Priority of Uganda's water resources utilization and development is given to domestic water supply for both human consumption and livestock. Rural domestic water demand is by far greater than the urban demand because about 80% of the country's population is rural. Currently the rural population is dependent on ground water development or rainwater harvesting to meet their needs. Given the MDG water targets and GoU water development programmes, Uganda will resort to the utilization of the watercourses and the shared water bodies to satisfy rural water demand sooner than later. The water supply sector is under expansion and small-scale irrigation is being promoted and may in the future be of increased importance, but rainfall, which is a major source of water, is on a decrease as water demand increases. Detailed assessments are thus necessary on ground water resources, groundwater recharge and water balance.

Francis N. W. Nsubuga, Edith N. Namutebi, Masoud Nsubuga-Ssenfuma (24,71)">24] . Major source of water for the country are the springs and wells. The new national strategy of bulk water supply is to meet rural settings demands including demands of water for livestock (WSS sector working group report, 2006). The biggest challenge is that this demand is not quantified, not forgetting that Uganda is both a downstream and an upstream riparian state.

REFERENCES

1. Kumar, R., Singh, R.D. and Sharma, K.D. (2005) Water Resources of India. Current Science, 89, 794-811.
2. Kundzewicz, Z.W., Mata, L.J., Arnell N.W., Döll, P., Kabat, P., Jiménez, B., Miller, K.A., Oki, T., Sen, Z. and Shiklomanov, I.A. (2007) Freshwater Resources and Their Management. Climate Change 2007: Impacts, Adaptation and Vulnerability. In: Parry, M.L., Canziani, O.F., Palutikof, J.P., van der Linden, P.J. and Hanson, C.E., Eds., Contribution of Working Group 11 to the Fourth Assessment Report of the Intergovernmental Panel on Climate Change, Cambridge University Press, Cambridge, 173-210.
3. WRMD (2004) The Year-Book of Water Resources Management Department (WRMD) 2002-2003. Entebbe.
4. UN-WWAP (2006) Uganda National Water Development Report; Prepared for the 2nd UN World Water Development Report "Water a Shared Responsibility" UN-WATER, WWAP/2006/9. World Water Assessment Programme (WWAP).

5. Asadullah, A., Mcintyre, N. and Kigobe, M. (2008) Evaluation of Five Satellite Products for Estimation of Rainfall over Uganda. Hydrological Sciences Journal, 53, 1137-1150. http://dx.doi.org/10.1623/hysj.53.6.1137
6. NEMA (2008) State of Environment Report for Uganda. National Environment Management Authority (NEMA), Kampala.
7. Anyah, R.O. and Semazzi, F.H.M. (2004) Simulation of the Sensitivity of Lake Victoria Basin Climate to Lake Surface Temperatures. Theoretical and Applied Climatology, 79, 55-69. http://dx.doi.org/10.1007/s00704-004-0057-4
8. FAO (2005) Uganda-AQUASTAT Survey, Irrigation in Africa in Figures. FAO Water Reports, Rome. http://www.fao.org/nr/water/aquastat/countries_regions/uganda/index.stm
9. NEMA (1998) State of Environment Report for Uganda. National Environment Management Authority (NEMA), Kampala.
10. DWRM (2011) A Case Study of River Rwizi, Lake Wamala, Lake Victoria Catchments and Representative Ground Water Monitoring Stations. Water Resources Monitoring and Assessment Division. Directorate of Water Resources Management (DWRM) Entebbe, Uganda.
11. Mölg, T., Georges, C. and Kaser, G (2003) The Contribution of Increased Incoming Shortwave Radiation to the Retreat of the Rwenzori Glaciers, East Africa, during the 20th Century. International Journal of Climatology, 23, 291- 303.http://dx.doi.org/10.1002/joc.877
12. Awange, J.L., Ogalo, L., Bae, K.-H., Were, P., Omondi, P., Omute, P. and Omullo, M. (2008) Falling Lake Victoria Water Levels: Is Climate a Contributing Factor? Climate Change, 89, 281-297. http://dx.doi.org/10.1007/s10584-008-9409-x
13. Mileham, L., Taylor, R.G., Todd, M., Tindimugaya, C. and Thompson, J. (2009) The Impact of Climate Change on Groundwater Recharge and Runoff in a Humid, Equatorial Catchment: Sensitivity of Projections to Rainfall Intensity. Hydrological Sciences Journal, 54, 727-738.
14. Taylor, R.G. and Howard, K.W.F. (1996) Groundwater Recharge in the Victoria Nile Basin of East Africa: Support for the Soil-Moisture Balance Method Using Stable Isotope and Flow Modelling Studies. Journal of Hydrology, 180, 31- 53. http:// dx.doi.org/10.1016/0022-1694(95)02899-4
15. Tate, E., Suitcliffe, J., Conway, D. and Farquharson, F. (2004) Water Balance of Lake Victoria: Update to 2000 and Climate Change Modelling to 2100. Hydrological Sciences Journal, 49, 563-574.
16. Tindimugaya, C. (2006) Overview of Groundwater Development in Uganda. Proceedings of the Workshop for Groundwater Professionals in Uganda, Kampala, 25 August 2006, Unpublished.
17. Nyenje, P.M. and Batelaan, O. (2009) Estimating the Effects of Climate Change on Groundwater Recharge and Baseflow in the Upper Ssezibwa Catchment, Uganda. Hydrological Sciences Journal, 54, 713-726. http://dx.doi.org/10.1623/hysj.54.4.713
18. Tindimugaya C (2005) Groundwater Resources Management in Urban Areas of Uganda: Experiences and Challenges. Conference Paper Maximizing the Benefits from Water and Environmental Sanitation. Proceedings of the 31st WEDC Conference, Kampala, 31 October-4 November 2005, 311-313.
19. Ministry of Water, Lands and Environment (1999) The National Water Policy. MWLE, Kampala.
20. NBCBN (2005) Small Scale Hydropower for Rural Development. Hydropower Development Research Cluster, Group 1, Nile Basin Capacity Building Network NBCBN-RE.
21. ERA (2009) Development and Investment Opportunities in Renewable Energy Resources in Uganda. Electricity Regulatory Authority, Kampala.
22. Umeme (2011) Umeme Annual Report and Financial Statements for the Year Ended December 2011. Kampala.
23. Kasita, I. (2011) Electricity Demand to Triple by 2023. The New Vision, Kampala.

24. MOFPED (2011) Annual Economic Performance Report 2010/11. Directorate of Economic Affairs, Ministry of Finance, Planning and Economic Development, Kampala.
25. UBOS (2009) Uganda Bureau of Statistics, 2009 Statistical Abstracts. Kampala.
26. African Development Bank (2010) The African Development Bank in Action, Activities in the Water and Sanitation Sector in Uganda: Overview and Key Elements of Interventions. Water and Sanitation Department of the African Development Bank. http://www.afdb.org/fileadmin/uploads/afdb/documents/Projects-and-Operations/9_AFDB_watsan_UG
27. African Development Bank (2012) AFDB Promotes Improved Access to Water, Sanitation and Health Services in Rural and Urban Uganda. African Development Bank Group. http://www.afdb.org/en/news-and-events/article/afdb-promotes-improved-access-to-water-sanitation-and-health-services-in-rural-and-urban-uganda-8771/
28. Kull, D. (2006) Connections between Recent Water Level Drops in Lake Victoria, Dam Operations and Drought. http://www.internationalrivers.org/files/attached-files/full_report_pdf.pdf
29. Hulme, M. (1992) Rainfall Changes in Africa: 1931-1960 to 1961-1990. International Journal of Climatology, 12, 685-699. http://dx.doi.org/10.1623/hysj.54.4.713
30. Hastenrath, S. (2001) Variations of East African Climate during the Past Two Centuries. Climatic Change, 50, 209- 217. http://dx.doi.org/10.1023/A:1010678111442
31. Mistry, V.V. and Conway, D. (2003) Remote Forcing of East African Rainfall and Relationships with Fluctuations in Levels of Lake Victoria. International Journal of Climatology, 23, 67-89. http://dx.doi.org/10.1002/joc.861
32. Nicholson, S.E., Yin, X. and Ba, M.B. (2000) On the Feasibility of Using a Lake Water Balance Model to Infer Rainfall: An Example from Lake Victoria. Hydrological Science Journal, 45, 75-95. http://dx.doi.org/10.1080/02626660009492307
33. Taylor, R.G., Mileham, L., Tindimugaya, C. and Mwebembezi, L. (2009) Recent Glacial Recession and Its Impact on Alpine River Flow in the Rwenzori Mountains of Uganda. Journal of African Earth Sciences, 55, 205-213. http://dx.doi.org/10. 1016/j.jafrearsci.2009.04.008
34. Kaser, G., Hardy, D.R., Mölg, T., Bradley, R.S. and Hyera, T.M. (2004) Modern Glacier Retreat on Kilimanjaro as Evidence of Climate Change: Observations and Facts. International Journal of Climatology, 24, 329-339. http://dx.doi.org/10.1002/joc.1008
35. MWE (2011) Uganda Water and Environment Sector Performance Report 2011, Ministry of Water, Lands and Environment, Kampala.
36. GWP (2000) Integrated Water Resources Management. Global Water Partnership (GWP) Technical Advisory Committee, Background Paper No.4.
37. Rogers, P. and Hall, A.W. (2003) Effective Water Governance. Technical Committee Background Papers No.7, Global Water Partnership (GWP).
38. Turton, A.R., Hattingh, H.J., Maree, G.A., Roux, D.J., Claassen, M. and Strydom, W.F., Eds. (2007) Governance as a Trialogue: Government-Society-Science in Transition. Springer-Verlag Berlin, Heidelberg. http://dx.doi.org/10.1007/978-3-540-46266-8
39. Sewagudde, S. (2011) Catchment Based Water Resources Management: Uganda's Journey to Taking the IWRM Concept to the Grass Root. Wash Watch, 2, 19-23.
40. Human Development Report, HDR (2006) Beyond Scarcity: Power, Poverty and Global Water Crisis. Palgrave Macmillan, New York.
41. NEMA (1996) State of the Environment Report for Uganda 1996. National Environment Management Authority (NEMA), Kampala.

CHAPTER 13

Recent Developments in Monitoring of Microbiological Indicators of Water Quality Across a Range of Water Types

Sandra Mesquita[1] and Rachel T. Noble[2]

[1] Center Centre for Marine Sciences from Algarve (CCMAR), Campus de Gambelas, Portugal

[2] Institute of Marine Sciences, University of Chapel Hill, North Carolina, Morehead City, USA

1. INTRODUCTION

As human pressure increases, so does exploitation of natural water sources for production of drinking water, public use and appreciation of recreational waters. Subsequently, the demand for faster and reliable monitoring methods and approaches has also been intensified. Ensuring that water is safe, whether for bathing or consumption is a critical complex process that requires the participation of multiple stakeholders [1, 2] and also distinct degrees of knowledge with respect to water management.

For quantifying microbial contaminants that are dangerous to public health, fecal indicator bacteria (FIB) are used worldwide as indicators of the potential presence of dangerous pathogens that can be found in water used for bathing, drinking and harvesting of seafood. The "indicator" system relies on the assumption that FIB are present concomitant to the presence of human bacterial, viral and parasitic pathogens of concern. Furthermore, the indicator system simplifies management, because no single pathogen can be singularly attributed to the majority of waterborne disease. If pathogens were measured for management of water quality, the list of candidates to be quantified would be many. The FIB system relies on thresholds that have been established using epidemiological studies and risk assessment frameworks (see Annapolis Protocol and WHO bathing water guidelines).While the standards for drinking water, recreational water and shellfish harvesting waters and meats are based upon similar approaches [3], different types of waters can have different acceptable thresholds for the same FIB because of differences in either acceptable risk, the relative use of the water (i.e. for consumption versus recreation), and a range of other factors. Therefore, acceptable thresholds for FIB in drinking water, recreational water and shellfish harvesting waters and meats vary by continent, region, state and locality.

By definition an indicator (in this case, FIB) should be, 1) able to be enumerated using standardized tests that are relatively cost-effective and user-friendly, 2) present in high concentrations in human fecal contamination, matching the high concentrations of other respective pathogens, and 3) well studied in relation to human health, with epidemiological studies to link FIB concentrations and adverse health outcomes such as gastrointestinal and respiratory human diseases [4, 5, 6]. These assumptions can be tenuous at times for all types of waters and for all of the FIB groups currently utilized for water quality management. However, it is generally accepted that regulation using FIB to protect public health is successful. For example, total coliforms do not exactly fit these requirements since they are not exclusively fecal in origin; however they are often used in conjunction with other FIB enumeration approaches to protect public health in drinking water [7]. Even though some of the assumptions listed above are violated periodically for the use of FIB to protect public health in the regulation of drinking water, an alternative approach is often unattainable. Human pathogens are heterogeneous, can be highly seasonal, and can be highly infectious at low doses. The range of enteric viruses alone that could be monitored in drinking water is vast, and these pathogens are difficult to detect [8, 9, 11]. A discussion of the pathogens of concern is beyond the scope of this chapter, and the reader is referred to a range of quality publications on the topic (e.g. 11, 12).The "indicator" system can be used successfully, particularly if new research and methodologies in the water quality arena are incorporated into management. As a community, we can attain improved means of management of public health in water, if attention is paid to utilization and incorporation of new monitoring tools, new models and molecular tools. Current advances in science include rapid quantitative PCR assays (QPCR) developed to track distinct microorganisms in multiple water sources [12, 13, 14, 15, 16, 7].

In this chapter, we highlight recent advancements in the management of water, and in monitoring of microbial water quality in drinking water production and recreational waters. First, we highlight the current regulatory landscape for microbiological indicators. Second, we will describe the application and evolution of molecular tools for rapid quantification of FIB (E. coli and *Enterococcus* sp.), in a range of water types. We will include some of the limitations and advantages of different types of culture based methods as compared to the recently developed rapid QPCR-based methods. We will present a small case study conducted along high use beaches along the Pacific Ocean of California, USA. Third, we highlight recent developments in the field of Harmful Algal Blooms (HABs) and concerns over cyanobacterial toxins in surface water sources for drinking water. Following the presentation of information presented on HABs, and comparing *E. coli* and *Enterococcus*sp. concentrations using culture based and molecular (qPCR) rapid methods, the authors will consider the application of newly developed source tracking markers for quantification of fecal contamination fecal contamination. These microbial source tracking (MST) tools are currently being used and expanded for the management of a range of water types. The chapter is intended to highlight recent advancements, while promoting the consideration of other successful approaches for managing water resources well into the 21st century with an eye toward improved protection of public health.

2. REGULATORY LANDSCAPE ACROSS MAJOR WATER TYPES: HIGHLIGHTING THE USE OF THE FECAL INDICATOR BACTERIA AS PART OF THE "INDICATOR" SYSTEM

Traditionally, the application of FIB for water quality monitoring included enumeration of members of the total coliform group. This group was originally described on the basis of lactose fermentation detection instead of upon the principles of systematic bacteriology [3; 17-26]. Total coliforms are still used widely for management of drinking water, but their sole use in recreational water quality monitoring has decreased in favor of enumeration of FIB that are specific to fecal contamination from warm-blooded animals (i.e. *E. coli* and *Enterococcus* sp.). Total coliforms are defined by their respective characteristics, i.e. they are Gram negative facultative anaerobes that do not form spores are rod-shaped bacteria, with lactose fermentation occurring with acid production (24-48h, 36°C), and are indole-negative. Coliforms belong to the family Enterobacteriaceae which includes *Escherichia, Enterobacter, Klebsiella* and *Citrobacter, Kluyvera, Leclercia*, and some members of the genus *Serratia* [27].

In all of the culture-dependent methods, cultivation conditions are choose in order to improve the growth of the target microorganism while simultaneously inhibiting the growth of other microorganisms. Balance amongst sensitivity and selectivity is the reason for different methods for sample processing (drinking water *vs* highly contaminated waters) [28] and selection of different approaches for quantification using traditional culture based methods. The three most widely used culture based methods for quantification of FIB in any water type are Defined Substrate Technology (DST, e.g. IDEXX Colilert-18®), multiple tube fermentation (MTF), and membrane filtration (MF). The first edition of "Standard Methods for the Examination of Water and Wastewater" was released in 1905 and *E. coli* was selected as the most suitable indicator organism for raw drinking water [29]. Therefore multiple detection methods for *E. coli* were available early in the century. As scientific knowledge evolved a broader group of microorganisms (still including total coliforms, fecal coliforms, and *E. coli*) have been selected as surrogate measures in a wide range of water sources and treated water [30]. Two key factors have led to the trend of using *E. coli* as the 'preferred' indicator for the detection of fecal contamination, not only in drinking water but in other matrices as well: first, the finding that some 'fecal coliforms' were non-fecal in origin, and second, the development of improved testing methods for *E. coli* [1-19]. Membrane filtration is commonly selected for FIB quantification along the drinking water treatment process since it is flexible to the amount of sample filtered and since the method permits specific quantification. That is, both MTF and DST-based methods are either reported in a presence/absence format or in a "most probable number" format. Colilert® and Colilert- 18®, however, tend to be used at the "end of pipe" and are user friendly methods, such that their popularity has increased in the past decade.

An example of current regulations for protection of public health in Europe, USA and other countries is summarized in Table 1 [20]. Note that many currently used methods both total coliform and *E. coli* enumeration together, as IDEXX DST kits such as Colilert-18® permit the simultaneous quantification of both groups. Furthermore, while guidelines have been issued in different countries, as highlighted in Table 1, the interpretation and implementation of monitoring typically falls in the hands of member states and provinces; this creates a wide range of monitoring approaches that are currently employed for management of drinking, recreational, and shellfish harvesting water quality. It is for this reason that only a table for drinking water is presented, as drinking water regulations are the most stringent and consistent across nations.

Table 1. International drinking water standards and guidelines from the World Health Organization (WHO) Modified from [31, 32 and 33]

Paramete r	**Canada**	**United States**	**United Kingdom**	**EU direct ive**	**WHO***	**Austr alia**
Total Coliforms	0/100mL in 90%	0/100mL in 95%	0/100mL	0/100 mL	*	*
Thermotol erant coliforms (fecal coliforms) or *E. coli*	*	0/100mL	*	*	0/100m L	*
E. coli	0/100mL	0/100mL	0/100mL	0/100 mL	0/100m L	*
Enterococc i	*	*	0/100mL	0/100 mL	*	*
Cryptospori dium parvum	*	99% removal or inactivati on	<1oocyte /10L	*	*	*
Cryptospori dium perfringens (including spores)	*	*	0/100mL	a	*	*
Pseudomon as aeruginosa	*	*	*	0/250 mL	*	*
Colony count 22°C	*	*	No abnormal change	No abnor mal chang e	*	*

Parameter	Canada	United States	United Kingdom	EU directive	WHO*	Australia
Colony count 37°C	<500CFU /mL[b]	<500CFU /mL[b]	No abnormal change	20/mL[c]	*	*
Microcystins _LR	1,5μg/L	*	*		1,0 μg/L (provisional)	Lifetime exposure 1,3μg/L

[i] - * spaces left in blank indicate parameters not specified;[a] Necessary only if the water originates from or is influenced by surface water;

[ii] - [b] HPC(35ºC for 48h) or <200 background coliforms on a total coliform membrane filter;[c]Necessary only in the case of water offered for sale in bottles or container

The World Health Organization (WHO) Guidelines for Drinking-water Quality [32] represent an overall international scientific consensus, based on a wide range of participants, of the health risks presented by microbes and chemicals in drinking water. The existence of international guidelines for cyanobacteria microcystins (1 μg/L microcystins –LR) in drinking water has been established by WHO (1998). Most countries rely upon this value and developed specific systems adapted for their own reality. For instance, in Brazil federal legislation requires more intensive monitoring programs including toxin analyses or toxicity testing if cyanobacteria exceed 10,000 cells/ml or 1 mm^3 biovolume in a given water sample. This includes a mandatory standard of 1 μg/L applied for microcystins (variants not specified), and recommendations limiting to saxitoxin 3μg/L.

Evaluation of water quality for drinking water production requires sampling collection at specific temporal intervals to ensure safety for the public. Usually it is recommended to sample across the process of drinking water treatment, permitting an analysis of bacteriological water quality throughout processing and distribution [6,7]. This permits rapid identification of a process or component in the distribution system that is failing, and rather than troubleshooting across the entire process, subsequent attention can be focused on specific components. This approach simultaneously prevents local contamination from developing undetected at cross-connections or breaks in the distribution lines or due to a drop in positive pressure [1, 34, 35].

The definition of a recreational water body represents a wide range of environments that include the ocean, hot springs, lakes, reservoirs, streams and rivers [34]. Similarly, the diversity of potential human pathogenic microorganisms includes all viruses, protozoans, and bacteria that could potentially be present in natural fresh and marine recreational waters particularly those contaminated by wastewater [9, 36]. In recreational water quality monitoring, monitoring approaches are often guided by use. Beaches that are used intensely by the public (such as those in Santa Monica Bay, California, USA) are monitored daily throughout the summer bathing season. In some states or areas with lower beach usage, during specific "off seasons", or areas that are used only for secondary contact recreation such as kayaking and sailing, monitoring for FIB might be conducted only once per month. Although a broad discussion is beyond the scope of this document, the use of FIB for monitoring shellfish harvesting waters also relies heavily on the use of traditional culture-based methods such as MF and MTF, but management differs by country. In some EU states, management of shellfish is by quantities of *E. coli* in shellfish meats (e.g. oysters), whereas, quantification in the United States is based upon fecal coliforms in the water column surrounding the shellfish beds.

Advantages connected to the application of culture-based methods include low equipment requirements, user-friendliness, and low cost, which makes it the most used approach for quantifying FIB in any water type. On the other hand, the enrichment stage takes time (typically from 18-24 hours) causing the notification of the public to be delayed during a contamination event [37].

This delay also causes vital revenue to be lost because contaminated waters are often still restricted or "closed" for use (in the case of beaches) after the contamination event has already passed. Incubation methods also suffer from the fact that they can promote the growth of false positive organisms [38]. Rapid molecular methods have recently been developed for application to protection of public health in recreational waters [39]. These methods require 3 hours from sample processing to results, and they are qPCR based.

They target specific organisms (*Enterococcus* sp. or *E. coli*). The use of such methods permits rapid public notification (e.g. [40]) resulting in improved protection of public health. One other advantage of such methods is the high specificity [28X]. Other rapid methods using qPCR have already been developed for quantification of microbial markers that are specific to sources of fecal contamination [44, 45].

3. NEWLY DEVELOPED RAPID MOLECULAR METHODS FOR RECREATIONAL WATER QUALITY MONITORING

Millions of dollars are expended each year to measure FIB to and assess whether swimming at recreational beaches posed a risk to public health. The monitoring programs are typically conducted using culture-based methods that require an 18 to 96 hour incubation period. The lengthy incubation step results in situations where beaches are managed inaccurately, by keeping beaches open that are contaminated and closing beaches that are potentially clean again [40]. Newer, more rapid methods, such as qPCR have been developed by EPA and others for use in recreational water quality monitoring [39, 41]. Several of these new rapid PCR-based methods for quantification of *Enterococcus* sp. have been found to be significantly related to human health outcomes in epidemiology studies [42, 43]. In some cases, the association between rapid methods and adverse human health outcomes has been stronger than that observed between traditional culture-based methods and adverse human health outcomes [43].

Quantitative PCR is a novel primer-based molecular technique that combines the specificity of traditional PCR with the quantitative measurement of fluorescence for quantification of presence of specific types of nucleic acid in environmental samples. Many of the recently developed methods are actually quite similar to one another; differing only in the type of qPCR fluorophore "chemistry" used [39,41]. Even though epidemiological studies to assess relationships between rapid FIB quantification methods and human health outcomes have been successful, the implementation or widespread use of the rapid qPCR-based methods has not been without hurdles. Noble et al., 2010 [41] successfully demonstrated the use of rapid molecular methods by water quality personnel that had no previous experience with molecular techniques. Griffith and Weisberg, 2011 [40] built upon this success, by training three water quality laboratories to conduct the rapid methods, and implementing the use of the rapid methods for active beach management decisions (posting and closing of beaches). They also demonstrated the use of real time notification technology at the beaches to rapidly convey real-time water quality results through the assistance of a non-profit organization called "MiOcean". Griffith and Weisberg (2011) reported excellent management agreement (>96%, i.e. the agreement in whether a beach would be posted or remain open) among the rapid QPCR based enumeration of *Enterococcus* sp. [41] and culture based methods such as EPA Method 1600 (membrane filtration based quantification for

Enterococcus sp.). Lavander and co-workers [46] have shown similar successes with quantification of *E. coli* at freshwater Great Lakes beaches, reporting management agreement of 98% between the rapid qPCR based quantification of *E. coli* and traditional culture based methods such as Colilert-18.

Even though success has been demonstrated in specific environments, it is clear that rapid qPCR-based methods for quantification of FIB are not appropriate for all places at all times. Three major issues to be resolved are ways to standardize assays, the approach used for quantification of the target organism, and inhibition of the qPCR.

Currently there are two modes of quantification being employed in the literature, both with advantages and disadvantages. One relies on quantifying the target cell, e.g. *Enterococcus* sp., where results are reported as cell equivalents/100 ml, where inhibition is assessed using a specimen processing control, but the final reported target concentrations are not modified quantitatively using inhibition data (see [40] for an example). The second approach reports target cell concentrations as calibrated cell equivalents/100 ml and relies on the use of cycle threshold ratios for both the target cell and the specimen processing control for final calculated quantities of the target cell concentration. The cell equivalent approach can, in certain circumstances, provide data that is more related to historical data reported for culture based methods [40,41, 46], but cannot be used at beaches were partial inhibition of the qPCR is common. On the other hand, the calibrated cell equivalent quantification approach can be more widely applied at a range of beaches, because inhibition is accounted for. However, the approach is imperfect because the specimen processing controls that are currently used are DNA-based, and they do not adequately predict inhibition of the amplification of FIB cells in real world samples [46].Dilution is another approach that can be used to effectively reduce inhibition, but this increases the limit of detection.

Figure 1 is a snapshot of results from a preliminary assessment of the performance of rapid qPCR-based methods as compared to membrane filtration (culture-based) methods for the quantification of Enterococcus sp. in water samples collected from an array of beach locations in Santa Monica Bay, California, USA. The rapid methods used here provide results within a few hours of sample collection. Quantification was conducted using the cell equivalent approach.

Inhibition of the qPCR has been discussed at length in other publications but is relevant to the implementation of rapid qPCR based methods for water quality assessment because of its unpredictable nature. Inhibition of qPCR can occur when high molecular weight compounds in the surface water (e.g., humic acids and other complex carbohydrates) combine with metal ions to sequester nucleic acids from polymerases and prevent amplification [47, 48, 49, 50]. Inhibition can sometimes be alleviated with the use of a commercial DNA extraction kit for sample purification. This requires extra time that is added to the total analysis time, and increases variability in the final results due to analyst error and variable DNA binding characteristics. Others have tried to deal with inhibition in analysis of surface and other water samples with the addition of adjuvants (e.g. bovine serum albumin, [49]. In recreational waters, many use salmon testes DNA to assess inhibition (e.g. 39). Cao et al. 2012 [10] assessed the use of dilution, salmon testes DNA, and internal controls for the assessment of inhibition in a range of water sample types, and no one approach clearly emerged as superior over the others. In fact, the internal controls and salmon testes DNA did not agree in their assessment of inhibition of the target qPCR [52]. It may be that implementation of the rapid methods, at least at first, follows the approach developed in southern California by [40]. They utilized the cell equivalent quantification approach, and for beaches that demonstrated inhibition of the QPCR, traditional culture based methods were utilized to quantify FIB concentrations for beach management purposes.

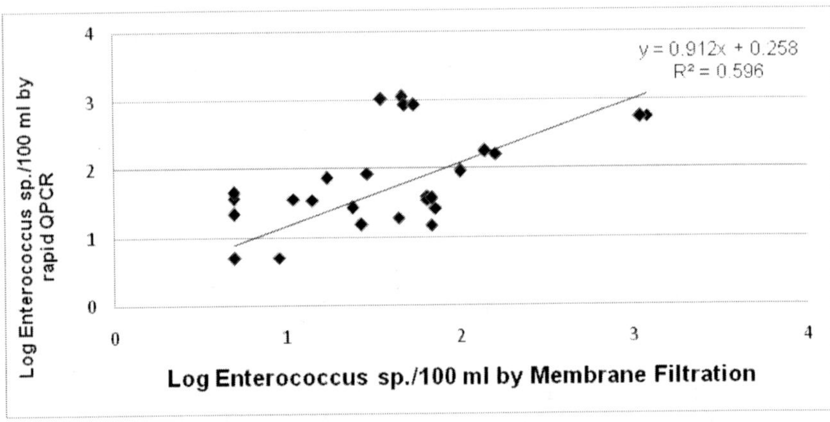

Figure 1. Example of rapid qPCR-based methods compared to membrane filtration (culture based) methods for the quantification of Enterococcus sp. in water samples collected from an array of beach locations in Santa Monica Bay, California, USA.

In the past years another new, rapid and reliable methodological approach has been tested to monitor microbial contamination, particularly in recreational marine and freshwater sources [53, 54]: the immunomagnetic separation / adenosine triphosphate (IMS / ATP) method. IMS-ATP was first developed by [55] and consists of the use of immune magnetic beads that selectively capture targeted bacteria using cell specific antibodies. A magnet is then used to separate the target bacteria from the rest of the sample with concentration determined by quantification of the ATP concentration in the sample using a bioluminescent assay [53]. While IMS/ATP has not been the subject of potential recommendation by the USEPA in 2012, there are advantages to its use. These advantages include the detection of only viable bacteria, therefore making it more directly related to quantification based upon other currently used culture based methods. Rapid quantification using IMS/ATP might be beneficial in areas receiving treated, disinfected wastewater for example, where the concentration of FIB-specific DNA is likely high, but the concentration of viable FIB cells would be lower. The capital costs of IMS/ATP are lower than those of the molecular methods. Another positive aspect of this technique is that it requires less than 4 hours. The major criticism of IMS/ATP concerns the availability of appropriate antibodies for quantification, particularly for the *Enterococcus* genus, which cannot be adequately captured using a single set of antibodies. However, improvements are being made in the development of cocktails of antibodies combined with the development of new bead based capture systems [e.g. 54].

In the coming years, there is the potential to dramatically improve the sensitivity, accuracy, precision, and speed of the QPCR assays used for water quality applications, while dramatically reducing the cost associated. One key step forward will be in the area of standardization, as standardized kits and cell standards and controls become available in ready-made kits, the analyst error and variability associated with the assays will decrease dramatically. The idea is to develop an elegant and powerful, yet simple, tool that can be applied to a wide variety of water sample types, to simultaneously determine the FIB concentration and enumerate bacterial tracers or pathogen types, necessary for preliminary source identification and/or protection of public health. Many key groups around the globe are working collaboratively to ensure this success.

4. HARMFUL ALGAE BLOOMS (HABS) IN DRINKING WATER TREATMENT PLANTS AND RECREATIONAL WATERS

In general, harmful algal blooms (HABs) are a cause of great concerns amongst stakeholders involved in water quality regulation and management. Since this chapter is specifically devoted to the problems caused by microbial contamination, we are specifically focusing our treatment of the subject on the closely related cyanobacteria, and their ability to form toxic blooms (referred to herein as CYANOHABs). There is an array of reports on the increased incidence of toxic blooms caused by Cyanobacteria. Their cyanotoxins are a serious concern for raw water sources that are exploited for drinking water production and recreational purposes [33]. Most countries around the world rely upon the exploration of surface freshwater sources for drinking water production. Efficient surface water treatment systems usually combine a series of treatment stages. Commonly employed water supply treatment includes multiple steps, generally including in order: pre-chlorination, flocculation and coagulation (addition of salts of iron or aluminium), sedimentation and post-chlorination [R]. One specific problem associated with toxic blooms of cyanobacteria is that specific stages of drinking water treatment are needed to remove the toxins. Not all of the drinking water treatment plants have these stages available or possess a flexible system [56]. The goal is to guarantee the production of drinking water according to national water quality guidelines, including microbial and chemical parameters that could potentially promote human disease and/or fatalities. In Table 2 some of the international guidelines for the presence of cyanobacteria toxins is presented. Cyanotoxins can be classified from to a toxicological perspective: hepatotoxins, cytotoxins, neurotoxins, irritant compounds as well as dermatotoxins [33]. Common methods of detecting and quantifying cyanotoxins include Enzyme-linked immunosorbent assay (ELISA), and chromatographic analysis [33, 57].

The main concern for public health associated with cyanobacterial blooms is exposure to cyanotoxins which can cause central nervous system or liver damage. There are multiple modes of exposure to cyanotoxins but most common are: ingestion of contaminated water, accidental ingestion by swimming on contaminated waters, and consumption of vegetables irrigated by contaminated waters or even inhalation (Table 3) [33,57]. A CYNOHAB's cause adverse effects to drinking water production plants. For instance there are physical impacts and the release of hazard by-products from the treatment stages such as chemical treatment. Other hazards come from oxygen depletion or pipe corrosion due to microbial activity within the pipes coming from bloom collapse. It is fundamental to have knowledge of source waters such as lakes that are prone to the development of toxic and non-toxic cyanobacterial blooms so that the overall population can be informed about public health risks. Several statistical approaches have been used to create models that could explain phytoplankton feed-back to nutrient enrichment and oxygen reduction, and these models may be adapted to other locations [56,58].

Whilst WHO's guidelines are based on total of enumerated cells (using microscopy techniques) and/or determination of chla (chlorophyll a) concentrations related to the amount of cells, there is a clear desire to develop faster quantification of cyanotoxins in environmental samples, especially large population areas. It is important to have rapid, reliable and accurate analysis i.e. high throughput molecular detection methods. In order to develop molecular assays, pure cultures from toxin producers need to be utilized to focus on genes specifically involved in the synthesis of a given toxin [AS].In Table 2, the guidelines from WHO for cyanobacterial levels and expected health effects due to exposure of contaminated recreational waters are shown.

Table 2. WHO guidance values for recreational waters with respect to cyanobacteria contaminates waters.(modified from [33]).

Guidance level or situation	**How guidance level derived**	**Health risks**
20000 cyanobacterial cells/mL 10μg chlorophyll-a/L Low impacts in human health	From human bathing epidemiological study	Skin irritations and gastrointestinal illness
100000 cyanobacterial cells/mL or 50μg chlorophyll-a/L moderate chances of affecting human health	From provisional drinking water guideline value and data concerning other cyanotoxins	Long-term illness and/or Short-term problems such as skin irritations and gastrointestinal illness
High probability of adverse effects Scum formation in areas where whole-body contact and/or risk of ingestion/aspiration occur	Inference from oral animal lethal poisonings	Potential for acute poisoning
	Actual human illness case histories	Potential for long-term illness with Adverse health problems such as skin irritations and gastrointestinal illness

Recently published molecular approaches for detection and quantification of toxigenic and non-toxigenic cyanobacteria are based on the application of multiplex approaches for targeting the genus (16S rDNA), and subsequently using other nested assays for sequences found in the non-ribosomal polypeptide synthetize complex (55KB in size) responsible for toxin production [56,57, 58]. Al-Tabrineh et al. (2012) recently developed and validated a quadruple target qPCR assay specifically designed for the detection and quantification of genes involved in the production of microcystins, nodularins, cylindrospermopsins and saxitoxins. There is a multiplicity of qPCR protocols published recently; however they typically focus on the quantification and detection of *Microcystis* spp. as a target due to the fact that it represents the most recurrent blooming cyanobacterial group in multiple water sources around the globe [59].

CYANOHAB related problems begin at the raw water source. Therefore, international guidelines such as the Water Frame Directive (Water Frame Directive, WFD; Directive 2000/60/CE of 23 October 2000) require a good ecological state in freshwater or highly modified freshwater sources (example water dams). Therefore, if a water body doesn't comply with these requirements it should be considered contaminated treated accordingly adding necessary steps to guarantee good finished water quality [60]. Thus, regular monitoring of freshwater sources for both recreational and drinking water purposes are necessary. Current modes of quantification starts with chla, but can extend to molecular quantification of specific phytoplankton groups, quantification of toxins (typically ELISA based), or pigment based quantification of phytoplankton groups [61].

Table 3. Examples of cyanobacterial toxins and their acute toxicity modified from [33]

<table>
<tr><th>Cyanotoxins</th><th>Mechanism of toxicity</th><th>Taxa toxin producer</th><th>LD_{50}(mouse) pure toxin</th></tr>
<tr><td>Protein-phosphatase blockers</td><td></td><td></td><td></td></tr>
<tr><td>General microcystins</td><td rowspan="3">Blocks protein phosphatases by covalent binding promoting liver haemorrhage damage could occur</td><td rowspan="3">Microcystis, Planktothrix, Anabaena, Oscillatoria, Nostoc, Anabaenopsis, Hapalosiphon, Nodularin</td><td>45->1000μg/Kg</td></tr>
<tr><td>Microcystin-LR</td><td>60(25-125)μg/Kg</td></tr>
<tr><td>Microcystin-YR</td><td>70 μg/Kg</td></tr>
</table>

Cyanotoxins	Mechanism of toxicity	Taxa toxin producer	LD_{50}(mouse) pure toxin
Microcystin-RR			300-600 μg/Kg
Nodularin			30-50 μg/Kg
Neurotoxins			
Anatoxin-a	Blocks postsynaptic depolarization	*Aphanizomenon, Oscillatoria, Anabaena, Cylindrospermum*	250 μg/Kg
Saxitoxins	Blocks sodium channels	*Aphanizomenon, Anabaena, Lyngbya, Cylindrospermopsis raciborskii*	10-30 μg/Kg
Cytotoxin			
Cylindrospermopsin	Blocks protein synthesis; substantial cumulative toxicity	*Cylindrospermopsis raciborskii*	200 μg/Kg/d

5. THE APPLICATION OF MICROBIAL SOURCE TRACKING IN WATER

5.1. MICROBIAL SOURCE TRACKING IN WATER

Microbial source tracking (MST) is the use of microbial markers (including bacteriophage, bacteria, viruses, protozoans, etc.) to determine the source of fecal pollution present in an aquatic system (i.e., human, pet, livestock). Some refer to the scientific discipline as bacterial source tracking (BST), but this term is older and has given way to MST, acknowledging the wider array of information that can be gleaned from targets other than bacteria. The

MST field continues to advance rapidly and much that has been published ten years ago is now obsolete and so it is generally better to refer to studies conducted since 2005 for the most updated information. For preliminary information on the evolution of the MST field, the reader should refer to the US Environmental Protection Agency's MST Guide (EPA 2005), article by [62]. MST methods were once categorized into two types — library dependent approaches and library independent approaches. Library dependent methods relied on matching the known source "fingerprints" or isolate patterns generated by molecular or phenotypic methods to unknowns collected from the environment. The most commonly used library dependent methods include, antibiotic resistant analysis (ARA), carbon utilization profiles (CUP), pulse field gel electrophoresis (PFGE), restriction- or amplified- fragment length polymorphism (RFLP or AFLP), random amplified polymorphic DNA (RAPD), repetitive extragenic palindromic PCR (rep-PCR) and ribotyping.

Library independent approaches are based upon detection (often presence/absence) or quantification of specific markers of fecal contamination, and are not dependent upon any sort of classification into categories. Library independent methods include F-specific and somatic coli phage serotyping and genotyping, and host-specific and marker-specific PCR and qPCR for a wide range of targets (see Table 3). Many older approaches utilized conventional PCR, but those have generally given way to the increased use of qPCR. Conventional PCR results are typically reported as presence/absence, or at best by using a serial dilution technique, can produce semi-quantitative information [63]. As opposed to conventional PCR, which is limited to a presence/absence result, qPCR provides for quantification over a wide dynamic range. The platforms currently being used for qPCR include hardware with proven rapid cycling, sensitive optics, and multiplex capabilities.

5.2. Bacteroides based microbial source tracking

One of the most fruitful organisms of focus in MST has been the Bacteroidales family. The Bacteroidales family has yielded a range of quantitative markers for human, dog, cow, horse, and other types of fecal contamination (e.g. 23, 71) The Bacteroides genus was suggested as early as 1967 as an alternative indicator due to the fact that *Bacteroides* spp. levels present in human and animal intestinal tracts are about 10 to 100 times above the other FIB [14, 16]. The high degree of difficulty for cultivation of *Bacteroides* spp. reduced its use in the late decades of the 20th century, but the development of molecular tools allowed the detection and quantification of members of the family Bacteroidales, without cultivation [64]. Consequently, there has been a dramatic increase in publications detailing development of qPCR assays for the quantification of

Bacteroidales in recreational waters as indicators of specific types of fecal contamination [14,65]. While several studies have demonstrated the value of the family Bacteroidales, some have also highlighted problems associated with cross-reactivity for certain markers, making testing and validation of assays particularly important [64, 65]. One of the most important advancements has been the development of the HF183 human specific markers that has been shown in a range of studies to be specific and relatively sensitive [66,67,68]. Acknowledging that no one marker can be used everywhere and in all circumstances, the HF183 marker has been atypical is that it has been successfully utilized more than any other single marker in application for detection and quantification of human fecal contamination in recreational waters. Recent work [69] using the HF183 marker has shown that data generated can be suitable for providing stakeholders and decision markers with valuable information for remediation projects [31].

Recently, another qPCR-based method has been developed by [70] to quantify fecal *Bacteroides* spp. The assay was specifically designed to select for the *Bacteroides* species that are most human-associated, but it also amplifies Bacteroides spp. from animals. The focus of the assay is on a cohort of *Bacteroides* species that are most prominent in the human gut. This method has been successfully used as a screening tool, with more specific methods used to quantify specific types of fecal contamination [71]. Another successful approach is reported by [71], where the authors developed an array of fecal *Bacteroides* based markers for a range of types of fecal contamination. They reported a quantitative universal *Bacteroidales* assay called BacUni-UCD that detected universal fecal *Bacteroidales*in all of the test samples (n=73) including human, cat, seagull, horse, dog and cow feces. The human assay BacHum-UCD presented by [68], successfully discriminated fully between human and cow stool samples and had slight cross-reactivity with dog stool, however, the method is highly quantitative when used in combination with human specific approaches outlined by [68]. All of the wastewater samples tested during the [68], study were positive for the BacHum-UCD marker, showing it to be 100 percent sensitive for human source identification. In 2009, Ahmed and co-workers [73] published a test of the host specificity of the five published sewage-associated *Bacteroides* markers. They tested 186 fecal samples from 11 animal species including humans. All of the human fecal samples were positive for all five markers indicating 100 percent sensitivity of the above referenced markers of [71], [68] and [72]. The HF183 marker has been found to have the capability to differentiate between human and animal feces with 99 percent accuracy [73]. It has been used in several mid-Atlantic stormwater studies to indicate the presence of human fecal contamination [74].

5.3. Virus based methods

Pathogen-based research for MST has focused almost solely upon pathogens that are carried via the fecal-oral route. The work that has been accomplished regarding enteric viruses has relied on the fact that high concentrations of these viral pathogens can be found in human sewage. High concentrations are a vital characteristic, given the limitations with concentration and isolation of pathogens in wastewater and complex aquatic matrices. In 2005, in a review written by Fong and Lipp [75], the authors recount the major developments in the past decades, including applications of human and animal viruses to source tracking. The authors detail the disadvantages and advantages of virus-based detection and quantification methods. In particular, they present a summary of information of the myriad of animal viruses applied to source tracking studies for human fecal contamination, namely human enteroviruses (echovirus and Coxsackie virus for example) and adenoviruses (Ad5, 7, 40, and 41). The review is completed by iterating the need for strong study designs with the use of virus detection and quantification methods for microbial source tracking applications, due to the limitations with the ability to detect viral targets, which can often occur in aquatic systems at low concentrations and high patchiness.

Rajal and co-workers [76] presented work on the quantification of human and animal pathogenic viruses in real world water samples. They used a known sample volume and tracer addition of surrogate viruses to calculate the limit of detection. This work was one of the first to detail carefully the quantification and recovery in a tightly designed study. For any type of human pathogenic virus quantification, recovery estimations are vital. This can typically be done through the use of spiked samples or by adding surrogate DNA or RNA to the sample, or with the use of a competitive internal positive control (CIPC). In 2006, Gregory et al.[78] reported the first use of a CIPC for full quantification of RNA viruses for water quality applications. The point of this internal control was to permit assessment of the efficiency of the combined reverse-transcriptase and PCR steps to enterovirus quantification and to provide a scenario for CIPC design approaches for other virus types. Similar approaches were developed by [79] and have resulted in successful quantification of important human pathogenic enteroviruses in recent years.

While not all viruses will be infectious in a water sample, accurate quantification of viruses in a water sample along with flow measurements can permit loading estimates of potentially pathogenic material, specific to source, thereby permitting partitioning specific sources of fecal contamination in complex environments. Also, the accurate quantification of the human pathogenic viruses is an important parameter in the successful determination of the potential public health risk and can be a key component to Quantitative Microbial Risk Assessment (QMRA).

Recently, human polyomaviruses (HPyV) have emerged as a useful human-specific viral marker. There has been a series of studies conducted that report the presence and quantification of these viruses in raw sewage samples and/or environmental waters in sites ranging from Florida [82], California [82, 83], Australia [66], Spain [84], Germany [85] and Japan [86]. These viruses are double-stranded DNA viruses frequently isolated from urine, and in some cases feces, of both healthy and immunocompromised individuals. In a recent study conducted by [87], both nested PCR and qPCR were used to assess the presence of HPyV in raw and treated sewage in Rio de Janeiro, Brazil. The study interestingly demonstrated the clustering of different types of the HPyV coming from distinct African and European lineages. While this study was focused on samples from a sewage treatment plant, it demonstrates the specificity of information that can be gleaned from viral pathogen-based investigations.

A very recent publication by [87] has demonstrated the use of a range of MST approaches at a beach in southern California that has been the site of a range of water quality investigations (e.g. [43, 88]). Doheny State Beach, in Orange County, California, USA is a high priority, intensely utilized beach location that has suffered from a range of beach water quality problems, In the past decade, for example, the beach has been given a poor grade several times on the "Heal the Bay Beach Report Card" (www.healthebay.org). This study demonstrated the utility of the HF183 human specific marker, adenovirus, HPyV, and *Methanobrevibacter smithii* as markers of fecal contamination. They found distinct correlations among the highly human specific molecular markers, i.e. adenovirus at Doheny State Beach was correlated to both HPyVs and the HF183 molecular marker. This study dramatically presents the benefits of application of a range of microbial source tracking approaches (i.e. the use of the toolbox approach), as opposed to a single marker. It also demonstrates the complexity of real-world water quality investigations and the importance of conducting studies over appropriate time and sampling scales.

6. CONCLUSIONS

Over the past decades, the use of culture-based FIB based quantification approaches has dominated the management of microbial contaminants for drinking water, recreational water, and shellfish harvesting water quality. While these approaches are useful and have resulted in a dramatic improvement in water quality management over that observed prior to the 1950's, there is an array of new molecular approaches that can be used to bolster water quality management and the process of public notification (i.e. water advisories, beach posting and closings, and shellfish harvesting water closures).

Here, highlights have been presented of three areas for immediate advancement of the science. Rapid qPCR based methods are poised for future use to improve the accuracy of beach water quality postings and beach closures. Molecular assays specific for toxin producing strains of cyanobacteria, and other HAB species can be utilized to improve the management of our vital raw surface waters. Microbial source tracking techniques are poised to improve all facets of water quality management; from site specific investigations of the predominant fecal sources to assessment of the efficiency of wastewater treatment systems.

This is an exciting time for researchers given the fast evolution of molecular approaches. The design of appropriate methods, that are versatile, fast and reliable, can permit information to be transmitted to managers in shorter periods of time. The molecular methods will only improve in their cost effectiveness and user-friendliness in the next few years. Finally, the successful application of predictive multivariate models and formalization of QMRA approaches will improve dramatically with the infusion of the scale appropriate molecular information on microbial contaminants, sources of contamination and rates of discharge and loading.

REFERENCES

1. Tallon P, Magajna B, Lofranco C and Leung KT. Microbial indicators of fecal contamination in water: a current perspective. Water, Air, and Soil Pollution, 2005; 166:139–166
2. Bláha L, Babica P, Blahoslav. Toxins produced in cyanobacterial water blooms – toxicity and risks Interdisc Toxicol, 2009; 2(2): 36–41
3. Standard methods for the examination of water and wastewater. Greenberg, A.E, Clesceri, L.S. and Eaton, A.D. (Eds). 18th edition; 1992;
4. Balarajan R, Raleigh VS, Yuen P, Wheeler D, Machin D, Cartwright R.(1991) Health risks associated with bathing in seawater. British Medical Journal, 1991; 303: 1444
5. Haile RW, Witte JS, Gold M, Cressey R, McGee C, Millikan RC, Glasser A, Harawa N, Ervin C, Harmon P, Harper J, Dermand J, Alamillo J, Barrett K, Nides M, Wang GY. The health effects of swimming in ocean water contaminated by storm drain runoff. Epidemiology, 1999; 10: 355-363
6. Dufour AP, Ballentine P. Ambient water quality criteria for bacteria. Technical Report EPAA440/5-84-002. U.S. Environmental Protection Agency, Washington DC, 18, 1986
7. Izbicki JA, Swarzenski PW, Reich CD, Carole Rollins C and Holden, PA. Sources of fecal indicator bacteria in urban streams and ocean beaches, Santa Barbara, California. Annals of Environmental Science, 2009; 3:139-178
8. World Health Organization. Report of WHO Scientific Group of human viruses in water, wastewater and soil. Technical Report Series, no. 639. World Health Organization, Geneva, Switzerland, 1979
9. Ikner LA, Gerba CP and Bright KR,. Concentration and Recovery of Viruses from Water: A Comprehensive Review on Food Environmental Virology, 2012;4:41–67

10. Cao Y, John F Griffith JF, Samuel Dorevitch S and Stephen B Weisberg SB. Effective- ness of qPCR permutations, internal controls and dilution as means for minimizing the impact of inhibition while measuring Enterococcus in environmental waters. Journal of Applied Microbiology 2012; 113(1):66-75
11. World Health Organization. Report of WHO Guidelines for drinking water quality, I-recommendations. World Health Organization, Geneva, Switzerland, 2004;
12. Figueras MJ and Borrego JJ.. New Perspectives in Monitoring Drinking Water Microbial Quality. Int. J. Environ. Res. Public Health, 2010; 7:4179-4202
13. U.S. ENVIRONMENTAL PROTECTION AGENCY. Ambient Water Quality Criteria for Bacteria. EPA-440/5-84-002, U.S. Environmental Protection Agency, Washington, D.C., 1986
14. Griffith JF, Cao Y, McGee CD, Stephen B. and Weisberg B. Evaluation of rapid methods and novel indicators for assessing microbiological beach water quality. Water Research, 2009; 43:4900 - 4907
15. Harmsen, H J. M, Raangs GC, He T, Degener JE, and Welling GW. Extensive set of 16S rRNA-based probes for detection of bacteria in human feces. Applied and Environmental Microbiology, 2002; 68:2982–2990
16. Leila Kahlisch l, Henne K, and Gröbe, L.. Assessing the Viability of Bacterial Species in Drinking Water by Combined Cellular and Molecular Analyses. Microb Ecol 2012; 63:383–397
17. Eichler S, Christen R, Höltje C, Westphal P, Bötel J, BrettarI,Mehling A. and Höfle MG. Composition and dynamics of bacterial communities of a drinking water supply system as assessed by RNA- and DNA-based 16S rRNA gene fingerprinting. Applied and Environmental Microbiology, 2006; 72:1858–1872
18. Paavola, J., Institutions and environmental governance: a reconceptualization. Ecological Economics 2007; 6, 1:93–103
19. Orin C. Shanks, O C, Kelty,CA, Sivaganesan, M, Varma M and Richard A. Haugland, RA Quantitative PCR for Genetic Markers of Human Fecal Pollution, Applied and Environmental Microbiology, 2009;.75, 17:5507–5513
20. U.S. ENVIRONMENTAL PROTECTION AGENCY, Ambient Water Quality Criteria for Bacteria. EPA-440/5-84-002, U.S. Environmental Protection Agency, Washington, D.C., 1986
21. Dafour, AP. Health Effects Criteria for Fresh Recreational Waters. EPA-600/1-84-004, U.S. Environmental Protection Agency, Research Triangle Park, N.C., 1984
22. Davies, CM, Long, JA, Donald, M and N. J. Ashbolt, NJ, Survival of fecal microorganisms in marine and freshwater sediments. Applied and Environmental Microbiology 1995; 61:1888–1896
23. Layton, A., McKay L,Williams D, Garrett V, Gentry, R and Sayler. G. Development of Bacteroides 16S rRNA gene TaqMan-based real-time PCR assays for estimation of to- tal, human, and bovine fecal pollution in water. Applied Environ Microbiology 2006; 72:4214–4224
24. Matsuki T, Watanabe K, Fujimoto J, Takada T, and Tanaka, R. Use of 16S rRNA gene-targeted group-specific primers for real-time PCR analysis of predominant bacteria in human feces. Applied and Environmental Microbiology, 2004; 70:7220–7228
25. Boehm, AB, Ashbolt, NJ. A sea change ahead for recreational water quality criteria. Journal of Water Health 2009; 7, 1: 9-20
26. Gentry, R W, Layton AC, McKay LD, McCarthy JF, Williams DE, Koirala SR, and Sayler GS. Efficacy of Bacteroides measurements for reducing the statistical uncertainty associated with hydrologic flow and fecal loads in a mixed use watershed. Journal of Environmental Quality, 2007; 36:1324–1330
27. Figueras MJ, Robertson W, Pike E and Borrego JJ. Sanitary inspection and microbiological water quality. In: Monitoring Bathing Waters. A Practical Guide to the Design and Implementation of Assessments and Monitoring Programs; Bartram J. and Rees G (Eds), E & FN Spon: London, UK, 2000; pp. 113-126
28. Fielda KG and Samadpourb M. Fecal source tracking, the indicator paradigm, and managing water quality. Water Research, 2007; 41: 3517-3538

29. Environmental Health Directorate – Health Protection Branch. Recreational Water Quality-Published by Authority of the Minister of National Health and Welfare, 1978;
30. Kornacki JL and Johnson JL. Enterobacteriaceae, Coliforms and *Escherichia coli* as quality and safety indicators'. In: F. Downs (Eds), Compendium of Methods for the Microbiological Examination of Foods, APHA. Washington DC, 2001;
31. World Health Organization. Guidelines for drinking-water quality, 3rd ed., vol. 1. World Health Organization, Geneva, Switzerland; 2008
32. WHO, Guidelines for Drinking-Water Quality, 2nd edition Geneva, 1993; ISBN 92 4 154460(Albinana-Gimenez et al. 2009), Germany (Hamza et al. 2009) and Japan (Haramoto et al. 2010)
33. Chorus I and Bartram J (Eds.). Toxic cyanobacteria in water: a guide to their public health consequences, monitoring and management. E&FN Spon, ISBN 0-419-23930-8, London, 1999;
34. Izbicki JA, Swarzenski PW, Reich CD, Carole Rollins C and Holden, PA. Sources of fecal indicator bacteria in urban streams and ocean beaches, Santa Barbara, California. Annals of Environmental Science, 2009; 3:139-178
35. Whitman RL, Przzybyla-Kelly K,Shivley DA and Byappanhali MN. Incidence of the enterococcal surface protein (esp) gene in human and animal fecal sources. Environ SciTechnol 2007; 41(17):6090–6095
36. Dafour, AP. Health Effects Criteria for Fresh Recreational Waters. EPA-600/1-84-004, U.S. Environmental Protection Agency, Research Triangle Park, N.C., 1984
37. Leecaster MK and Weisberg SB. Effect of Sampling Frequency on Shoreline Microbiology Assessments. Marine Pollution Bulletin, 2003;42(11): 1150-1154.
38. Pisciotta J, et al. The role of neutral lipid nanospheres in *Plasmodium falciparum* crystallization. Biochem J, 2007, 402:197–204.
39. Haugland RA, Siefring SC, Wymer LJ, Brenner KP and Dufour AP. Comparison of Enterococcus density measurements by quantitative polymerase chain reaction and membrane filter culture analysis at two freshwater recreational beaches. Water Research, 2005; 39:559-568.
40. Griffith JF and Weisberg SB. Challenges in Implementing New Technology for Beach Water Quality Monitoring: Lessons from a California Demonstration Project. Marine Technology Society Journal 2011; 45:65–73
41. Noble RT, Blackwood AD, Griffith JF, McGee CD and Weisberg SB. Comparison of Rapid Quantitative PCR-Based and Conventional Culture-Based Methods for enumeration of *Enterococcus* spp. and *Escherichia coli* in Recreational Waters. Applied and Environmental Microbiology, 2010; 76(22):7437-7443
42. Wade TJ, Calderon RL, Brenner KP, Sams E, Beach M, Haugland R, Wymer L and A. P. Dufour AP. High Sensitivity of Children to Swimming-Associated Gastrointestinal Illness Results Using a Rapid Assay of Recreational Water Quality. Epidemiology, 2008; 19: 375-383
43. Colford JM, Schiff K, Griffith JF, Yau V, Arnold BF, Wright C, Gruber J, Wade T, Burns S, Hayes S, McGee C, Gold M, Noble RT and Weisberg, SB. Using rapid indicators for Enterococcus to assess the risk of illness after exposure to urban runoff contaminated marine water. Water Research, 2012; 1-11.
44. Scott TM, Jenkins TM,, Lukasik J and Rose JB. Potential use of a host-associated molecular in *Enterococcus faecium* as an index of human fecal pollution. Environmental Science and Technology, 2005; 39 (1):283–287.
45. McQuaig SM, Scott TM, Harwood VJ, Farrah SR and Lukasik JO. Detection of hu- man-derived fecal pollution in environmental waters by use of a PCR-based human polyomavirus assay. Applied and Environmental Microbiology, 2006; 72(12):7567–7574
46. Lavender JS and Kinzelman JL. A cross comparison of QPCR to agar-based or de- fined substrate test methods for the determination of *Escherichia coli* and enterococci in municipal water quality monitoring programs. Water Research 2009; 43(19): 4967-4979
47. Tsai YL and. Olson BH. Rapid method for separation of bacterial DNA from humic substances in sediments for polymerase chain reaction. Applied and Environmental Microbiology, 1992; 479(58):2292-2295

48. De Boer SH, L.J. Ward LJ, X. Li X and S. Chittaranjan. S. Attenuation of PCR in- hibition in the 411 presence of plant compounds by addition of BLOTTO. Nucleic Acids Research, 1995; 23:2567-2568
49. Kreader CA. Relief of amplification inhibition in PCR with bovine serum albumin or T4gene 32 protein. Applied and Environmental Microbiology, 1996; 62:1102-1106
50. Watson RJ and B. Blackwell. Purification and characterization of a common soil component which inhibits polymerase chain reaction. Canadian Journal of Microbiology, 2000; 46:633-642
51. Thurman EM, Aiken GR, Ewald M, Fischer WR, Forstner U, Hack AH, Mantoura RFC, Parsons JW, Pocklington R, Stevenson FJ, Swift RS and SzpakowskaB.Isolation of soil and aquatic humic substances. pp. 31-43 in: F.H. Frimmel and R.F. Christman (ed.), Humic Substances and Their Role in the Environment. John Wiley and Sons, Ltd. New York, NY, 1988
52. Cao Y, Griffith JF and Weisberg SB. Evaluation of optical brightener photo decay characteristics for detection of human fecal contamination. Water Research, 2009; 49 (8):2273-227
53. Bushon RN, Brady AM, Likirdopulos CA and Cireddu JV. Rapid detection of *Escherichia coli* and enterococci in recreational water using an immunomagnetic separation/ adenosine triphosphate technique Journal of Applied Microbiology, 2009;106: 432–441
54. Lee CM, Griffith JF, Kaiser W and Jay, JA. Covalently linked immune magnetic separation/ adenosine triphosphate technique (Cov-IMS/ATP) enables rapid, in-field detection and quantification of *Escherichia coli* and Enterococcus spp. in freshwater and marine environments. Journal of Applied Microbiology, 2009; 109:324–333
55. Lee JY. and Deininger RA. Detection of *E-coli* in beach water within 1 hour using immunomagnetic separation and ATP bioluminescence. Luminescence 2004; 19:31–36
56. Wiegand C and Pflugmacher S. Ecotoxicological effects of selected cyanobacterial metabolites a short review. Toxicology and Applied Pharmacology, Vol. 203, (March 2005); pp. 201-218, ISSN 0041-008X
57. Al-Tebrineh J, Leanne A. Pearson LA, Serhat A. Yasar SA and Brett A. Neilan BA. A multiplex qPCR targeting hepato- and neurotoxigenic cyanobacteria of global significance. Harmful Algae, 2012; 15:19–25
58. Sivonen K and Jones G. Cyanobacterial toxins. In: Toxic cyanobacteria in water: a guide to public health significance, monitoring and management. Chorus I and Bar- tram (Eds).The World Health Organization, ISBN 0-419-23930-8, E and FN Spon, London, UK, 41-111pp, 1999
59. Conradie KR and Barnard S. The dynamics of toxic *Microcystis* strains and microcys- tin production in two hypertrofic South African reservoirs. Harmful Algae 2012; in press
60. Henderson, Chips M, Cornwell, Hitchins P, Holden B, Hurley S, Parsons SA, Wetherill A and Jefferson B. Experiences of algae in UK waters: a treatment perspective. Water and Environment Journal, 2008; 22:184–192
61. Pinckney JL, Richardson TL, Millie DF, Paerl HW. Application of photopigment biomarkers for quantifying microalgal community composition and in situ growth rates. Organic Geochemistry, 2001; 32: 585–595
62. Simpson JM, Santo Domingo JW and Reasoner DJ. Microbial source tracking: state of the science. Environmental Science and Technology 2002; 36:5279–5288
63. Noble RT and J. A. Fuhrman, JA. Enteroviruses detected in the coastal waters of Santa Monica Bay, CA: Low correlation to bacterial indicators. Hydrobiologia, 2001;460:175-184
64. Gentry, R W, Layton AC, McKay LD, McCarthy JF, Williams DE, Koirala SR, and Sayler GS. Efficacy of Bacteroides measurements for reducing the statistical uncertainty associated with hydrologic flow and fecal loads in a mixed use watershed. Journal of Environmental Quality, 2007; 36:1324–1330
65. Boehm, AB, Ashbolt, NJ. A sea change ahead for recreational water quality criteria. Journal of Water Health, 2009; 71: 9-20

66. Ahmed W, Stewart J, Powell D and Gardner T. Evaluation of Bacteroides markers for the detection of human faecal pollution. Letters in Applied Microbiology, 2008; 46(2):237-242
67. Ahmed W, Goonetilleke A, Powell D, Chauhan K and Gardner T. Comparison of molecular markers to detect fresh sewage in environmental waters. Water Research, 2009;43:4908-4917
68. Seurinck S, Defoirdt T, Verstraete W and Siciliano SD.Detection and quantification of the human-specific HF183 Bacteroides 16S rRNA genetic marker with real-time PCR for assessment of human faecal pollution in freshwater. Environmental Microbiology, 2005; 7(2):249-59
69. Mieszkin S, Furet JP, Corthier G, and M. Gourmelon M. Estimation of pig fecal contamination in a river catchment by real-time PCR using two pig-specific Bacteroi- dales 16S rRNA genetic markers. Applied and Environmental Microbiology 2009; 75:3045–3054
70. Converse RR, Blackwood AD, Kirs M, Griffith JF and Noble RT, Rapid QPCR-based assay for fecal Bacteroides spp. as a tool for assessing fecal contamination in recreational waters. Water Resources, 2009; 43:4828–4837
71. Kildare BJ, Leutenegger CM, McSwain SM, Bambic DG, Rajal VB and Wuertz S. 16S rRNA-based assays for quantitative detection of universal, human-, cow-, and dog-specific fecal Bacteroidales: A Bayesian approach. Water Research, 2007; 41:3701-3715
72. Bernhard A and Field, KG. Identification of nonpoint sources of fecal pollution in coastal waters by using host-specific 16S ribosomal DNA genetic markers from fecal anaerobes. Applied and Environmental Microbiology, 2000; 66:1587–1594
73. Ahmed W, Goonetilleke A, Powell D, Chauhan K and Gardner T. Comparison of molecular markers to detect fresh sewage in environmental waters. Water research, 2009; 43:4908-4917
74. Parker JK, McIntyre D and Noble RT. Characterizing fecal contamination in storm-water runoff in coastal North Carolina, USA. Water Resources, 2010; 44(14):4186-4194
75. Fong TT and Lipp EK. Enteric viruses of humans and animals in aquatic environments: health risks, detection, and potential water quality assessment tools. Micro Mol Biol Rev, 2005; 69(2):357-371
76. Rajal, McSwain BS, Thompson DE, Leutenegger CM, Wuertz, S. Molecular quantitative analysis of human viruses in California stormwater. Water Research, 2007; 41(19):4287-4298
77. Wong M, Kumar L, Jenkins TM, Xagoraraki I, Phanikumar MS and Rose JB, (2009) Evaluation of public health risks at recreational beaches in Lake Michigan via detection of enteric viruses and a human-specific bacteriological marker. Water re- sources, 2009; 43:1137-1149
78. Gregory JB, Litaker RW and Noble RT. Rapid One-Step Quantitative Reverse Tran-scriptase PCR Assay with competitive internal positive control for detection of En-teroviruses in environmental samples. Applied and Environmental Microbiology, 2006; 72(6):3960-3967
79. Fuhrman JA, Liang X and Noble RT. Rapid Detection of Enteroviruses in Small Vol- umes of Natural Waters by Real-Time Quantitative Reverse Transcriptase PCR Ap- plied and Environmental Microbiology, 2005; 71(8): 4523–4530
80. Baums L, Goodwin KD, Kiesling T, Wanless D, Diaz MR, Fell,]W. Luminex detection of fecal indicators in river samples, marine recreational water, and beach sand. Marine Pollution Bulletin, 2007; 54 (5):521-536
81. Carson CA, Christiansen JM, Yampara-Iquise H, Benson VW, Baffaut C, Davis JV, Broz RR, Kurtz WB, Rogers WM and Fales WH. Specificity of *Bacteriodes thetaiotao- micron* marker for human feces. Applied and Environmental Microbiology, 2005; 71(8):4945-4949
82. McQuaigSM , Scott TM , Lukasik JO , Paul JH , Harwood VJ . Quantification of human polyomaviruses JC virus and BK virus by TaqMan quantitative PCR and comparison to other water quality indicators in water and fecal samples .Applied and Environmental Microbiology 2009; 75: 3379-3388

83. Rafiquea and Jiang, CS. Genetic diversity of human polyomavirus JCPvV in Southern California wastewater. Journal of Water Health, 2009; 6:533-538
84. Albina-Gimenez N, Clemente-Casares P, Calgua B, Huguet, JM, Courtois S and Gi- rones R. Comparison of methods for concentrating human adenoviruses, polyomavi- rus JC and noroviruses in source waters and drinking water using quantitative PCR. J Virol Methods, 2009; 158: 104-109
85. Hamza IA, Jurzik L, Stang K, Uberla K and Wilhem M. Detection of human viruses of a densly-populated area in Germany using virus adsorption method optimized for PCR analyses. Water Resources, 2009; 43:2657-2668
86. Haramoto E, Kitajima M, Katayama H and Ohgaki S. Real-time pCR detection of adenoviruses, polyomaviruses and torque viruses in river water in Japan. Water resources, 2010; 44:1747-1752
87. Fumian TM, Guimarães FR, Pereira Vaz BJ, da Silva MT, Muylaert FF, Bofill-Mas S, Gironés R, Leite JP and Miagostovich MP. Molecular detection, quantification and characterization of human polyomavirus JC from waste water in Rio De Janeiro, Brazil, 2010; 8(3):438-45
88. McQuaig S, Griffith J and Harwood VJ. The Association of Fecal Indicator Bacteria with Human Viruses and Microbial Source Tracking Markers at Coastal Beaches Impacted by Nonpoint Source Pollution Applied and Environmental Microbiology AEM.00024-12; published ahead of print 6 July 2012

CHAPTER 14

Assessing the Governance Capacity of Cities to Address Challenges of Water, Waste, and Climate Change

S. H. A. Koop[1,2], L. Koetsier[1] A. Doornhof[1], O. Reinstra[3], C. J. Van Leeuwen[1,2], S. Brouwer[2], C. Dieperink[1], & P. P. J. Driessen[1]

[1]*Copernicus Institute of Sustainable Development Utrecht University, Heidelberglaan 2, Utrecht 3584 CS, The Netherlands*

[2]*KWRWatercycle Research Institute, Groningenhaven 7, Nieuwegein 3430 BB, The Netherlands*

[3]*Waternet, Korte Ouderkerkerdijk 7, 1096 ACAmsterdam, The Netherlands*

ABSTRACT

The challenges of water, waste, and climate change in cities are overwhelming and underpin the importance of overcoming governance issues impeding adaptation. These "*governance challenges*" typically have fragmented scopes, viewpoints, and responsibilities. As there are many causes leading to this uncertainty and disagreement, there is no single best approach to solve these governance challenges. In fact, what is necessary is iterative and requires governance capacity to find dynamic long-term solutions that are supported by flexible interim targets, so as to anticipate emerging barriers and changing situations. The literature contains a plethora of governance gaps, barriers, and capacities, which sometimes overlap, are contradictory and case-specific, and reflect disciplinary scopes.

We argue that a balanced set of well-developed conditions is needed, to obtain the governance capacity that enables effective change. Therefore, we aim to obtain deeper understanding of the key conditions determining the urban water governance capacity, by developing an integrated empirical-based approach that enables consistent city comparisons and facilitates decision-making. We propose a governance capacity framework focusing on five governance challenges: 1) water scarcity, 2) flood risk, 3) wastewater treatment, 4) solid waste treatment and 5) urban heat islands. Nine governance conditions, each with three indicators, are identified and empirically assessed using a Likert-type scoring method. The framework is illustrated by a case study on Amsterdam, the Netherlands. We conclude our approach shows great potential to improve our understanding of the key conditions determining the governance capacity to find solutions to the urban challenges of water, waste, and climate change.

Keywords: Governance capacity, Water governance, Water management, Adaptive governance, City Blueprint, Social learning

1 INTRODUCTION

1.1 Governance Challenges

Cities accross the globe face huge challenges of water, waste and climate change in a rapidly urbanizing world (World Economic Forum 2016; Koop and Van Leeuwen 2016). By 2030, the world will be experiencing an estimated 40% freshwater shortage (2030 WRG 2009). Sea-level rise, soil subsidence, and extreme river discharges pose risks to around 15% of the global population, mostly in urban areas, including almost all the world's mega-cities (Ligtvoet et al. 2014). Cities are particularly vulnerable to flooding by storm events, which are increasing in magnitude and frequency (EEA 2012). Furthermore, streams, rivers, and oceans are heavily polluted by insufficiently treated wastewater and solid waste (Derraik 2002; Ligtvoet et al. 2014), which is leading to losses of valuable resources such as phosphate (European Commission 2015). Finally, global warming is exacerbating heatwaves that pose serious health risks to people living in Urban Heat Islands (UHI): for example, it is estimated that in 2003 heatwaves were responsible for 70,000 deaths in Europe (Baccini et al. 2008). In order to combat UHI, vegetation and water bodies need to be intertwined in the urban infrastructure.

Altogether, urban areas face five main interrelated water challenges: 1) water scarcity, 2) flood risk, 3) wastewater treatment, 4) solid waste treatment, and 5) urban heat islands (EEA 2012, 2016; UCCRN 2011). Meeting these challenges requires good governance, because it entails managing long-term, complex, uncertain, and imperfectly known risks that can have large impacts. Typically, multiple governance layers (OECD 2011) and a variety of stakeholders, sectors, and policies are involved, each with different time horizons and agendas (Segrave et al. 2013). Because there are many causes that can lead to the complexity, uncertainty, and disagreement, there is no single best approach to address these *governance challenges*. In fact, what is necessary is an iterative process that requires governance capacity to find integrated long-term solutions that are supported by flexible intermittent targets to anticipate changing situations and adapt to emerging barriers. Hence, it is essential to manage governance challenges in an integrative long-term approach (Patterson et al. 2013).

1.2 Knowledge Gap

The need for integrated approaches is reflected in two main concepts: 1) Integrated Water Resources Management (IWRM), and 2) Adaptive Management (AM). IWRM aims to reshape institutional structures and redefine problems, to broaden scope. AM focuses on changing the way responsible authorities perceive and act, by emphasizing experimentation and the value of learning to adapt to changing and uncertain challenges such as climate change (Medema et al. 2008). Despite their important contributions, both approaches appear to be difficult to implement, as they are not very concrete and their explanations differ as they refer to largely undefined, complex and unpredictable processes. Moreover, they tend to be focused on more technical solutions with lesser attention to governance processes (Medema et al. 2008; Rahaman and Varis 2005; Gregory et al. 2006). Therefore, these approaches sometimes fail to provide clear and effective guidelines for their concrete application (Gregory et al. 2006).

With respect to water management, the City Blueprint Trends and Pressure Framework and the City Blueprint Performance Framework provide integrated and strategic insights into the management performance of local authorities (Koop and Van Leeuwen 2015a, b). Furthermore, the OECD (2015a) principles of good water governance provide for an integrated network analysis of the fragmented water sector. They include analyses of international, national, and sub-national indicators and learning practices but do not focus on urban water governance.

A few worthy attempts have been made to compare organizations and institutions (Gupta et al. 2010; Ballard 2008; Engle and Lemos 2010). However, the rich literature on governance and transformation processes has a theoretical nature and only a few approach have been reported as providing tangible results to facilitate decision-making (Kersberger and Waarden 2004). At the same time, there is a seemingly endless list of social factors and conditions that impede or enhance our ability to respond proactively to future changes (Biesbroek et al. 2013). They often reflect a disciplinary scope and focus on specific case studies, which limits their usefulness and learning value (Measham et al. 2011; Plummer et al. 2012). Hence how socioeconomic factors actually influence urban governance processes largely remains an open question (Biesbroek et al. 2013) that can only be explored by applying a coherent framework that assesses different contexts consistently, provides empirical-based understanding of underlying governance processes, and searches for transferable lessons that enhance overall governance effectiveness.

1.3 Research Aim

In many cases, organizations or institutions are taken as the point of departure for governance assessment, whereas in practice, organizations often tend to focus on sub-tasks, lacking a full notion of their interdependencies, and underestimating the need for cooperation in addressing shared goals (Emerson et al. 2012). We argue that this is a missed opportunity, because interaction and collaboration between all relevant actors is critical. On the contrary, overly comprehensive, multi-layer governance structures that address multiple issues across different jurisdictions, proof to be difficult to analyse (Medema et al. 2008). Interestingly, the urban scale is increasingly recognized as having a crucial role in finding the most suitable solutions to address context-specific climate vulnerabilities (Measham et al. 2011; OECD 2015b). So far, little effort has been made to consistently assess the urban water governance capacity. However, such an assessment creates great potential to understand specific (local) issues and underlying processes, can provide recommendations for stakeholders, and shape learning alliances in and between cities. Here we aim to provide a deeper, integrated, and empirically-based understanding of the most important enabling conditions that determine the governance capacity needed to continuously solve governance challenges of water, waste, and climate change in urban networks. To do so, we develop a cohesive, comprehensive, and applicable Governance Capacity Framework (GCF) for cities, which can:

1) compare cities in order to develop a deeper empirical-based understanding of the key enabling governance conditions and identify transferable lessons.
2) reveal the limiting conditions and thereby formulate pathways for effective and efficient increase in the local governance capacity.

The paper is structured as follows. Section 2 provides a literature-based overview that frames and defines governance capacity. This definition is operationalized into a comprehensive framework in section 3. In section 4 a Likert-type method to score governance capacity is described. Section 5 provides an illustration of the first GCF pilot study in the city of Amsterdam, the Netherlands. Finally, section 6 presents the major discussion points and conclusions about the framework.

2 FRAMING GOVERNANCE CAPACITY

Governance capacity has a rich literature in the fields of environmental governance, climate adaptation, capacity building, public administration, and water governance. An important component of governance capacity is institutional capacity: this generally focuses on how institutional setting, rules, and regulations enable actors to collaborate and address shared problems (UNDP 2008; Dang et al. 2016). Governance capacity is broader, also including resources and the role of discourses (Engle and Lemos 2010; Pahl-Wostl 2009). As it is widely recognized that governance capacity is context-dependent, definitions diverge considerably. Some emphasize integration (Emerson et al. 2012), others cooperation (Dang et al. 2016), yet others focus on flexibility (Termeer et al. 2015). We take the position that governance capacity is about enabling effective change.

The kind of change that is effective is context-dependent. For example, integration is needed whenever scopes are fragmented, whereas adaptive approaches are needed to address inflexibility, and anticipatory governance is required when responses are reactive (Segrave et al. 2016). Nevertheless, a few communalities regarding governance capacity can be identified. First, this capacity is about the ability of actors to continuously identify and jointly act on collective problems (Dang et al. 2016). Second, the capacity is determined by actors' interactions formed by social-institutional settings and allocation of resources (Pahl-Wostl 2009). Third, actors' frame of reference, including their interests, values, and culture, shape interactions and influence collective problem-solving (Adger et al. 2009). Therefore, no single governance condition is decisive. On the contrary, governance capacity is determined by a balanced set of conditions that need to be well developed. Importantly, the nature of actors' interactions is complex, unpredictable, and susceptible to external social-ecological developments. Hence, governance capacity per see does not lead to efficacious change, but rather is a precondition or enabler for effective change. Accordingly, we define water governance capacity as "*the key set of governance conditions that should be developed to enable change that will be effective in finding dynamic solutions for governance challenges of water, waste, and climate change in cities".*

3 CONSTRUCTING THE GOVERNANCE CAPACITY FRAMEWORK

Policy actions often appear to be ineffective, and a wide range of governance gaps or barriers have been suggested as reasons for this (e.g. Biesbroek et al. 2013; Eisenack et al. 2014). The literature has identified normative principles (e.g. OECD 2015a) and enabling or adaptive capacities to overcome barriers (e.g. UNDP 2008; Ford and King 2015). We make use of this rich knowledge base by selecting and redefining key conditions and their indicators that enable effective change, exploring their interrelations, assessing their relevance for urban water governance, and reformulating them into a well-balanced framework. We have also studied existing frameworks that analyze social processes that have inhibited or stimulated effective change in the field of urban water governance (Brown and Farralley 2009), river basins (Engle and Lemos 2010), water systems (Van Rijswick et al. 2014), or have a multi-level perspective (OECD 2011, 2015a; Pahl-Wostl et al. 2010). Other frameworks that have been applied in the past have concerned the adaptive capacity of institutions (Gupta et al. 2010) and organizations (Ballard 2008), or have been scale-independent (e.g. Moser and Ekstrom 2010; Ford and King 2013).

We identified key conditions for good urban water governance and classified them into three dimensions: knowing, wanting, and enabling. The "knowing" dimension refers to the need to be fully aware, understand, and learn the actual or possible risks and impacts of actions, policy, and strategic choices. We created the "wanting" dimension because actors need to commit to cooperate, express, and act upon ambitions, and apply their skills and capabilities to find solutions. Finally, the "enabling" dimension was created because actors need to have the network, resources, and instruments to enable them to implement their ambitions. The resulting framework has nine governance conditions; an in-depth literature study for each condition yielded three indicators. The results are shown in Table 1, below the findings from the literature research are described.

3.1 Condition 1: Awareness

Awareness is a prerequisite to enable effective change. It refers to a more profound understanding of the causes, impact, and risks of governance challenges (Raaijmakers et al. 2008). Awareness is both cognitively and emotionally felt by individuals, organizations, and society (Ballard 2008) and forms the base for learning and action (Adger et al. 2009). *Awareness* is assessed by the indicators 1.1 *community knowledge,* 1.2 *local sense of urgency* and 1.3 *behavioral internalization*. *Community knowledge* refers to the extent to which different stakeholders possess relevant knowledge about the challenges. This is the first step in achieving conscious behavior (Gifford 2011). *Local sense of urgency* reflects the perception of importance of the governance challenge, which may or may not result in actions and policies (O'Connor et al. 1999). Finally, *behavioral internalization* indicates that a higher level of knowledge affects actors' problem- framing, goals, values, and perceptions, changing their behavior and increasing their commitment to sustainable approaches (Gifford 2011).

Table 1 The water Governance Capacity Framework (GCF). The GCF consists of nine conditions, each defined by three indicators. For each indicator, a Likert-type scoring scale has been developed, which ranges from very encouraging (++) to very limiting (−−) to the governance capacity

Dimensions	Conditions	Indicators
Knowing	1 Awareness	1.1 Community knowledge
		1.2 Local sense of urgency
		1.3 Behavioral internalization
	2 Useful knowledge	2.1 Information availability
		2.2 Information transparency
		2.3 Knowledge cohesion
	3 Continuous learning	3.1 Smart monitoring
		3.2 Evaluation
		3.3 Cross-stakeholder learning
Wanting	4 Stakeholder engagement . process	4.1 Stakeholder inclusiveness
		4.2 Protection of core values
		4.3 Progress and variety of options
	5 Management ambition	5.1 Ambitious and realistic management
		5.2 Discourse embedding
		5.3 Management cohesion
	6 Agents of change	6.1 Entrepreneurial agents
		6.2 Collaborative agents
		6.3 Visionary agents
Enabling	7 Multi-level network. . potential	7.1 Room to manoeuvre
		7.2 Clear division of responsibilities
		7.3 Authority
	8 Financial viability	8.1 Affordability
		8.2 Consumer willingness to pay
		8.3 Financial continuation
	9 Implementing capacity	9.1 Policy instruments
		9.2 Statutory compliance
		9.3 Preparedness

3.2 Condition 2: Useful Knowledge

The field of information science distinguishes between data, information and knowledge (Zins 2007). Data in itself is not necessarily informative, as useful knowledge can only be obtained by data interpretation and analysis (Zins 2007; Rowley 2007; Van Leeuwen 2007). *Useful knowledge* consists of 2.1 *information availability*, 2.2 *information transparency*, and 2.3 *knowledge cohesion*. *Information availability* refers to the extent that reliable knowledge is available. A lack of knowledge inhibits informed decision-making (Rowley 2007; Van Rijswick et al. 2014). Many cities authorities recognize the lack of knowledge of how future trends, such as urbanization and climate change, will affect them (Amundsen et al. 2010). *Information transparency* refers to the effective communication and sharing or co-creation of knowledge with all interested stakeholders. The information needs to be good quality, credible, understandable, and accessible for non-experts, in order to prevent miscommunication, knowledge gaps, and fragmented policy (Lemos et al. 2012; Füssel 2007). Finally, *knowledge cohesion* refers to the conformity of knowledge across actors, sectors, and administrative layers.

3.3 Condition 3: Continuous Learning

Continuous learning is required, in order to adapt to changing situations with many uncertainties, complexities, and unknowns (Folke et al. 2005). *Continuous learning* is assessed by 3.1 *smart monitoring*, 3.2 *evaluation* and 3.3 *cross-stakeholder learning*. *Smart monitoring* is a precondition for learning and may serve as tool for identifying alarming situations, clarifying underlying processes, and predicting future developments (Van Leeuwen 2007). Regular monitoring and evaluation are imperative for continuous learning and enhance preparedness for uncertain futures. In order to conceptualize *evaluation*, the theory of triple-loop learning is used, which has three levels: 1) single-loop learning which is incremental learning to refine current management and policy; 2) double-loop learning refers to the critical investigation of assumptions and key relationships, which reframes problems; 3) triple-loop learning questions underlying norms and values and can transform the wider social and institutional structure (Pahl-Wostl 2009). Finally, the third indicator, *cross-stakeholder learning* is crucial for learning in a public policy context, as the interaction among actors and their understanding of different perspectives lead to a more comprehensive, if not consensual, evaluation (Emerson et al. 2012). Furthermore, this can prevent overly limited scopes or path-dependencies (Termeer et al. 2015; Brown and Farrelly 2009).

3.4 Condition 4: Stakeholder Engagement Process

The importance of stakeholder engagement is widely recognized from a normative, substantive, and instrumental rationale (Glucker et al. 2013; OECD 2015a; UNDP 2008). Stakeholder engagement may lead to a more complete problem-framing and widely accepted optimized solutions for all parties involved (Pahl-Wostl 2009; Carlsson and Berkes 2005). Active stakeholder engagement is generally more time-consuming than unilateral decision-making. However, this can be more than compensated for by time gains in the implementation phase (Ridder et al. 2005). The *stakeholder engagement process* consists of 4.1 *stakeholder inclusiveness*, 4.2 *protection of core values* and 4.3 *progress and variety of options. Stakeholder inclusiveness* refers to the extent to which the representatives are able to speak and decide on behalf of all relevant stakeholders in clear and transparent engagement processes (Ford and King 2015; Ridder et al. 2005). *Protection of core values* refers to the importance of ensuring that all stakeholders feel confident that their core values are not harmed, in order to create a safe environment for trust relationships (Ridder et al. 2005; Pahl-Wostl et al. 2011). Therefore, it is essential that stakeholders become actively involved and commit to the process, rather than the outcome is predetermined or intermediate decisions are made early on (Folke et al. 2005). Moreover, stakeholders' contribution should influence the end-result. *Progress and variety of options* encompasses the prospect of gain for each stakeholder, which is ensured by clear and realistic procedures. Stakeholders should co-produce and, at the end of the process, select from a variety of options, to ensure learning and authoritative decisions (Ridder et al. 2005).

3.5 Condition 5: Management Ambition

Management ambition is a measure of the extent to which sustainable management and policy is interwoven with historical, cultural, normative, and political context. This is measured by assessing the sustainability ambitions within policies. *Management ambition* is assessed by 5.1 *ambitious and realistic management*, 5.2 *discourse embedding*, and 5.3 *management cohesion. Ambitious and realistic goals* need to be long-term, with intermittent measurable targets, all provided with sufficient resources and flexible mechanisms to deal with changing situations (Brown and Farrelly 2009). *Discourse embedding* is important, as management ambitions need to match the dominant values, discourses, and principles, in order to be successful (Van Rijswick et al. 2014). Hence, the degree to which the challenges of water, waste, and climate change are embedded in the dominant discourse, strongly determines the effectiveness of ambitious management and policy. *Management cohesion* assesses the level of integration between different sectoral policies and strategies, across governance levels, and between organizations. Often, the over-fragmentation of roles and responsibilities means that no single agency is in charge of water policy, and opportunities to create co-benefits are not seized (OECD 2011, 1015a; Head and Alford 2015).

3.6 Condition 6: Agents of Change

The concept of agents of change is often described in the fields of organizational change, AM, and innovation studies, although different terminology is used (e.g. leaders, policy entrepreneurs, institutional entrepreneurs; Pahl-Wostl et al. 2011; Ballard 2008; Brouwer and Biermann 2011). "Agents of change" refers to the intrinsic motivation of people, their willingness to take risks, and the support given to these efforts to change current approaches. The concept is therefore not limited to people in leading positions (Brouwer and Biermann 2011; Head and Alford 2015; Schultz and Fazey 2009). For this condition, three types of agents of change are distinguished: 6.1 *entrepreneurial agents*, who have the means and skills to gain access to resources, seek opportunities, and manage risks; 6.2 *collaborative agents*, who have the skills to build bridges and coalitions between actors; and 6.3 *visionary agents*, who envision long-term adaptive approaches and are able to steer current policy and actions (Brouwer and Huitema 2017; Termeer et al. 2012; Gupta et al. 2010; Ford and King 2015).

3.7 Condition 7: Multi-Level Network Potential

Flexible and dynamic networks are important, in order to deal with governance challenges with different interests and perspectives, and with stakeholders acting at different levels (Pahl-Wostl 2009; Gupta et al. 2010; Moser and Ekstrom 2010). *Multi-level network potential* consists of 7.1 *room to manoeuvre*, 7.2 *clear division of responsibilities*, and 7.3 *authority*. *Room to manoeuvre* assesses the opportunity that actors have to explore different alternative pathways, develop knowledge, and put ideas into practice. This also involves the possibility and autonomy of actors to form new fit-for-purpose partnerships that can address unconventional and emerging challenges (Gupta et al. 2010; Folke et al. 2005). *Clear division of responsibilities* refers to the accurate and clear division of tasks and roles for which stakeholders can be held accountable (Mees et al. 2014). *Authority* refers to the presence of legitimate forms of authority (e.g. embedded in policy or law), regulations, and policy networks that promote the necessity to address water-related challenges (Van Rijswick et al. 2014).

3.8 Condition 8: Financial Viability

Addressing urban water-related challenges requires the assurance of long-term financial support (OECD 2015b; UNECE 2009), as short budgetary cycles prevent long-term thinking (Ford and King 2015) and will most likely substantially increase overall cost (UNEP 2013; Koop and Van Leeuwen 2016). Two important aspects of financial viability are the costs and benefits of measures: e.g., who is affected, who benefits, and, therefore, who should pay (UNECE 2009). *Financial viability* is characterized by 8.1 *affordability*, 8.2 *consumer willingness to pay*, and 8.3 *financial continuation*. *Affordability* of water and climate adaptation services is assessed with a focus on the poor and marginalized groups (OECD 2011; UNDP 2008). *Consumer willingness to pay* assesses how expenditure and risks are perceived. Often, trust in local authorities and their accountability, as well as the sense of urgency or worry, are key (Raaijmakers et al. 2008). Finally, *financial continuation* is needed for solving long-term challenges and avoiding resources being squandered as a result of uncoordinated investments (Adger et al. 2005).

3.9 Condition 9: Implementing Capacity

Most studies mention policy implementation as crucial (Adger et al. 2005; Ekstrom et al. 2011; Van Rijswick et al. 2014). *Implementing capacity* is substantiated through 9.1 *policy instruments*, 9.2 *statutory compliance* and 9.3 *preparedness*. *Policy instruments* can be used to stimulate desired behavior and discourage undesired activities (Mees et al. 2014). Examples are the inclusion of the user-pays and polluter-pays principles in pricing. Continuous monitoring, evaluation, and adjustments are needed, to check and improve the effectiveness of instruments. *Statutory compliance* ensures that stakeholders respect and understand agreements, objectives, and legislation, which contributes to the accountability of authorities. *Preparedness* increases the implementation capacity, as the existence of action plans, procedures, and scripts supports policy and prepares the city for both gradual and sudden changes, events, and calamities (Gupta et al. 2010; Raaijmakers et al. 2008; Runhaar et al. 2016).

4 DETERMINING LEVELS OF GOVERNANCE CAPACITY

Despite the rich literature on governance capacity which provides many clues, it remains a puzzle to identify gradual levels of increasing governance capacity and only a few studies explicitly described them (e.g. Gupta et al. 2010; Ballard 2008). Hence, a better understanding of these gradual levels of governance capacity is needed to provide valuable insights into key governance processes.

It should also be noted that a scaling system provides cities with a better and more nuanced indication of where they are, and what steps to take to improve their capacity. For each of the twenty-seven indicators we therefore developed a Likert-type scoring system, with scores ranging from very encouraging to the overall governance capacity (++) to very limiting to the overall governance capacity (--). The indicator levels were determined from a wide-ranging perusal of the literature, including AM theory (e.g. Folke et al. 2005; Engle and Lemos 2010; Gupta et al. 2010) in combination with specific theory for each condition. In addition, we included practical indicators from governance assessments and policy documents (BAGroep 2016; KING 2016).

Table 2 illustrates the scoring methodology. Each of the twenty-seven indicators are scored by answering a predefined question, which are illustrated for the indicators belonging to condition 4 *stakeholder engagement process*. Next, the Likert-type scoring scale for indicator 4.2 *protection of core values* is provided. The Likert-type scoring levels, together with predefined questions, and the five main literature sources are available for each indicator at the EIP Water website (EIP Water 2017). The Likert-type scoring scale for indicator 4.2 *protection of core values* is based on three main aspects which together ensure that stakeholders feel confident that their core values are not harmed (Ridder et al. 2005):

1) Stakeholders need to be asked to commit to the process rather than to a predetermined outcome or intermediate decisions early in the process.
2) The existence of clear rules and procedures that have been agreed upon before the start of the engagement process, in order to ensure a sound environment in which trust relationships can be developed.
3) The actual influence stakeholders have on the end-result is important. It is largely determined by the type of stakeholder interaction, which can be conceptualized into three layers. The first layer (*information supply*) indicates one-way communication. The second layer (*consultation*) indicates that stakeholders can give feedback on developed plans. The third layer (*active involvement*) actively involves stakeholders throughout the policy-making and implementation process (CIS Working Group 2.9 2003).

Table 2 Illustrative overview of the GCF scoring methodology. First, an overview of the predefined questions for the indicators belonging to condition 4 *stakeholder engagement process* are given. Second, the indicator 4.2 *protection of core values* Likert-type scoring is provided as an illustration. The predefined questions, the Likert-type scoring scale and a literature overview for each of the twenty-seven indicators are summarized at the EIP Water website (EIP Water 2017)

Predefined questions for condition 4 *stakeholder engagement process*	
Indicator	**Predefined question**
4.1 Stakeholder inclusiveness	To what extent do stakeholders interact in the decision-making process interaction (i.e., are merely informed, are consulted, or are actively involved)? Are their engagement processes clear and transparent? Are stakeholders able to speak and decide on behalf of a group?
4.2 Protection of core values	To what extent 1) is commitment focused on the process instead of on early end-results? 2) do stakeholders have the opportunity to be actively involved? 3) are the exit procedures clear and transparent? (All 3 ensure that stakeholders feel confident that their core values will not be harmed.)
4.3 Progress and variety of options	To what extent are procedures clear and realistic, are a variety of alternatives co-created and thereafter selected from, and are decisions made at the end of the process in order to secure continued prospect of gain and thereby cooperative behavior and progress in the engagement process?

Likert-type indicator levels for indicator 4.2 *Protection of core values*		
Level	**Description**	
Very encouraging (++)	**Maximal protection of core values**	Stakeholders are actively involved and co-create the end-result. There are clear exit possibilities and clear process procedures. All relevant stakeholders are engaged and a variety of options are assessed. The final options are chosen at the end of the engagement process
Encouraging (+)	**Demand for commitment to early output**	Stakeholders are actively involved and expected to commit to early process outcomes. Hence some relevant stakeholders are discouraged from committing, as not all options are being assessed and at this stage the stakeholders contribution might be small. The stakeholders have influence on the end-result
Indifferent (0)	**Suboptimal protection of core values**	Stakeholders are consulted or actively engaged for short periods. The number of options considered and influence on the end-result are limited. Exit rules are vague. Decisions mainly comply with the interests of the initiating party
Limiting (-)	**Low influence on end-result**	Stakeholders are kept informed or consultation meetings are taking place for already partly or fully elaborated plans. The influence on the end-result is small and resistance may be evoked
Very limiting (--)	**Ignorance of core values**	Stakeholders are hardly engaged, not informed or only informed after decisions have already been made. Resistance to implementation often occurs, as do distrust and lack of stakeholder participation, and no clear communication

5 ILLUSTRATING THE FRAMEWORK: GOVERNANCE CAPACITY IN AMSTERDAM

5.1 Assessing the Governance Capacity

We applied the GCF to the five water governance challenges faced by the city of Amsterdam. Amsterdam has a complex hydrological setting: large areas are below sea level, many canals and sluices regulate the different water levels, and there is a sophisticated system for supplying drinking water, which involves infiltrating surface water into the nearby dunes.

A triangular method was applied to score indicators according to the Likert-type method. First, an analyses of policy documents and reports provided preliminary scores. Second, fifteen interviewees, three for each of the five governance challenges, were selected. To this end, the most relevant stakeholders were identified and their interdependencies were plotted, and key persons from different levels of decision-making were selected (Reed et al. 2009). As explained above, there were twenty-seven predefined questions that the research needed to answer: one for each indicator and each asked separately with respect to the five governance challenges. In this way, a consistent assessment approach was applied that enables basic comparisons to be made between, on the one hand, governance challenges, and, on the other hand, between cities. The interview questions were aimed to gather all the information needed to answer the predefined questions.

They were open and non-technical, with follow-up questions to target specific elements, or to achieve further clarification. Finally, after the interviews, the participants received the predefined questions with the preliminary indicator scoring and were asked to provide constructive feedback and additional information to be included in the final scoring. The assessment was fully transparent, as the Likert-type scales, twenty-seven predefined questions, and the full list of references are publicly available.

6 RESULTS

Each of the governance challenges was separately assessed and scored: from very encouraging (++) to very limiting (--) to the governance capacity (Table 3).

Table 3 Overview of the twenty-seven governance indicator scores for each of the five water-related governance challenges for the city of Amsterdam. Scores range from very encouraging (++) to very limiting (--) to the city's governance capacity to find dynamic solutions

	Water scarcity	Flood risk	Wastewater treatment	Solid waste treatment	Urban heat islands
1.1 Community knowledge	-	0	-	0	-
1.2 Local sense of urgency	-	++	-	0	-
1.3 Behavioral internalization	+	++	++	+	-
2.1 Information availability	++	++	0	0	0
2.2 Information transparency	0	0	0	-	-
2.3 Knowledge cohesion	+	0	0	0	+
3.1 Smart monitoring	++	++	++	++	-
3.2 Evaluation	++	++	++	++	-
3.3 Cross-stakeholder learning	+	++	++	0	-
4.1 Stakeholder inclusiveness	++	++	++	0	++
4.2 Protection of core values	++	++	0	+	-
4.3 Progress and variety of options	++	++	0	0	0
5.1 Ambitious and realistic management	+	+	+	0	--
5.2 Discourse embedding	++	+	0	++	-
5.3 Management cohesion	++	+	++	+	-
6.1 Entrepreneurial agents	++	+	++	+	-
6.2 Collaborative agents	+	+	+	0	-
6.3 Visionary agents	+	++	+	0	+
7.1 Room to manoeuvre	+	+	0	+	0
7.2 Clear division of responsibilities	+	++	+	0	--
7.3 Authority	+	++	+	++	--
8.1 Affordability	+	+	+	+	+
8.2 Consumer willingness to pay	-	+	0	0	0
8.3 Financial continuation	+	+	+	+	0
9.1 Policy instruments	+	+	+	+	0
9.2 Statutory compliance	++	++	++	+	0
9.3 Preparedness	+	++	+	+	--

This first assessment suggests that Amsterdam excels in flood risk governance, and most indicators regarding water scarcity and wastewater treatment are encouraging. However, more governance capacity needs to be developed to address Urban Heat Islands (UHI), since the number of tropical days (>30 °C) in the Netherlands is predicted to rise substantially, from 4 days at present to 7–13 days by 2050 and 8–21 by 2100 (KNMI 2014). Interestingly, cities in Northern Europe appear to be most affected by the predicted rise in high temperatures, because here, temperatures will strongly exceeded the usual seasonal conditions (EEA 2012). Despite this, in the Netherlands, no separate policy on UHI has been developed so far. Consequently, Amsterdam lacks specific targets and policies regarding UHI. The lack of policy may explain the low multi-level network potential (condition 7) to address UHIs. Averaging the scores of the five challenges for each indicator yields a more general overview of Amsterdam's water governance capacity (Fig. 1). It suggests that the knowledge level of communities (indicator 1.1) and the access to understandable information for non-experts (indicator 2.2) may slightly limit local sense of urgency regarding water challenges (indicator 1.2) and consumer willingness to pay (indicator 8.2). These results are in line with the OECD (2014) analyses of the Dutch water governance, which conclude that Dutch citizens take water services for granted and that this "awareness gap" tends to decrease public involvement and the willingness to pay for water services. Therefore, the most feasible way for Amsterdam to further enable effective policy change is to focus on improving the indicator scores found to be limiting (–) or very limiting (--) to the city's governance capacity.

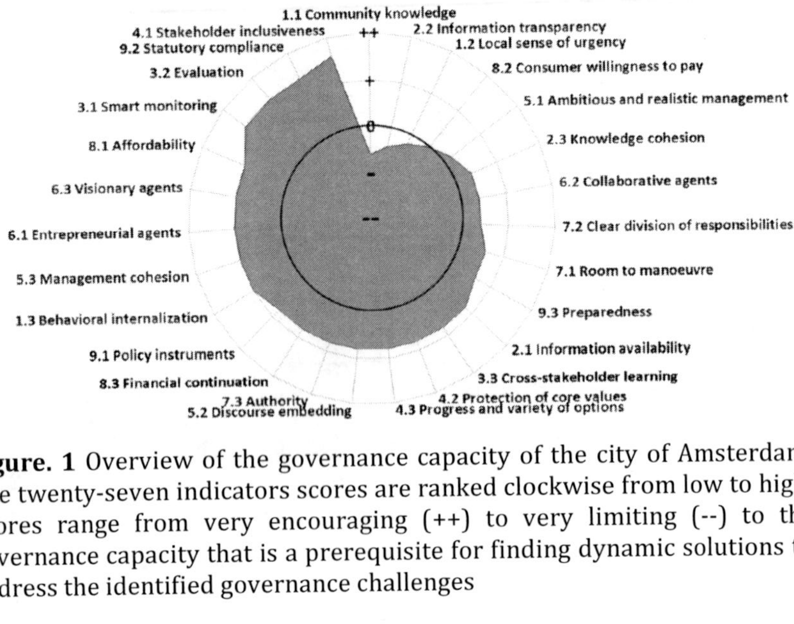

Figure. 1 Overview of the governance capacity of the city of Amsterdam. The twenty-seven indicators scores are ranked clockwise from low to high. Scores range from very encouraging (++) to very limiting (--) to the governance capacity that is a prerequisite for finding dynamic solutions to address the identified governance challenges

7 DISCUSSION AND CONCLUSION

First, we will briefly discuss some limitations of the GCF. Next, we will outline the main contributions of the GCF to theory and practice and, more specifically, its role in connecting science, policy, and implementation.

7.1 Framework Discrepancies and Reproducibility

As governance processes are often interconnected, some of the conditions and indicators identified in our framework are inherently interrelated. Although each indicator is designed to provide an independent score, inevitably, a hypothetical "ideal" situation will not always result solely in very encouraging (++) indicator scores. For example, entrepreneurial, collaborative, and visionary agents of change (condition 6) are all relevant but their importance is context-dependent (Patterson et al. 2013): visionary agents, for example, may be more necessary in times of crisis, whereas collaborative agents are more valuable in initiating new joint activities in established collaborative networks, and entrepreneurial agents operate best in open governance networks that are ambitious and flexible (Brouwer and Huitema 2017). Hence, situations may exist in which entrepreneurial and collaborative agents of change are very encouraging (++) while, as a consequence, visionary agents are less prominently active. Another important interrelation is between indicators 5.1 *ambitious and realistic goals* and 9.2 *statutory compliance*, as it is easier to comply with non-ambitious goals. Furthermore, indicators 6.1 *entrepreneurial agents of change* and 7.3 *room to manoeuvre* are reinforcing, and their scoring is interrelated. Their main difference is that entrepreneurial agents of change focuses on actors' ability and skills to create and seize opportunities, whereas the score for room to manoeuvre represents the degree of freedom and existing opportunities that actors can utilize.

7.2 Bridging Theory and Practice

Different time frames, reward structures, process cycles, epistemologies, and goals impede the effective use of scientific knowledge in practice (Hegger et al. 2012). Scientific knowledge is often fragmented, as it is intertwined with values, discourses, disciplinary scopes, and traditions that are often context-specific. Therefore, existing knowledge often fails to provide applicable insights that can help decision-makers achieve their intended goals and objectives. The gap between science, policy, and implementation has been widely acknowledged in water governance (OECD 2011; Medema et al. 2008; Patterson et al. 2013). Our work provides three important contributions to improve the connection between scientific knowledge, policy, and implementation in the field of water governance, climate adaptation, and beyond:

1) **Integration**: The GCF is one of the first attempts to integrate the plethora of contradicting, overlapping, and fragmented governance gaps, barriers, and capacities with respect to prevailing urban water challenges. It may reveal more effective and efficient pathways for cities to increase their governance capacity. At present, our understanding of underlying interconnections and relations is often insufficient to provide overarching pragmatic insights that facilitate decision-makers.

2) **Communication**: The GCF is designed to be easy to understand and transparent, and has been developed with the end-users (who include decision-makers, stakeholders, and citizens) in mind. Information

needs to be understandable for them. This is essential, to facilitate constructive discussions, knowledge co-production, and cooperation.

3) **City comparison**: The GCF provides a framework for comparison cities and the accumulation of empirical data that can improve our understanding of underlying governance processes that limit or encourage governance capacity. At the same time, it provides a practical framework for cities to exchange learning experience, knowledge, and good practices.

Because the framework is embedded in the literature of governance and transformation processes, it is possible to assess a broader range of issues that involve processes of change in multi-organizational networks. An empirical database is currently being developed, as the GCF is being used to assess the cities of Melbourne (Australia), Quito (Ecuador; Schreurs et al. 2017), New York (USA) and Ahmedabad (India; EIP 2017). Moreover, the results of the city of Quito (Ecuador) are about to be published (Schreurs et al. 2017). The GCF has the potential to act as a portal of communication for constructive cross-city, cross-sector, and cross-stakeholder discussions, learning and theory building. Finally, the framework provides the basis for common understanding and action, by revealing the most effective and efficient pathways for increasing the governance capacity needed to address the challenges of water, waste, and climate change.

ACKNOWLEDGEMENTS

Our gratitude goes to the Amsterdam water network for their participation: Roelof Kruize, Jos Ketelaars, Ingrid Heemskerk, Alice Fermont, Peter Simoës, Mark Nijman, Lex Lelijveld, Kees van der Drift, Rolf Steenwinkel, Jeroen Ponten, Martine Lodewijk, Ed Cousin, Gerard Korrel, Geertje Wijten, Maarten Claassen, Rob Koeze and Wiegert Dulfer. We thank Joy Borrough for her editorial contribution. The Governance Capacity Framework is part of City Blueprint Approach developed at KWR Watercycle Research Institute in the context of Watershare® (http://www.watershare.eu). The City Blueprint Action Group is part of the European Innovation Partnership on Water of the European Commission (http://www.eip-water.eu/City_Blueprints). The European Commission is acknowledged for Funding POWER in H2020-Water Under Grant Agreement No. 687809.

REFERENCES

1. Adger WN, Arnell NW, Tompkins EL (2005) Successful adaptation to climate change across scales. Glob Environ Chang 15:77–86
2. Adger WN, Dessai S, Goulden M, Hulme M, Lorenzoni I, Nelson DR, Naess LO, Wolf J, Wreford A (2009) Are there social limits to adaptation to climate change? Clim Chang 93:335–354
3. Amundsen H, Berglund F, Westskogh H (2010) Overcoming barriers to climate change adaptation-a question of multilevel governance? Environ Plan C 28:276–289
4. Baccini M, Biggeri A, Accetta G, Kosatsky T, Katsouyanni K, Analitis A, Anderson HR, Bisanti L, D'Ippoliti D, Danova J, Forsberg B, Medina S, Paldy A, Rabczenko D, Schindler C, Michelozzi P (2008) Heat effects on mortality in 15 European cities. Epidemiology 2:711–719
5. BAGroep (2016) Governance capacity measurement: How is your municipality, province or water boardfunctioning? IOP Publishing Physics Web. http://www.bagroep.nl. Accessed 23 Jan 2017
6. Ballard A (2008) Adaptive capacity benchmarking: a handbook and toolkit. Berkshire, Hungerford
7. Biesbroek GR, Klostermann JEM, Termeer CJAM, Kabat P (2013) On the nature of barriers to climate change adaptation. Reg Environ Chang 13:1119–1129
8. Brouwer S, Biermann F (2011) Towards adaptive management: examining the strategies of policy entrepreneurs in Dutch water management. Ecol Soc 16:5
9. Brouwer S, Huitema D (2017) Policy entrepreneurs and strategies for change. Reg Environ Chang. doi:10.1007/s10113-017-1139-z
10. Brown RR, Farrelly MA (2009) Delivering sustainable urban water management: a review of the hurdles we face. Water Sci Technol 59:839–846CrossRef
11. Carlsson L, Berkes F (2005) Co-management: concepts and methodological implications. J Environ Manag 75:65–76
12. CIS Working Group 2.9 (2003) Common implementation strategy for the water framework directive. Guidance document, Luxembourg: Official Publications of the European Communities. https://circabc.europa.eu/sd/a/0fc804ff-5fe6-4874-8e0dde3e47637a63/Guidance%20No%208%20-%20Public%20participation%20%28WG%202.9%29.pdf
13. Dang TKP, Visseren-Hamakers IJ, Arts BJM (2016) A framework for assessing governance capacity: an illustration from Vietnam's forestry reforms. Environ Plann C Gov Policy 34:1154–1174
14. Derraik JGB (2002) The pollution of the marine environment by plastic debris: a review. Mar Pollut Bull 44:842–852
15. EEA (2012) European Environment Agency: urban adaptation to climate change in Europe. Challenges and opportunities for cities together with supportive national and European policies. ISBN 978-92-9213-308-5
16. EEA (2016) European Environmental Agency: Urban adaptation to climate change in Europe 2016. Transforming cities in a changing climate. Publication office of the European Union, Luxembourg. ISBN 978-92-9213-742-7
17. EIP Water (2017). European innovation partnership on water. Indicators of the Governance Capacity Framework. http://www.eip-water.eu/City_Blueprints Accessed 21 November 2016
18. Eisenack K, Moser SC, Hoffmann E, Klein RJT, Oberlack C, Pechan A, Rotter M, Termeer CJAM (2014) Explaining and overcoming barriers to climate change adaptation. Nat Clim Chang 4:867–872
19. Ekstrom JA, Moser SC, Torn M (2011) Barriers to climate change adaptation: a diagnostic framework. Project Report. California Energy Commission, Berkely
20. Emerson K, Nabatchi T, Balogh S (2012) An integrative framework for collaborative governance. Public Adm Res Theory 22:1–29

21. Engle NL, Lemos MC (2010) Unpacking governance: building adaptive capacity to climate change of river basins in Brazil. Glob Environ Chang 20:4–13
22. European Commission (2015) Closing the loop – an EU action plan for the circular economy. Brussels, Belgium
23. Folke C, Hahn T, Olsson P, Norberg J (2005) Adaptive governance of social-ecological systems. Annu Rev Environ Resour 30:441–473
24. Ford JD, King D (2015) A framework for examining adaptation readiness. Mitig Adapt Strateg Glob Chang 20:505–526
25. Füssel H (2007) Adaptation planning for climate change: concepts, assessment approaches, and key lessons. Sustain Sci 2:265–275
26. Gifford R (2011) The dragons of inaction: psychological barriers that limit climate change mitigation and adaptation. Am Psychol 66:290–302
27. Glucker A, Driessen PPJ, Kolhoff A, Runhaar HAC (2013) Public participation in environmental impact assessment; why, who and how? Environ Impact Assess Rev 43:104–111
28. Gregory R, Ohlson D, Arvai J (2006) Deconstructing adaptive management: criteria for application to environmental management. Ecol Appl 16:2411–2425
29. Gupta J, Termeer C, Klostermann J, Meijerink S, Van Den Brink M, Jong P, Nooteboom S, Bergsma E (2010) The adaptive capacity wheel: a method to assess the inherent characteristics of institutions to enable the adaptive capacity of society. Environ Sci Pol 13:459–471
30. Head BW, Alford J (2015) Wicked problems: implications for public policy and management. Adm Soc 47:711–739
31. Hegger D, Lamers M, Van Zeijl-Rozema A, Dieperink C (2012) Conceptualising joint knowledge production in regional climate change adaptation projects: success conditions and levers for action. Environ Sci Pol 18:52–65
32. Kersberger KV, Waarden FV (2004) Governance' as a bridge between disciplines: cross-disciplinary inspiration regarding shifts in governance and problems of governability, accountability and legitimacy. EJPR 43:143–171
33. KING (2016) Quality Institue. Dutch municipalities (in Dutch) http://www.waarstaatjegemeente.nl
34. KNMI (2014) Royal Dutch Meteological Institute: KNMI'14 climate cenario's for the Nederlands (in Dutch). Ministry of infrastructure and Environment. www.climatescenarios.nl
35. Koop SHA, Van Leeuwen CJ (2015a) Assessment of the sustainability of water resources management: a critical review of the City Blueprint approach. Water Resour Manag 29:5649–5670
36. Koop SHA, Van Leeuwen CJ (2015b) Application of the improved City Blueprint framework in 45 municipalities and regions. Water Resour Manag 29:4629–4647
37. Koop SHA, Van Leeuwen CJ (2016) The challenges of water, waste and climate change in cities. Environ Dev Sustain 19:385–418
38. Lemos MC, Kirchhoff CJ, Ramprasad V (2012) Narrowing the climate information usability gap. Nat Clim Chang 2:789–794
39. Ligtvoet W, Hilderink H, Bouwman A, Puijenbroek P, Lucas P, Witmer M (2014) Towards a world of cities in 2050. An outlook on water-related challenges. Background report to the UN-Habitat Global Report. PBL Netherlands Environmental Assessment Agency
40. Measham TG, Preston BL, Smith TF, Brooke C, Gorddard R, Withycombe G, Morrison C (2011) Adapting to climate change through local municipal planning: barriers and challenges. Mitig Adapt Strateg Glob Chang 16:889–909
41. Medema W, McIntosh BS, Jeffrey PJ (2008) From premise to practice: a critical assessment of integrated water resources management and adaptive management approaches in the water sector. Ecol Soc 13:29
42. Mees HLP, Dijk J, Van Soest D, Driessen PPJ, Van Rijswick MHFMW, Runhaar H (2014) A method for the deliberate and deliberative selection of policy instrument mixes for climate change adaptation. Ecol Soc 19

43. Moser SC, Ekstrom JA (2010) A framework to diagnose barriers to climate change adaptation. Proc Natl Acad Sci U S A 107:22026–22031
44. O'Connor RE, Bord RJ, Fisher A (1999) Risk perceptions, general environmental beliefs, and willingness to address climate change. Risk Anal 19:461–471
45. OECD (2011) Organization for Economic Cooperation and Development: water governance in OECD countries: a multi-level approach. OECD Studies on Water, Paris
46. OECD (2014) Organization for Economic Cooperation and Development: water governance in the Netherlands. Fit for the future? OECD studies on water, OECD publishing
47. OECD (2015a) Organization for Economic Cooperation and Development: OECD principles on water governance. OECD Ministerial Council Meeting, Paris
48. OECD (2015b) Organization for Economic Cooperation and Development: water and cities: ensuring sustainable futures. OECD Studies on Water, Paris
49. Pahl-Wostl C (2009) A conceptual framework for analysing adaptive capacity and multi-level learning processes in resource governance regimes. Glob Environ Chang 19:354–365
50. Pahl-Wostl C, Holtz G, Kastens B, Knieper C (2010) Analyzing complex water governance regimes: the management and transition framework. Environ Sci Pol 13:571–581
51. Pahl-Wostl C, Nilsson C, Gupta J, Tockner K (2011) Societal learning needed to face the water challenge. Ambio 40:549–553
52. Patterson J, Smith C, Bellamy J (2013) Understanding enabling capacities for managing the 'wicked problem' of nonpoint source water pollution in catchments: a conceptual framework. J Environ Manag 128:441–452
53. Plummer R, Crona B, Armitage DR, Olsson P, Tengo M, Yudina O (2012) Adaptive comanagement: a systematic review and analysis. Ecol Soc 17:11
54. Raaijmakers R, Krywkow J, Van Der Veen A (2008) Flood risk perceptions and spatial multi-criteria analysis: an exploratory research for hazard mitigation. Nat Hazards 46:307–322
55. Rahaman MM, Varis O (2005) Integrated water resources management: evolution, prospects and future challenges. Sustain Sci Pract Policy 1:15–21
56. Reed MS, Graves A, Dandy N, Posthumus H, Hubacek K, Morris J, Prell C, Quinn CH, Stringer LC (2009) Who's in and why? A typology of stakeholder analysis methods for natural resource management. J Environ Manag 90:1933–1949
57. Ridder D, Mostert E, and Wolters HA (2005) Learning together to manage together. HarmoniCOP, Osnabrück: University of Osnabrück
58. Rowley J (2007) The wisdom hierarchy: representations of the DIKW hierarchy. J Inf Sci 33:163–180
59. Runhaar HAC, Uittenbroek CJ, Van Rijswick HFMW, HLP M, PPJ D, Gilissen HK (2016) Prepared for climate change? A method for the ex-ante assessment of formal responsibilities for climate adaptation in specific sectors. Reg Environ Chang 5:1389–1400
60. Schreurs E, Koop S, Van Leeuwen K (2017) Application of the City Blueprint approach to assess the challenges of water management and governance in Quito (Ecuador). Environ Dev Sustain. doi:10.1007/s10668-017-9916-x
61. Schultz L, Fazey I (2009) Effective leadership for adaptive management. A Practitioner's Guide, Adaptive Environmental Management, pp 295–303
62. Segrave AJ, van der Zouwen MW, Vierssen W (2013) Water planning: from what time perspective? Technol Forecase Soc 86:157–167
63. Segrave A, Brouwer S and Frijns J (2016) Understanding anticipatory water governance. Global Water Forum
64. Termeer C, Biesbroek R, Van Den Brink M (2012) Institutions for adaptation to climate change: comparing National Adaptation strategies in Europe. EPS 11:41–53
65. Termeer CJAM, Dewulf A, Breeman G, Stiller SJ (2015) Governance capabilities for dealing wisely with wicked problems. Adm Soc 47:680–710

66. UCCRN (2011) Urban climate change research network: climate change and cities. First assessment Report of the Urban Climate Change Research Network. Cambridge University Press
67. UNDP (2008) United Nations Development program: governance principles, institutional capacity and quality. New York, USA
68. UNECE (2009) United Nations Economic Commission for Europe: guidance on water and adaptation to climate change. Geneva
69. UNEP (2013) United Nations environmental program: City-level decoupling. Urban resources flows and the governance of infrastructure transition. A report of the working group on cities of the international resource panel
70. Van Leeuwen CJ (2007). Introduction. In: Van Leeuwen, CJ and Vermeire TG (eds) Risk Assessment of Chemicals. An Introduction, 2nd edn. Springer, Berlin, 1–36
71. Van Rijswick M, Edelenbos J, Hellegers P, Kok M, Kuks S (2014) Ten building blocks for sustainable water governance: an integrated method to assess the governance of water. Water Int. doi:10.1080/02508060.2014.951828
72. World Economic Forum (2016) The Global Risk Report 2016. 11the Edition. Cologny/Geneva, Switzerland
73. WRG (2009) 2030 water resources group: charting our water future. Economic frameworks to inform decision-making. West Perth, USA
74. Zins C (2007) Conceptual approaches for defining data, information, and knowledge. J Am Soc Inf Sci Technol 58:479–493

CHAPTER 15

Virtual Water Trade as a Strategy to Water Resource Management in Iran

F. Mohammadi-Kanigolzar*, J. Daneshvar Ameri, N. Motee

Department of Agriculture Development and Management, Faculty of Agriculture Economic and Development, University of Tehran, Tehran, Iran

ABSTRACT

Agriculture has historically played a central role in the economy, life and culture of the Iranian population. Nowadays, this sector is facing the reality that its natural fresh water resources become fully utilized. Considering the climate conditions and limitation of using new water resources and the necessity of increasing agricultural product as a result of population growth, there is a general doubt about Iran's ability to maintain this level of production amid the mounting water challenges, among other obstacles. Therefore, the evaluation of virtual water and water footprint can provide new indicators for informing water policy decisions. So, in order to study the situation at the national level, we estimated virtual water consumption in term of virtual water theory as well water footprint in agriculture sector of Iran. Data from 2001-2008 were used to account for yearly. The results of this study show that Iran has water import dependency and also net water import is 12.7 billion m^3 averages. So Iran country saved 12.7 billion m^3 from their domestic water resource for utilization in other sectors. Finally, it should be concluded that virtual water trade as a policy measure to water resources management will be provided to a great extent in order to reach both significant water saving and environmental sustainability.

Keywords: Virtual Water; Footprint; Water Resource Management; Environmental; Iran

1. INTRODUCTION

There are many studies concerning the increasing threat of water scarcity and vulnerability of water resources at regional and global scales [1-3]. With no significant room to expand cultivation areas in water scarce regions, increased demand for food will have to be met through sustainable agricultural production, which entails improved management of the available resources and development of crop production strategies [4], as well as import from outside [5,6]. Since in water scarce countries, agriculture sector has the highest rate of consumption water among the various sectors, the main focus of most water scarcity studies is on the impact on agricultural and food security.

Iran is located in dry area, large parts of which are characterized as arid and semi-arid environment. According to predictions, Iran's population will reach 100 million by 2021 and this will result in a drastic drop in the water per capita. Self-sufficiency in wheat production, self-reliance in the production of some strategic goods, and great investment in the implementation of infrastructural water projects are needed to meet this population's water and food demands. On the other hand, the need for industrial development, eradication of poverty, the creation of job opportunities and the construction of economic infrastructure in order to achieve the minimum economic and social indices, and a move in coordination with the world economy, have created problems as far as meeting financial demands is concerned. Issues such as high water consumption in the agricultural sector (more than 90% of clean water) and the low output of irrigation clearly show that these two sectors have to be strengthened in the country. However, there is a general doubt about Iran's ability to maintain this level of production amid the mounting water challenges, among other obstacles.

Virtual water, however, can be used in water-scarce countries such as Iran, so that they can use it as a way to relieve the pressure on water resources, and accordingly, as an alternative in the demand management [7]. Allan elaborated on the idea of using virtual water import (coming along with food imports) as a tool to release the pressure on the scarcely available domestic water resources. Virtual water import thus becomes an alternative water source, next to endogenous water sources. Imported virtual water has also been called "exogenous water" [8]. Measuring virtual water is a useful concept in assessing water resource management as it permits the comparison of crops and livestock from the perspective of embedded water [9]. A number of studies have recognized the usefulness of the concept of virtual water for analyzing production patterns and associated water flows [10]. For example, the average global volume of virtual water flows related to the international trade in agricultural products was 1263 Gm^3/yr in the period 1997-2001 by Chapagain and Hoekstra [11]. This estimate is based on the virtual water content of the products in the exporting countries. It would be interesting to see the volume of virtual water trade internationally based on the virtual water content of the products in the importing countries. In addition, in many countries, international trade in agriculture products effectively reduces domestic water demand. These countries import commodities that are relatively water intensive while they export commodities that are less water intensive. In the period 1997- 2001, Japan, the largest (net) importer of water-intensive goods in the world, annually saved 94 billion m^3 from its domestic water resources. This volume of water would have been required, in addition to its current water use, if Japan had produced all imported products domestically. In a similar way, Mexico annually saved 65 billion m^3, Italy 59 billion m^3, China 56 billion m^3 and Algeria 45 billion m^3 [12].

Besides, there are plenty studies about virtual water trade. Some researchers have been working on a specific crop farming [12]. Some have looked at the trade aspects from the national or local perspectives, for example, Earle et al. [13] & Guan and Hubacer [14]. Others have tried to make the analysis of international trade and virtual water put together, for example, Hoekstra and Yang [15] & Chapagain and Hoekstra [11].

So the evaluation of virtual water and water footprint can provide new indicators for informing water policy decisions. Water footprint of a country or region is the total volume of water resources required of the production of the goods and services consumed by a defined population during a given period time [16]. With similar ecological footprint, the virtual water trade (VWT) that introduced by Allan [6] has been seen as one of the ways to improve water use efficiency and to mitigate water scarcity at the regional level and interregional food trade [16,17]. At the national level, food self-sufficiency has been a desired objective of the Iranian government; nevertheless, large amounts of food are imported into the country in drought years. This is partly due to the lack of water for expanding agricultural production. Wheat import during the drought years of 1999 to 2001 accounted for 80% of the country's total domestic wheat supply, making Iran one of the largest wheat importers of the world at the time [18]. Given the close relationship between water and food, a systematic assessment of water resources availability with high spatial and temporal resolution is essential in Iran for strategic decision making on food security.

The present study evaluates the agricultural water footprint and virtual water "flows" through primary crops in Iran differentiating the green and blue water components. This paper is motivated, in part, by a few but crucial methodological issues that question current virtual water and water footprint evaluations for Iran [11]. Therefore, the main objectives of the present work are 1) to obtain new evaluations of both water footprint and virtual water "flow" at a lower spatial scale and for different years; 2) to evaluate water scarcity in light of the evaluations of water footprint and virtual water "flows"; and 3) to distill water policy lesson from these results. Data from 2001-2008 were used to account for yearly.

1.1 Study Area

Geographically, Iran is located between 25 - 40 degrees north latitude and 44 - 63 degrees east longitude with total area of about 1.648 million km^2. Climatic conditions of Iran are mostly typical of arid and semi-arid regions of the Middle East. Nevertheless, the country has a wide spectrum of climatic, physiographic, edaphic, and hydrological conditions. The country-wide average precipitation is about 252 $mm \cdot year^{-1}$. The per capita freshwater availability for the country was estimated at around 2000 $m^3 \cdot capita^{-1} \cdot year^{-1}$ in the year 2000 and expected to go below 1500 $m^3 \cdot capita^{-1} \cdot year^{-1}$ by 2030 due to population growth [17]. Winter temperatures of less than −20°C in high altitude of the country and summer temperatures of more than 50°C in some areas are recorded [19]. Of total land area of the country 12% is under cultivation (arable land, orchards and vineyards). About 9 million hectares of this land are irrigated using traditional and modern techniques, around 6.5 million hectares are rain fed, and the rest is fallow every year.

2. METHODOLOGY

In this study, virtual water of 34 important agricultural products (m^3/ton) was estimated based on crop water requirements and yields. These products categorized in different groups such as legumes, fruits, industrial crops, oil seeds, dried fruit, summer crops, selected fodder crops, Tea and Saffron. However, to calculation of virtual water and virtual water trade flow process was used of the methods provided by Chapagain et al. [16], Chapagain and Hoekstra [11]. Furthermore, the processing water requirement is calculated as follows Chapagain and Hoekstra [11]. For water footprint calculation we used the methodology described by Hoekstra and Hung [15] and Van Oel et al. [20]. Water footprint of a country or region is the total volume of water resources required of the production of the goods and services consumed by a defined population during a given period time [2006]. To calculate the water footprint offered two method of Top-down and Bottom-up, but in this approach, to asses water footprint (Top-down method), we subtracted virtual water export and added virtual water import to total national water withdrawal. National water scarcity, water dependency on foreign water resources and sustainability of national consumption (water self-sufficiency) indexes had calculated to explain relation between water scarcity and virtual water import of country according to Hoekstra and Hung [21] and Arabi et al. [22].

The information about yield of crops, water requirement for each crop, crop productivity and water resource is derived from Power and Agriculture ministry and also FAOSTAT. Required data about Iranian trade have been taken from Agriculture ministry, FAO and Statistics of exports and imports of agricultural commodities and food industries, published by the Iran Chamber of Commerce, Industries, Mines and agriculture for during 2000-2009. Information for adaptive capability collected from Agriculture ministry, FAO, Statistical Center of Iran and the World Bank.

3. ANALYSIS OF THE RESULTS

Table 1, illustrates the mean water requirement, yield, amount of virtual water and water productivity of the products that has been evaluated for during the 2001- 2008 in Iran. As can be seen, among the crops, the average cane and date palm have the greatest water requirements and also the tobacco and kiwi have the least. Plant water demand represents the minimum amount of water required for the activity of physiology and evapotranspiration.

Virtual water content of each crop is different due to climate change and yield in the period of survey. So the average virtual water for products was calculated in 2001-2008. Based on the results, among the imported products respectively, sunflower, lentils, cotton and soybean and also in the exports crops saffron, pistachios, almonds, walnuts and chickpeas have the maximum amount of virtual water. In addition, average annual virtual water for exports and imports crops represented that the virtual water content (especially water demand) of products have been exported more than import products. In fact, in the current circumstances the amount of imports (in terms of volume) is much more than the amount of export (in terms of volume).

The averages of virtual water content ($m^3 \cdot ton^{-1}$) in separated categories of agricultural products in 2001-2008 are shown in Figure 1. Grains, summer crops, fruits, fodder crops and tea have virtual water content more than the virtual water standard limit as 1000 ($m^3 \cdot ton^{-1}$). So, they are included water-intensive products. Virtual water content of each type of crop depends on crop water requirements, climate conditions, and crop productivity and irrigation efficiency. Of course, the Saffron (export crop) is in those products which have very high virtual water this is because the low yield per hectare.

3.1. Virtual Water Trade Flow in Agricultural Products

Tables 2 & 3 show the virtual water trade related to agricultural crop trade in import and export groups respectively. The production trade causes virtual water flow which import/export water in virtual form to/from nation. It is found that the largest volume of products which are shared in import and export in 2001-2008.

From the Table 2, it can be seen that in during the period studied the productions such as corn, wheat and sugarcane, respectively, with the 2.894, 2.581 and 2.003 (billion cubic meters per year) has the highest amount of virtual water import. These results due to high level of imports of its crops were expected. In addition, Crude oil, soya and sunflower in the ranking for the next. The results were expected since the high amount of virtual water in raw product and import value. Based on this data, the total virtual water import into the country in 2008 (Greatest imports in this year due to low rainfall in the country) with a 22.813×10^9 ($m^3 \cdot yr^{-1}$) had the largest share of the total imports of virtual water but in other hands, in 2004 the amount of 9.144×10^9 ($m^3 \cdot yr^{-1}$) the lowest proportion of the total virtual water import into the country has had.

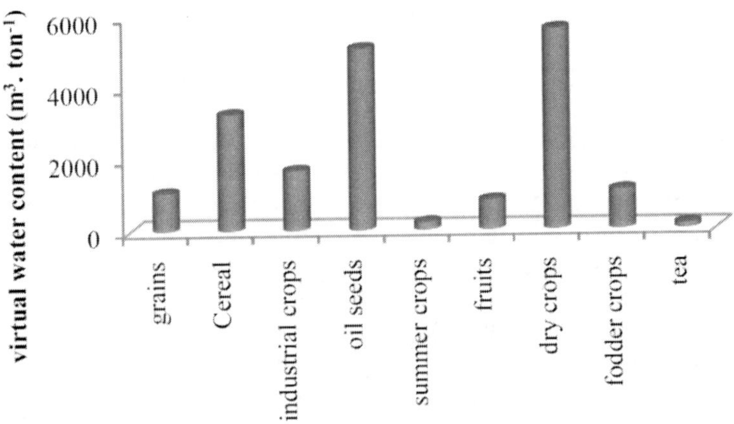

Figure 1. The average of virtual water content ($m^3 \cdot ton^{-1}$) in different groups of Agricultural products in 2001-2008.

Table 1. Yield, water requirement (CWR), virtual water crop (VWC), water productivity (CWP), 2001-2008.

Crops	Yield (kg/ha)	CWR ($m^3 \cdot ha^{-1}$)	VWC ($m^3 \cdot ton^{-1}$)	CWP ($ton \cdot m^{-3}$)
Wheat	3536	3615	1022.37	0.98
Barley	2987	2964	992.31	1.01
Rice	4310	4275	991.81	1.01
Maize	6790	8208	1208.86	0.83
Lentil	1107	3870	3496.90	0.29
Cotton	2427	7644	3149.74	0.32
Tobacco	1464	2353	1607.45	0.62
Cane	83062	18924	227.82	4.39
Soybean	2327	3145	1351.47	0.74
Banana	26278	9404	357.86	2.79
Sunflower	574th	5061	8820.72	0.11
Pea	1027	3872	3769.82	0.27
Beans	1957	4880	2493.09	0.40
Potato	24720	6145	248.58	4.02
Onions	34501	7579	219.67	4.55
Tomatoes	33434	4863	145.45	6.88
Watermelon	26653	5917	222.00	4.50
Cucumber	20776	5248	252.59	3.96
Alfalfa	8196	8897	1085.49	0.92
Orange	17071	6532	382.62	2.61
Date palm	5251	13794	2626.80	0.38
Kiwi	24177	2779	114.94	8.70
Tea	19869	2776	139.71	7.16
Saffron	3	3000	867804	0.00
Apple	16387	6732	410.80	2.43
Fresh grapes	11448	6256	546.47	1.83
Pistachio	671	5174	7710.34	0.13
Fig	7289	9730	1334.91	0.67
Almonds	1317	7333	5565.96	0.18
Walnut	2522	8971	3556.54	0.28

Table 2. The amount of virtual water imported in 2001-2008. Unit: billion cubic meters per year.

product	2001	2002	2003	2004	2005	2006	2007	2008	average
wheat	8.015	2.86	0.73	0.161	0.1	1.112	0.18	7.493	2.581
barely	1.056	0.024	0.175	0.773	1.175	0.352	0.187	1.687	0.678
rice	0.773	0.948	0.785	1.175	1.025	1.26	1.055	1.433	1.056
maize	2.28	2.078	4.568	2.337	2.401	3.216	2.825	3.452	2.894
lentil	0.026	0.002	-	-	0.007	0.052	0.032	0.168	0.047
cotton	-	0.002	0.014	0.076	0.075	0.122	0.21	0.211	0.101
tobacco	0.008	0.032	0.049	0.063	0.064	0.073	0.079	0.072	0.055
cane	2.045	1.419	0.412	0.289	1.361	4.798	1.994	3.708	2.003
soybean	1.112	0.436	1.096	1.066	0.938	1.384	1.058	0.991	1.010
banana	0.04	0.058	0.086	0.088	0.154	0.592	0.205	0.2	0.177
sunflower	-	-	0.001	0.001	0.009	0.001	0.005	0.009	0.004
Soybean meal	2.004	1.479	1.161	1.376	0.255	1.295	1.563	1.54	1.334
crude soybean oil	3.233	1.935	1.935	1.454	1.78	1.749	1.568	0.821	1.809
crude sunflower oil	1.132	1.829	3.442	0.285	0.569	1.887	2.694	1.028	1.608
total	21.724	13.102	14.454	9.144	9.913	17.893	13.655	22.813	

Table 3. The amount of virtual water exported in 2001-2008. Unit: billion cubic meters per year.

product	2001	2002	2003	2004	2005	2006	2007	2008	average
Pea	0.673	0.566	0.3	0.267	0.249	0.23	0.001	0.038	0.291
Beans	0.011	0.015	0.003	0.003	0.011	0.013	0	0.001	0.007
Potato	0.035	0.018	0.01	0.01	0.053	0.044	0.012	0.094	0.035
Onion	0.02	0	0.017	0.009	0.005	0.033	0.048	0.021	0.019
Tomatoes	0.003	0.963	0.008	0.003	0.004	0.013	0.015	0.012	0.128
Watermelon	0.023	0.021	0.019	0.02	0.035	0.042	0.063	0.05	0.034
Cucumber	0.001	0.002	0.003	0.009	0.011	0.022	0.038	0.064	0.019
Alfalfa	-	-	0.002	0	0.006	0.014	0.002	0.017	0.007
Orange	0.004	0.002	0.002	0.011	0.005	0.008	0.003	0.001	0.005
Date palm	0.324	0.323	0.291	0.234	0.298	0.415	0.318	0.282	0.311
Kiwi	0.001	0.001	0.002	0.003	0.003	0.003	0.003	0.005	0.003
Tea	0.003	0.002	0.001	0.01	0.004	0.011	0.015	0.001	0.006
Saffron	0.1	0.084	0.143	0.154	0.144	0.092	0.066	0.313	0.137
Apple	0.041	0.037	0.068	0.029	0.079	0.075	0.097	0.147	0.072
Fresh grape	0.001	0.002	0.004	0.003	0.004	0.008	0.007	0.004	0.004
Pistachio	1.546	0.934	1.189	1.252	1.073	1.071	1.3	1.078	1.180
Fig	0.013	0.01	0	0.012	0.01	0.009	0.012	0.007	0.009
Almond without skin	0.199	0.077	0.017	0.021	0.012	0.015	0.013	0.015	0.046
Walnut	0.003	0.003	0.004	0.001	0.002	0.003	0.002	0.002	0.003
Currant	0.259	0.27	0.303	0.287	0.293	0.339	0.309	0.247	0.288
Total	3.261	3.328	2.386	2.337	2.302	2.461	2.321	2.397	

Virtual water import of various groups of products is shown in Figure 2. As can be seen in grains groups "wheat, barley, rice and corn" are as the most important group of virtual water import in during the studied period. This can be due to high imports of grains (especially wheat) was discussed during the survey. Overall share of different groups of virtual water import value during the period are as follows: grains 46.8 percent, oil seeds 37.3, Industrial Products 14.3, fruit 1.14 percent and legumes 0.34 percent.

3.2. Virtual Water Export of Agricultural Products

Table 3 is shown the virtual water export of various products. As it can be seen that the average of during the period, respectively, pistachio, date palm, currant and peas, respectively, with 1.180, 0.311, 0.291 and 0.288 (billion cubic meters per year) has the maximum amount of virtual water exports. However, while virtual water content (especially demand water) of products likes nuts and dates was high. In other hands, although the grapes crop as the product of basic for the preparation of currant haven't high virtual water, however, the cause of currant has a significant export, hence virtual water exports content was high.

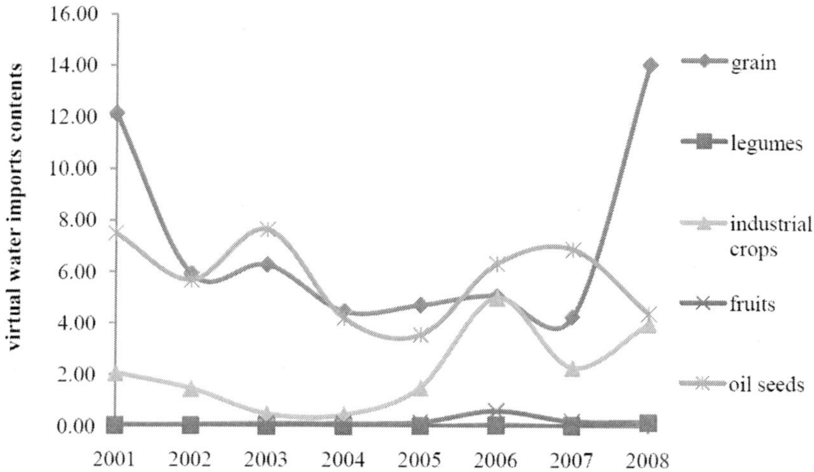

Figure 2. The virtual water imports (billion cubic meters per year) of various products Group during the 2001-2008 in Iran.

In addition, the total virtual water exports from the country in 2001 with a value of 3.328×10^9 ($m^3 \cdot yr^{-1}$) accounted the highest share of total exports of virtual water in during the studied period. Also in 2008 with a value of 2.302×10^9 ($m^3 \cdot yr^{-1}$) has the lowest share of virtual water exports.

Figure 3 illustrates the virtual water exports (billion cubic meters per year) of various products Group during the 2001-2008 in Iran. However, dried fruit Group (pistachios, walnuts, almonds and currant) with 57.8 percent was at the largest export amount of virtual water, so from the virtual water exports was as the most important groups in during the studied period. Furthermore, share of different groups of virtual water exports value during the period are as follows: fruits (oranges, date palm, kiwi, apples and grapes) 16.2 percent, legumes (peas and beans) 10.9 percent, vegetable and melon crops (potatoes, onions, tomatoes, tomatoes, watermelons, and cucumbers) 9.3%, saffron 5.2%, fodder crops 0.25% and tea 0.21% have been in the next category of virtual water export crops. In general it can be said that the export of virtual water in legumes had a decreasing trend, of course a reason for this can to be low commercial performance in this product.

3.3. Net virtual Water Import

Net virtual water content in relation to import and export products illustrates in Figure 4. The positive amounts mean that these groups are more import than export virtual water and the negative amounts mean that these groups are more export than import virtual water. This graph shows Iran imported net virtual water about 18.56 (bcm) in 2001 to 20.72 (bcm) in 2008, due to importing agricultural products and exported 1.6 bcm water in virtual form. The net virtual water of Iran in 2006 was about 11.64 (bcm). Average of this index for the period study was 13.7 (bcm). The maximum amount of net virtual water import appeared in 2008, and the lowest in 2004. Hence, it can be said that net virtual water import most effect of virtual water import.

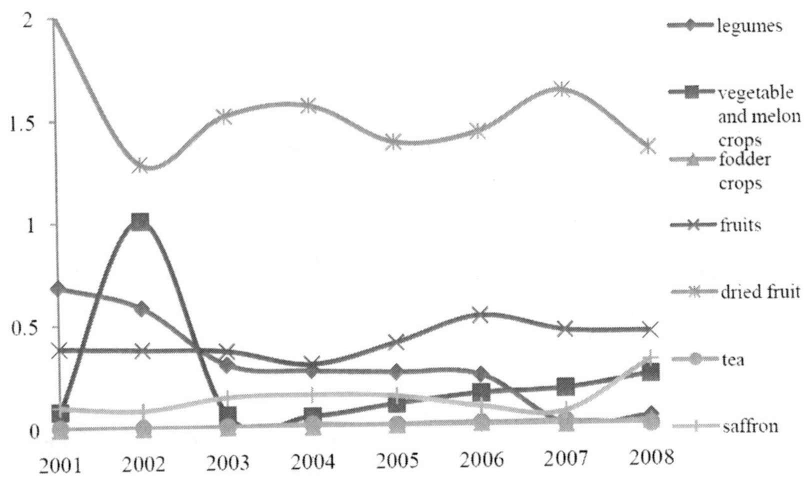

Figure 3. The virtual water exports (billion cubic meters per year) of various products Group during the 2001-2008 in Iran.

Generally, the net virtual water imports during the studied period were more affected of grains, especially wheat. Such that in the years the levels of import grain was high, net imports high and vice versa (These studies, like Hoekstra and Hung [21], Chapagain and Hoekstra [11], Ramirez and Rogers [2004] and Arabi et al. [22], have earned a positive net virtual water import).

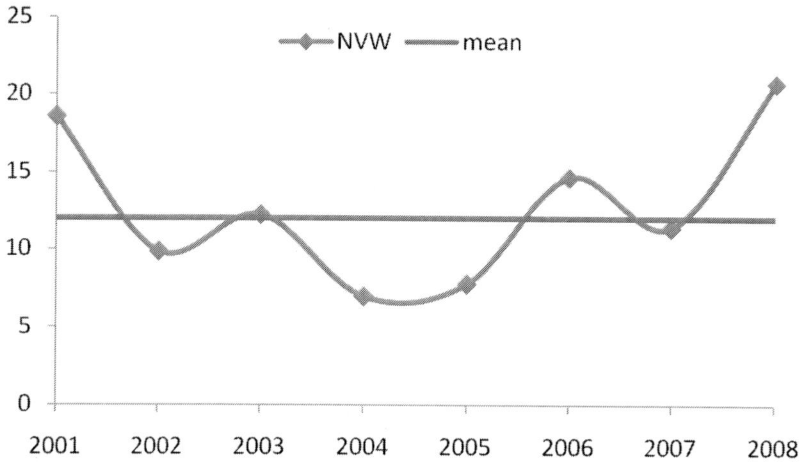

Figure 4. The net virtual water import of different agricultural products in 2001-2008.

Table 4 illustrates the founding about water footprint, Water Scarcity, virtual water dependence and water Selfsufficiency indicators. Survey the condition of water Footprint Index indicated that this index has been uneven over the years studied. Since this index is components of two parts; total domestic water use and net virtual water import. Thus, increasing net virtual water import will be increase water footprint. It has been expected that this increase will reduce pressure on domestic water resources. However, it is clear that the amount of water footprint in the years that the country is as importer is high, for both 2008 and 2001 (the lowest rainfall in these years, diagram 1).

Water Scarcity which increase with the rising net virtual water imports, in other words, there is a Positive correlation between of net virtual water import and water scarcity. The study of virtual water import dependency of the study shows that the level of index has increased. The average amount of it in during the years of period was 12.7 billion cubic meters per year. This means that if a country production domestically crops than imports agricultural products, should be added 12.7 bcm of internal water resource for production that crops.

The water self-sufficiency index (is as opposed index, the virtual water import dependency, that increase in this index will be decreased dependency) is defined as a proportion of domestic water footprint to the total water footprint of a country or region. Self-sufficiency is 100% if all the water needed is available and indeed taken from within the own territory. Water self-sufficiency approaches zero if the demands of goods and services in a country are heavily met with gross virtual water imports, i.e. it has relatively large external water footprint in comparison to its internal water footprint. The water self-sufficiency index shows a decreasing trend of this index in during the period 2001-2008. In other words, the demand for a product or service into country has been met plenty of by imports of virtual water.

Table 4. The index amounts of water Self-sufficiency (WSS), water Dependence (WD), Water Scarcity (WS), Water Footprint (WF).

Year	WF	WS	WD	WSS
		%	%	%
2001	111.11	85	16.71	83.29
2002	102.67	80	9.61	90.39
2003	105.31	81	11.6	88.40
2004	100.19	78	6.95	93.05
2005	100.95	77	7.68	92.32
2006	107.74	82	13.59	86.41
2007	104.20	80	10.94	89.06
2008	113.21	87	18.31	81.69

However, due to population growth and other sectors competitiveness with the agriculture in water consumption, expected that indicators of water scarcity and dependence are moving into the third (High water scarcity and water dependency increases).

4. CONCLUSIONS

In this paper, we estimated virtual water consumption in term of virtual water theory in agriculture sector of Iran. Furthermore, national water scarcity, water dependency on foreign water resources, and sustainability of national consumption (water self-sufficiency) index, have been calculated to explain relation between water scarcity and virtual water import of Iran. In addition, water resources utilization of this sector by the method of water footprint was analyzed. The findings in this paper are the following:

Calculations of the virtual water content of selected products indicate that exported products had been more specific water demand than imported products. In other words, the pattern of agricultural export during the period has been towards the export of water. Also, the results show that in the equal volume of trade, export of virtual water will be several times more than imports. These are notable cases in determining the pattern of virtual water trade.

The investigation showed that virtual water trade of the country has not been identified due to water conditions. Meanwhile, some of the goods that require less water and have high economic efficiency advantage have been placed on imported products, while many exported goods have a high specific water demand, that is to say, irrational use of water resources development in Iran. It seems that it is necessary to review water resource management policies and incorporate the development of virtual water resources based on comparative advantage, with the cooperation of all national and international sectors. During recent years, Iran has been one of the major importers of virtual water. Undoubtedly, the consideration of virtual water in the country's international and national water policies is in need as follows:

1) However, it seems that some of the decision makers have no reluctance to accept the role of virtual water in agricultural trade, but, in order to achieve the sustainable use of water resources, maximum use of existing capacities in this section is required.

2) A model of cultivation with a focus on virtual water and water resources in each region should be developed, because the current approach of self-sufficiency in local levels up to international level would be the loss of potential and water resource due to the lack of such a calculation.

3) Label installation of virtual water on the agricultural production, such as energy labels installed on industrial products, can be effective to promote knowledge and also optimize the use of agricultural products.

4) Fluctuation in production due to the lack of uniformity of rainfall is one of the features of Iran, which is harmful to the stability and security of food supplies. Iran has the large share of imports, so fluctuation in production may lead to increasing volatility in imports which can also affect the stability of global food markets. Changing the economic system to virtual water would allow the country to guarantee food security, and the production of low-cost desalinated seawater is able to support the domestic and industrial needs.

REFERENCES

1. S. L. Postel, G. C. Daily and P. R. Ehrlich, "Human appropriation of renewable freshwater," Science, Vol. 271, No. 5250, 1996, pp. 785-788. http://dx.doi.org/10.1126/ science. 271.5250.785

2. C. J. Vörösmarty, P. Green, J. Salisbury and R. B. Lammers, "Global Water Resources: Vulnerability from Climate Change and Population Growth," Science, Vol. 289, No. 5477, 2000, pp. 284-288. http://dx.doi.org/10.1126/science.289.5477.284
3. T. Oki and S. Kanae, "Global Hydrological Cycles and World Water Resources," Science, Vol. 313, No. 2790, 2006, pp. 1068-1072. http://dx.doi.org/10.1126/science.1128845
4. M. Qadir, B. R. Sharma, A. Bruggeman, R. Choukr and F. Allah Karajeh, "Nonconventional Water Resources and Opportunities for Water Augmentation to Achieve Food Security in Water Scarce Countries," Agricultural Water Management, Vol. 87, No. 1, 2007, pp. 2-22. http://dx.doi.org/10.1016/j.agwat.2006.03.018
5. H. Yang, L. Wang, K. C. Abbaspour and A. J. B. Zehnder, "Virtual Water Trade: An Assessment of Water Use Efficiency in the International Food Trade," Hydrology and Earth System Sciences, Vol. 10, 2006, pp. 443-454. http://dx.doi.org/10.5194/hess-10-443-2006
6. J. A. Allan, "Virtual Water: A Long Term Solution for Water Short Middle Eastern Economies," Occasional Paper 3, School of Oriental and African Studies (SOAS), University of London, 1997.
7. V. Esther, "Water Trade in Andalusia. Virtual Water: An Alternative Way to Manage Water Use," Ecological Economics, Vol. 63, No. 1, 2006, pp. 201-208.
8. M. J. Haddadin, "Exogenous Water: A Conduit to Globalization of Water Resources, Virtual Water Trade," In: A. Y. Hoekstra, Ed., Proceedings of the International Expert Meeting on Virtual Water Trade, Value of Water Research Report Series No. 12, 2003.
9. L. R. Brown, C. Scott, A. M. Craig, L. L. Joanna, M. Frank, P. Elizabeth, R. S. Arnold, L. N. Frederick, P. Hilda and S. José, "The Relationship of Built Environment to Perceived Social Support and Psychological Distress in Hispanic Elders: The Role of 'Eyes on the Street'," The Journals of Gerontology Series B: Psychological Sciences and Social Sciences, Vol. 64, No. 2, 2009, p. 234. http://dx.doi.org/10.1093/geronb/gbn011
10. M. Zeitoun, J. A. Allan and Y. Tony, "Virtual Water 'Flows' of the Nile Basin, 1998 - 2004: A First Approximation and Implications for Water Security," Global Environmental Change, Vol. 20, No. 2, pp. 229-242. http://dx.doi.org/10.1016/j.gloenvcha.2009.11.003
11. K. Chapagain and A. Y. Hoekstra, "Water Footprints of Nations," Value of Water Research Report Series, Vol. 1-2, UNESCO-IHE, Delft, 2004.
12. Y. Hoekstra, "The Role of Virtual Water Transfer in Water Management," International Workshop on European Water Scenarios: From the Mediterranean Sea to Central Asia, Brussels, 2003.
13. Earle and A. Turton, "The Virtual Water Trade amongst Countries of the SADC, Virtual Water Trade," In: A. Y. Hoekstra, Ed., Proceedings of the International Expert Meeting on Virtual Water Trade, Value of Water Research Report Series No. 12, 2003.
14. D. Guan and K. Hubacer, "Analysis: Assessment of Regional Trade and virtual Water Flows in China," Ecological Economics, Vol. 63, No. 1, 2007, pp. 159-170.
15. Y. Hoekstra and P. Q. Hung, "Globalization of Water Resources: International Virtual Water Flows in Relation to Crop Trade," Global Environmental Change, Vol. 15, No. 1, 2005, pp. 45-56. http://dx.doi.org/10.1016/j.gloenvcha.2004.06.004
16. K. Chapagain, A. Y. Hoekstra and H. G. Savenije, "Water Saving through International Trade of Agricultural Products," Hydrology and Earth System Sciences, 2006, Discuss 2. www.Copernicus.org/EGU/hess/hessd/2.
17. H. Yang and A. Zehnder, "Virtual Water: An Unfolding Concept in Integrated Water Resources Management," Water Resources Research, Vol. 43, No. 12, 2007, Article ID: W12301. http://dx.doi.org/10.1029/2007WR006048
18. Food and Agriculture Organization of United Nations, 2006.http://faostat.fao. org/site/342/default.aspx
19. NCCO, "Initial National Communication to United Nations Framework Convention on Climate Change," Iranian National Climate Change Office at Department of Environment, Tehran, Iran, 2003.
20. P. R. Van Oel, M. M. Mekonnen and A. A. Hoekstra, "The External Water Footprint of the Netherlands: Quan-tification and Impact Assessment," Value of Water Research Report Series No. 33, UNESCO-IHE, Delft, 2008.

21. Y. Hoekstra and P. Q. Hung, "Virtual Water Trade: A Quantification of Virtual Water Flows between Nations in Relation to International Crop Trade," Value of the Water Research Report Series No. 11, UNESCO-IHE, Delft, 2002.
22. Arabi, A. Alizadeh Y. Vahab Rajee, K. Jam and N. Niknia, "Study on Ecological Water Footprint in Agricultural Section of Iran," Journal of Water Resource and Protection, Vol. 4, 2012, pp. 318-324.

CHAPTER 16

The Need for Agricultural Water Management in Sub-Saharan Africa

Kbrom Ambachew Gebrehiwot*, Mehari Gidey Gebrewahid

Department of Water Resources and Irrigation Engineering, Institute of Technology, Haramaya University, Haramaya, Ethiopia

ABSTRACT

This paper discusses the current and future conditions that affect water resources and the constraints of water (agricultural) management in sub-Saharan Africa and suggests remedial measures to be considered by policy makers. The pressure on the quantity and quality of water resources is rising in sub-Saharan Africa due to the increased demand of water for agriculture and other purposes as a result of increase in population and food demand. The availability of water is also under threat from changing climate and as a result, water scarcity is expected in many countries in sub-Saharan Africa. On the other hand, the availability of water for agriculture is expected to further shrink due to the increasing demand of water for other purposes like industry, manufacturing and environmental requirement. The current poor efficiency rate of irrigation systems and massive expansion of irrigated area is expected to further exacerbate the water scarcity. Hence, a water management policy focused on maximization of water use efficiency and water productivity should be prioritized in order to meet the food demand of the growing population and cope with water scarcity problems. Engineering and management intervention integrated with strong society awareness and participation is considered very crucial in enhancing water use efficiency and crop water productivity.

Keywords: Water Scarcity, Water Management, Water Use Efficiency, Water Productivity, Sub-Saharan Africa

1. INTRODUCTION

World population has risen steadily from a total of 2.5 billion in 1950, to 7.4 billion in 2015 [1]. 16% of this growth comes from sub-Saharan Africa as the region has grown from 180 million to 962 million people from 1950 to 2015. This is about 12 million people a year for the past 65 years or approximately 782 million people in 65 years. According to the 2015 UN Revision, the current world population of 7.4 billion is expected to increase by 2.3 billion people within the next thirty five years, reaching 9.7 billion in 2050, and to further increase to 11.2 billion by 2100. Not surprisingly, by 2050 and 2100, the population of sub-Saharan Africa could be as large as 2.1 and 3.9 billion. This means 77% of the world population growth from 2015 to 2100 will be from sub-Saharan Africa.

This population growth would inevitably pose additional demand for food [2] [3] . As shown in Figure 1 and Figure 2, majority of countries with higher population growth rate, for example, Ethiopia, Kenya, Uganda and Nigeria, are precisely those showing high level of undernourishment as a result of inadequate food consumption. Nonetheless, these countries are on the verge of development and the food consumption pattern is expected to increase in the near future. The high population growth rate coupled with increased per capita food consumption will highly increase the food demand of the region which cannot be fulfilled simply with the existing production level.

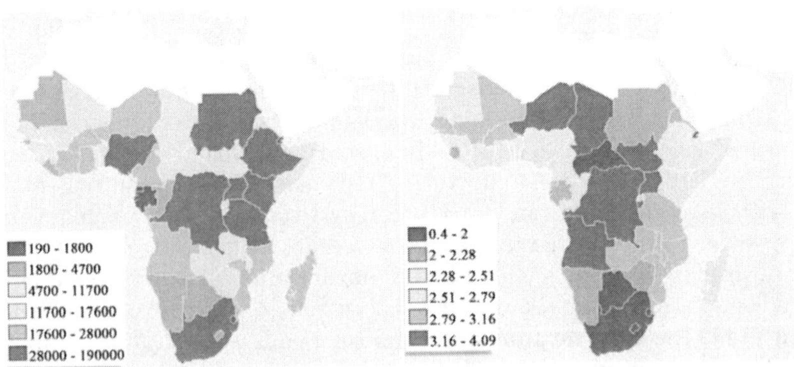

Figure 1. Total population (thousands) (left) and population growth rate (%) (right) in 2015 (source: United Nations) [1] .

Figure 2. Number of people who are undernourished in 2015 (Source: World Bank [4]).

In sub-Saharan Africa, food consumption of cereals such as maize, millet, sorghum, and teff (in Ethiopia) is the mainstay of diets [4]. As presented in Table 1, the demand for cereals is higher than the production in many of the sub-Saharan Africa countries. The gap between production and consumption would bemore pronounced if other food groups, like starchy roots, vegetables, meat, egg, and others, are considered. To fill this gap and ensure food security dramatic increase in agricultural production is inevitable.

Approximately 40% of the world's food comes from 17% of the world's cropland that is irrigated which consumes more than two-third of the world's developed water supplies [5] - [8] . The increase in population and the food demand as well as the changing consumption patterns toward a higher nutrition will be posing challenges to agricultural production [9] . A FAO analysis [10] of 93 developing countries expects agricultural production to increase over the period 1998-2030 by 49% in rain fed systems and by 81% in irrigated systems. According to World Bank [4] , agricultural production could nearly triple in sub-Saharan Africa by 2050.

In this regard, arable land expansion will remain an important factor in the growth of crop production in the region. In developing countries, it is expected to observe a projected net increase in the arable area of some 107 million ha (from 968 in 2005/07 to 1075 in 2050) of which the 51 million (48%) is expected to take place in sub-Saharan Africa [4] .

Table 1. Food (cereals) production and consumption in sub-Saharan Africa (Source: FAO food security data by food groups/items, June 2012).

Country	Production (kg/person/year)			Consumption (Kg/person/Year)			Deficit		
	1997	2002	2007	1997	2002	2007	1997	2002	2007
Angola	32	41	45	59.9	72.3	81.0	−27.9	−31.3	−36.0
Benin	130	136	130	108.0	104.0	116.1	22.0	32.0	13.9
Botswana	42	16	18	123.7	109.5	109.5	−81.7	−93.5	−91.5
Burkina Faso	215	233	242	223.7	219.7	233.6	−8.7	13.3	8.4
Burundi	43	37	35	30.3	28.8	35.8	12.7	8.2	−0.8
Cameroon	86	84	95	89.4	96.7	108.8	−3.4	−12.7	−13.8
CAR	35	45	53	41.6	48.9	57.3	−6.6	−3.9	−4.3
Chad	121	128	181	108.8	118.3	144.5	12.2	9.7	36.5
Congo	4	5	6	44.2	75.2	77.0	−40.2	−70.2	−71.0
Comoros	24	21	21	74.5	74.5	75.2	−50.5	−53.5	−54.2
DRC	30	27	23	34.7	32.9	39.8	−4.7	−5.9	−16.8
Djibouti	0	0	0	111.3	125.9	138.7	−111.3	−125.9	−138.7
Eritrea	31	31	74	127.8	120.8	134.0	−96.8	−89.8	−60.0
Ethiopia	145	132	168	123.0	138.3	142.7	22.0	−6.3	25.3
Gabon	25	20	23	94.2	101.8	109.5	−69.2	−81.8	−86.5
Gambia	85	120	117	139.8	142.7	147.5	−54.8	−22.7	−30.5
Ghana	95	87	81	82.5	87.6	97.5	12.5	−0.6	−16.5
Guinea	143	167	212	127.4	127.0	127.0	15.6	40.0	85.0
Kenya	101	93	100	120.1	117.5	120.5	−19.1	−24.5	−20.5
Lesotho	103	97	52	216.8	221.9	227.4	−113.8	−124.9	−175.4
Liberia	34	33	35	97.8	89.1	97.5	−63.8	−56.1	−62.5
Madagascar	137	121	145	111.7	117.9	129.6	25.3	3.1	15.4
Malawi	164	167	177	147.8	142.4	143.1	16.2	24.7	33.9
Mali	205	205	274	181.0	187.6	199.7	24.0	17.4	74.3
Mauritania	78	43	50	164.3	159.5	169.0	−86.3	−116.5	−119.0
Mauritius	0	0	1	152.2	155.1	161.0	−152.2	−155.1	−160.0
Mozambique	79	76	66	91.6	92.7	101.5	−12.6	−16.7	−35.5
Niger	207	247	281	200.0	201.1	197.5	7.0	45.9	83.5
Nigeria	185	156	181	136.9	133.6	145.6	48.1	22.4	35.4
Rwanda	31	33	39	33.9	32.1	43.4	−2.9	0.9	−4.4
Senegal	100	87	86	154.0	152.6	164.3	−54.0	−65.6	−78.3
Sierra Leone	78	54	136	107.3	113.2	112.1	−29.3	−59.2	23.9
Sudan	134	115	157	138.3	123.0	130.7	−4.3	−8.0	26.3
Swaziland	138	81	50	133.6	124.1	127.8	4.4	−43.1	−77.8
Tanzania	132	130	133	106.2	106.9	104.0	25.8	23.1	29.0
Togo	142	141	137	116.1	118.6	123.7	25.9	22.4	13.3
Uganda	80	88	84	59.5	58.8	63.5	20.5	29.2	20.5
Zambia	127	90	116	157.3	142.0	136.9	−30.3	−52.0	−20.9
Zimbabwe	192	141	120	148.6	144.5	157.0	43.4	−3.5	−37.0
South Africa	272	280	227	178.5	186.2	186.9	93.5	93.9	40.1
Namibia	68	59	64	130.7	132.1	131.4	−62.7	−73.1	−67.4
Côte d'Ivoire	70	62	58	96.0	89.4	93.4	−26.0	−27.4	−35.4
Guinea-Bissau	111	99	113	139.4	146.7	146.7	−28.4	−47.7	−33.7

In sub-Saharan Africa, the water use efficiency estimated for the year 2005/07 and projected for the year 2050 is 22 and 25 percent respectively, which is by far less than the world average which is 44 and 46 percent respectively [11] . Although, the current water withdrawal for agriculture and other purposes is lower than the total annual renewable water resource, this low water use efficiency will undoubtedly challenge the sustainable use and management of water in the region. Thus, expanding the irrigated area under the present agricultural water management characterized with low irrigation efficiency rates will result in severe water scarcity in the coming decades.

From Figure 3, it is clear that countries with larger total renewable water resources are also having a huge population, resulting in a lower total per capita water availability. In 2014, 16 SSA countries were already under water stress, water scarcity and absolute scarcity conditions according to the "Falkenmark indicator" or "water stress index". Based on the per capita usage, the water conditions in an area can be categorized as: no stress (> 1700 m^3 per capita), stress (1000 - 1700 m^3 per capita), scarcity (500 - 1000 m^3 per capita), and absolute scarcity (<500 m^3 per capita) [12] . As shown in Table 2, the per capita renewable water resource was declining with time due to the increase in population. As predicted by Falkenmark [13] , most of the East and some of the South African countries have already been water stressed. The projection by Falkenmark [13] indicates that eight of the East and some of the South African countries will arrived at absolute scarcity by 2025 and five of the West African countries will become at least water stressed in the early decades of the twenty first century.

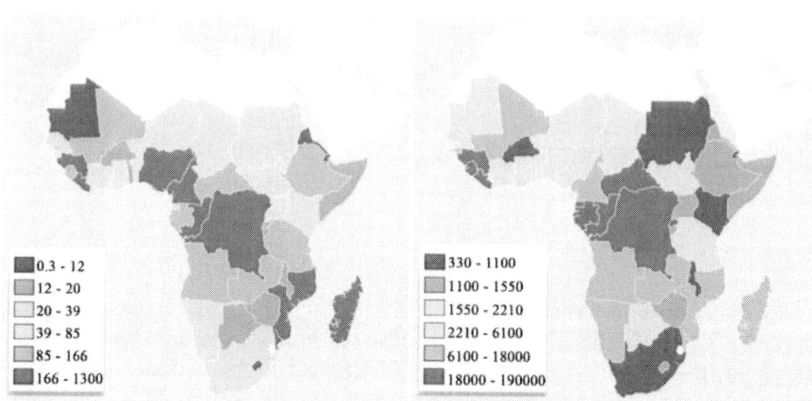

Figure 3. Total renewable water resources, BCM, (left) and total renewable water resources per capita, m^3/person/yr, (right) in 2014 (Source: FAO Aquastat 2015).

Not surprisingly, the water resources of some countries had already been declined and changed from no stress to stress and scarcity levels. For instance, in 2007, Comoros, Nigeria and Uganda were under no stress condition. However, in 2012 and 2014, their water resource was declined and changed to the stress category. Similarly,

Table 2. Per capita renewable water resources (Source: FAO Aquastat 2015).

Country	Water resources (m^3/per/yr)		
	2007	2012	2014
Djibouti	375.50	348.80	339
Kenya	813.20	711.00	674
Burkina Faso	948.40	820.20	775
South Africa	1035.00	980.20	966
Malawi	1260.00	1086.00	1027
Sudan	No data	1016.00	975
Rwanda	1340.00	1161.00	1099
Eritrea	1404.00	1193.00	1119
Burundi	1506.00	1273.00	1196
Ethiopia	1517.00	1330.00	1264
Lesotho	1545.00	1473.00	1440
Zimbabwe	1570.00	1457.00	1370
Somalia	1650.00	1442.00	1360
Comoros	1896.00	1671.00	1596.00
Nigeria	1944.00	1695.00	1603.00
Uganda	1956.00	1654.00	1547.00

South Africa's water resources status was changed from stress category in 2007 to scarcity category in 2014. Nevertheless, the attention given to water management in sub-Saharan Africa is still inadequate. As a result, over exploitation of water resources by different sectors is shadowing the efforts made to ensure sustainable water use in the region. Considering the global figure, withdrawal of water by about a quarter of the irrigated systems is above the regeneration rate [14]. This coupled with the poor property rights specifications on water resources and inefficient irrigation practices is creating a potential for serious water shortages. Moreover, with the growing demand of water for industry, manufacturing and urban sectors, combined with environmental problems, water available for agriculture will decrease in the future [15] - [18] . On the other hand, as irrigation (agriculture) is vitally important in meeting the food and fiber needs for a rapidly expanding world population, the demand for agricultural water is continuously growing [3] [7] [19] . Additionally, the rapid growth in population and income requires large water investments and financing that result in further pressure on the quantity and quality of water resources [20] . The problem of providing the food for a much greater world population, therefore, becomes focused on the area of growing more on existing water and land resources [21].

2. WHAT ARE THE AREAS OF IMPROVEMENT IN WATER MANAGEMENT?

Many sub-Saharan Africa countries must develop more than twice the amount of water they currently use to meet reasonable future requirements [21] . For instance, Ethiopia must double its cereal production by 2025 to meet the food needs of its rapidly growing population and secure food self sufficiency which in its turn escalate its agricultural water demand [22] . Considering the fact that 90% of staple food for sub Saharan Africa comes from rain-fed farming systems, [23] cited in [24] recommended that much greater emphasis will have to be given to increasing the productivity of global rain-fed agriculture. However, rain-fed agriculture remains risky because of spatial and temporal variability in rainfall, water scarcity, droughts, soil erosion by wind and water, low investment, or high population pressure [25] . Moreover, there is a significant difference in crop yields between irrigated and rain-fed agriculture. In developing countries, grain yields from rain-fed agriculture are 1.5 t/ha compared to 3.1 t/ha from irrigated agriculture on average [26] cited in [9] . Hence, due attention will have to be given to better management of water resources in both irrigated and rain-fed agriculture. In this regard, instead of trying to supply the growing demand for water with new sources, improvement in the water-use efficiency should be the main focus in water-management policies [18] [27] . To increase water-use efficiency, several practices can be adopted. Water-saving techniques, advances in irrigation technology (e.g., drip irrigation) and management, leaving crop residues on the soil surface and planting cover crops and adequate tillage practices are considered as potential measures to improve water use efficiency and adapt to climate change impacts [9] [28] - [30] . However, with the limited and shrinking water availability, the increasing water requirement, due to expansion of irrigable lands as a result of growing population and food demand, cannot be reversed by improving the water use efficiency only. Hence, increasing water productivity deserves much attention as well. Increasing water productivity means either to produce the same yield with less water resources or to obtain higher crop yields with the same water resources [31] which would enable growing more food and hence feeding the ever increasing population or gaining more benefits with less water [32] . In its broadest sense, it reflects the objectives of producing more food, income, livelihood and ecological benefits at less social and environmental cost per unit of water consumed [33] .

Sub-Saharan Africa countries, often characterized with low yield levels and large on-farm water losses resulting in poor water use efficiencies, have a promising potential for improvements in water productivity through introduction of agronomic or engineering measures [15] [33] . An important means of achieving increased water productivity will be to get more from the amount of water that is beneficially depleted by agriculture which can be achieved by introducing different agronomic, engineering and management technologies such as changing crop varieties, crop substitution, deficit, supplemental or precision irrigation [7] [16] . The next sections discuss small holder drip irrigation and water pricing as examples of the engineering and management techniques to improve water productivity and, of course, water use efficiency.

2.1. Small Holder Drip Irrigation System

Drip irrigation's combination of water savings and yield increases typically produces at least a doubling of water productivity, yield per unit water, and makes it a leading technology in the global challenge of boosting crop production in the face of serious water constraints [34] cited in [35] . However, due to its high capital cost, particularly for the poor farmers, the conventional drip irrigation is generally not widely used on a global scale and accounts for less than 0.1% of irrigated land worldwide [36] . Nonetheless, smallholders drip irrigation system attempts to retain the benefits of conventional systems while removing the factors that prevent their adoption by poor smallholders: purchase cost, the need for a pressurized supply, the associated pumping costs, and the complexity of operation and maintenance. It can help solve water management problems faced by smallholder farmers by making it easier and simpler to supply the right amount of water to their crops at the right place and at the right time [37]. Drip irrigation used about 35% of the water used by the surface irrigation systems thus giving much higher water use efficiencies [38] . It was also concluded by [38] that low cost drip systems achieved water saving of more than 50% compared to surface irrigation systems. Further results from field trials demonstrate that affordable smallholders drip systems easily pay for themselves in one growing season, and stimulate shifts to more intensive agricultural practices by small farmers [39] . Implementation of this technology might help improve livelihoods of individual farmers or small communities and is mentioned promising in the context of poverty alleviation which is the number one enemy of many sub-Saharan Africa countries [36]

Despite the higher potential for smallholder drip irrigation and other small scale irrigation technologies as supplementary irrigation for millions of people and to achieve household food security, the use of the technology has hardly begun in many sub-Saharan Africa countries. With proper marketing and proper agronomic and technical support low cost irrigation technologies can be adopted by smallholder farmers to change family lifestyles, increase people's incomes, create employment and go a long way towards food security and improved nutrition [38] . However, recent literatures argue that the success of the technology is interpreted through development agencies lenses and with the intention of continuing involvement in future projects [40] [41] . Hence, it is recommended to get the technology tested for its efficiency, productivity and social acceptance by farmers and independent researchers before its introduction and distribution to the vast agrarian population.

2.2. Water Pricing

Water pricing has become a key issue in both the developed and developing world. It is designed taking into account various objectives which are oftentimes more and more difficult to achieve due to environmental constraints, increased water demand, climate change and economic crisis which impacts all incomes (households, farmers, industries). The three main objectives of water pricing are balancing water budget, allowing water access, and inciting users to save water [42] .Achieving these objectives directly leads to a better water resource management which could help us in coping with the scarce and still shrinking water resource. [20] reported that pricing of water and water services are an acceptable practice that may result in better conservation and sustainable management of water. This is because higher prices for water use would encourage its more rational use by agricultural users and provide funds for investing in infrastructure which are two necessary conditions to avoid risk of water shortage [43]

Although, water pricing has long been practiced in some developed countries, it has been rarely or hardly implemented in sub Saharan Africa. The question here would be "how to introduce water pricing for the first time in an area where water has been accessed for free?" and "on what basis to effect the pricing?". In principle, there are multiple pricing patterns for irrigation water. The most common pricing systems are based on volume, area, electricity consumption (for groundwater based irrigation), type of irrigation or crop and block rate pricing structure [44] . The experience of the countries which have already implemented different pricing systems can be taken as a bench mark to choose better pricing approach in a particular area. The area pricing system, modified according to the crop or irrigation techniques, does not encourage water saving for a given choice of crop or irrigation technique, but it does have more effect on the choice of which crops to irrigate or which irrigation technique to adopt [45] . Pricing systems based on volume, volumetric pricing, and block rate pricing structure are known for their effect on high water savings. [46] cited in [47] estimated a 40% reduction in irrigation water demand associated with the increasing block rate structure.

Water pricing alone will not always be a sufficient incentive for users to enhance water saving and improve irrigation performance [45] . Thus, due consideration is needed to be given to water distribution and control. Rational water use has to be achieved by careful control of distribution and by allocations that broadly meet crop requirements. In this regard, it is required to enhance understanding of water scarcity among irrigators and encourage more efficient water use. It is also worthy to ensure better control over the supply of water as it can reduce wastage of water associated with excess amounts of water flowing through uncontrolled canals and un- gated turnouts onto fields and into drainage channels [48] .

3. CONCLUSION

Water scarcity is becoming severe and the availability of water for agriculture is dwindling due to increasing population and food demand, changing climate, expansion of industries and manufacturing. In order to sustain their economy and feed their population, sub-Saharan Africa countries should implement sound, feasible and implementable water management policy that can help in obtaining more production from the existing resources, land and water, through maximization of water use efficiency and crop water productivity. It is, thus, wise to develop a set of engineering and management interventions such as water-saving techniques, improved irrigation management, supplemental or precision irrigation, water pricing and better agronomy. Before implementation, however, the intervention measure should be tested through independent and participatory research. Above all, farmers, government and non-governmental organizations should cautiously evaluate the feasibility of the intervention measures and proper awareness should be given to the farming community.

REFERENCES

1. United Nations (2015) World Population Prospects: The 2015 Revision. DVD Edition, Department of Economic and Social Affairs, Population Division. United Nations, New York.
2. Rockström, J., Barron, J. and Fox, P. (2003) Water Productivity in Rain-Fed Agriculture: Challenges and Opportunities for Smallholder Farmers in Drought-Prone Tropical Agroecosystems. In: Kijne, J.W., Barker, R. and Molden, D, Eds., Water Productivity in Agriculture: Limits and Opportunities for Improvement, CABI, Wallingford, 145-162.http://dx.doi.org/10.1079/9780851996691.0145
3. Playan, E. and Mateos, L. (2006) Modernization and Optimization of Irrigation Systems to Increase Water Productivity. Agricultural Water Management, 80, 100-116.http://dx.doi.org/10.1016/j.agwat.2005.07.007
4. World Bank (2016) Health Nutrition and Population Statistics. World Bank, Washington DC.
5. Shiklomanov, I.A. (2000) Appraisal and Assessment of World Water Resources. Water International, 25, 11-32. http://dx.doi.org/10.1080/02508060008686794
6. Molden, D. (1997) Accounting for Water Use and Productivity. SWMI Paper, International Water Management Institute, Colombo.
7. Howell, T.A. (2001) Enhancing Water Use Efficiency in Irrigated Agriculture. Agronomy Journal, 93, 281-289.http://dx.doi.org/10.2134/agronj2001.932281x
8. Gleick, P.H. (2002) Water Management: Soft Water Paths. Nature, 418, 373. http://dx.doi.org/10.1038/418373a
9. Heumesser, C., et al. (2013) Current State and Future Potential of Global Food Production and Consumption. In: Rengel, Z., Ed., Improving Water and Nutrient-Use Efficiency in Food Production Systems, John Wiley & Sons, Inc., Hoboken,, 1-19. http://dx.doi.org/10.1002/9781118517994.ch1
10. Bruinsma, J. (2003) World Agriculture: Towards 2015/2030. An FAO Perspective. Earthscan Publications Ltd., London
11. Bruinsma, J. (2009) The Resource Outlook to 2050. Paper Presented at the Expert Meeting on How to Feed the World in 2050, Rome, 24-26 June 2009.
12. Falkenmark, M., Lundqvist, J. and Widstrand, C. (1989) Macro-scale Water Scarcity Requires Micro-Scale Approaches. Natural Resources Forum, 13, 258-267. http://dx.doi.org/10.1111/j.1477-8947.1989.tb00348.x
13. Falkenmark, M. (1989) The Massive Water Scarcity Now Threatening Africa: Why Isn't It Being Addressed? Ambio, 18, 112-118.
14. Ghanem, H. (2008) The State of Food Insecurity in the World, 2008: High Food Prices and Food Security: Threats and Opportunities. Food and Agriculture Organization of the United Nations (FAO), Rome.
15. Kijne, J.W., Barker, R. and Molden, D. (2003) Improving Water Productivity in Agriculture: Editors' Overview. In: Kijne, J.W., Barker, R. and Molden. D., Eds., Water Productivity in Agriculture: Limits and Opportunities for Improvement, CABI, Wallingford, xi-xix.
16. Molden, D. and Sakthivadivel, R. (1999) Water Accounting to Assess Use and Productivity of Water. International Journal of Water Resources Development, 15, 55-71. http://dx.doi.org/10.1080/07900629948934
17. De Fraiture, C., et al. (2007) Looking Ahead to 2050: Scenarios of Alternative Investment Approaches. In: Molden, D., Ed., Water for Food, Water for Life: A Comprehensive Assessment of Water Management in Agriculture, Earthscan/IWMI, London/Colombo, 91-145.
18. Garnett, T.P. and Rebetzke, G.J. (2013) Improving Crop Nitrogen Use in Dryland Farming. Improving Water and Nutrient-Use Efficiency in Food Production Systems, Chap. 8, 123-144. http://dx.doi.org/10.1002/9781118517994.ch8
19. Kumar, M.D., Sivamohan, M. and Bassi, N. (2012) Water Management, Food Security and Sustainable Agriculture in Developing Economies. Routledge, Abingdon-on-Thames.

20. Dinar, A., Pochat, V. and Albiac-Murillo, J. (2015) Water Pricing Experiences and Innovations. Global Issues in Water Policy. Springer International Publishing, Gewerbestrasse. http://dx.doi.org/10.1007/978-3-319-16465-6
21. Seckler, D.W. (1998) World Water Demand and Supply, 1990 to 2025: Scenarios and Issues. Vol. 19, Iwmi, Colombo, Sri Lanka.
22. Awulachew, S.B., et al. (2005) Experiences and Opportunities for Promoting Small-Scale/Micro Irrigation and Rainwater Harvesting for Food Security in Ethiopia. International Water Management Institute, Addis Ababa.
23. Rosegrant, M.W., Cai, X. and Cline, S.A. (2002) World Water and Food to 2025: Dealing with Scarcity. International Food Policy Research Institute.
24. Cooper, P., Dimes, J., Rao, K.P.C., Shapiro, B., Shiferaw, B. and Twomlow, S. (2008) Coping Better with Current Climatic Variability in the Rain-Fed Farming Systems of Sub-Saharan Africa: An Essential First Step in Adapting to Future Climate Change? Agriculture, Ecosystems & Environment, 126, 24-35. http://dx.doi.org/10.1016/j.agee.2008.01.007
25. Wani, S.P., Rockström, J. and Oweis, T.Y. (2009) Rainfed Agriculture: Unlocking the Potential. Vol. 7, CABI, Wallingford. http://dx.doi.org/10.1079/9781845933890.0000
26. Rosegrant, M.W., et al. (2002) The Role of Rainfed Agriculture in the Future of Global Food Production. Environment and Production Technology Division, International Food Policy Research Institute.
27. Gleick, P.H. (2003) Global Freshwater Resources: Soft-Path Solutions for the 21st Century. Science, 302, 1524-1528. http://dx.doi.org/10.1126/science.1089967
28. Hoanh, C.T., Johnston, R. and Smakhtin, V. (2015) Climate Change and Agricultural Development: A Challenge for Water Management. In: Hoanh, C.T., Johnston, R. and Smakhtin, V., Eds., Climate Change and Agricultural Water Management in Developing Countries, CABI, Wallingford, 1.
29. Fuchs, M. (2007) Impact of Research on Water Use for Irrigation in Israel. Irrigation Science, 25, 443-445. http://dx.doi.org/10.1007/s00271-006-0050-z
30. Stanhill, G. (1992) Irrigation in Israel: Past Achievements, Present Challenges and Future Possibilities. Water Use Efficiency in Agriculture: Proceedings of the Binational China-Israel Workshop, Beijing, 22-26 April 1991, 63-77.
31. Zwart, S.J. and Bastiaanssen, W.G. (2004) Review of Measured Crop Water Productivity Values for Irrigated Wheat, Rice, Cotton and Maize. Agricultural Water Management, 69, 115-133. http://dx.doi.org/10.1016/j.agwat.2004.04.007
32. Yihun, Y.M., Haile, A.M., Schultz, B. and Erkossa, T. (2013) Crop Water Productivity of Irrigated Teff in a Water Stressed Region. Water Resources Management, 27, 3115-3125. http://dx.doi.org/10.1007/s11269-013-0336-x
33. Molden, D., et al. (2009) Improving Agricultural Water Productivity: Between Optimism and Caution. Agricultural Water Management, 97, 528-535. http://dx.doi.org/10. 1016/j.agwat.2009.03.023
34. Postel, S. (1999) Pillar of Sand: Can the Irrigation Miracle Last? W.W. Norton & Company, New York.
35. Postel, S., Polak, P., Gonzales, F. and Keller, J. (2001) Drip Irrigation for Small Farmers. Water International, 26, 3-13. http://dx.doi.org/10.1080/02508060108686882
36. Mehari Haile, A., Depeweg, H. and Stillhardt, B. (2003) Smallholder Drip Irrigation Technology: Potentials and Constraints in the Highlands of Eritrea. Mountain Research and Development, 23, 27-31. http://dx.doi.org/10.1659/0276-4741(2003)023[0027:SDIT]2.0.CO;2
37. Stillhardt, B., Ghebru, B. and Mehari, A. (2003) Small-Scale Micro Irrigation in Eritrea. A Feasibility Study on the Introduction of Affordable Micro Irrigation Technology in Eritrea. Environment (CDE), University of Bern, Hochschulstrasse.
38. Maisiri, N., Senzanje, A., Rockstrom, J. and Twomlow, S.J. (2005) On Farm Evaluation of the Effect of Low Cost Drip Irrigation on Water and Crop Productivity Compared to Conventional Surface Irrigation System. Physics and Chemistry of the Earth, Parts A/B/C, 30, 783-791. http://dx.doi.org/10.1016/j.pce.2005.08.021

39. Postel, S. (2000) Entering an Era of Water Scarcity: The Challenges Ahead. Ecological Applications, 10, 941-948. http://dx.doi.org/10.1890/1051-0761(2000)010[0941:EAEOWS]2.0.CO;2
40. Wanvoeke, J., Venot, J.-P., De Fraiture, C. and Zwarteveen, M. (2015) Smallholder Drip Irrigation in Burkina Faso: The Role of Development Brokers. The Journal of Development Studies, 52, 1019-1033.
41. Venot, J.-P. (2016) A Success of Some Sort: Social Enterprises and Drip Irrigation in the Developing World. World Development, 79, 69-81. http://dx.doi.org/10.1016/j.worlddev.2015.11.002
42. Montginoul, M., Loubier, S., Barraqué, B. and Agenais, A.-L. (2015) Water Pricing in France: Toward More Incentives to Conserve Water. In: Dinar, A., Pochat, V. and Albiac-Murillo, J., Eds., Water Pricing Experiences and Innovations, Springer International Publishing, Gewerbestrasse, 139-160. http://dx.doi.org/10.1007/978-3-319-16465-6_8
43. De Andrade Resende Filho, M., Correa, J.S.O. and de Oliveira Torres, M. (2015) Water Pricing in Brazil: Successes, Failures, and New Approaches. In: Dinar, A., Pochat, V. and Albiac-Murillo, J., Eds., Water Pricing Experiences and Innovations, Springer International Publishing, Gewerbestrasse, 41-61. http://dx.doi.org/10.1007/978-3-319-16465-6_3
44. Che, Y. and Shang, Z. (2015) Water Pricing in China: Impact of Socioeconomic Development. In: Dinar, A., Pochat, V. and Albiac-Murillo, J., Eds., Water Pricing Experiences and Innovations, Springer International Publishing, Gewerbestrasse, 97-115. http://dx.doi.org/10.1007/978-3-319-16465-6_6
45. Chohin-Kuper, A., Rieu, T. and Montginoul, M. (2003) Water Policy Reforms: Pricing Water, Cost Recovery, Water Demand and Impact on Agriculture. Lessons from the Mediterranean Experience. Proceedings of the Water Pricing Seminar, Agencia Catalana del Agua and World Bank Institute, Barcelona, 30 June -2 July 2003, 1-9.
46. Pike, T., et al. (2007) Does Pricing Water Reduce Agricultural Demand? An Example from British Columbia. Sustainable Development Briefing Note. Government of Canada, Policy Research Initiative. http://www.policyresearch.gc.ca.
47. Renzetti, S. and Dupont, D.P. (2015) Water Pricing in Canada: Recent Developments. In: Dinar, A., Pochat, V. and Albiac-Murillo, J., Eds., Water Pricing Experiences and Innovations, Springer International Publishing, Gewerbestrasse, 63-81. http://dx.doi.org/10.1007/978-3-319-16465-6_4.
48. Molle, F. and Berkoff, J. (2007) Water Pricing in Irrigation: The Lifetime of an Idea. In: Molle, F. and Berkoff, J., Eds., Irrigation Water Pricing: The Gap between Theory and Practice, CABI, Wallingford, 1-20.

CHAPTER 18

How Subsurface Water Technologies (SWT) can Provide Robust, Effective, and Cost-Efficient Solutions for Freshwater Management in Coastal Zones

Koen G. Zuurbier[1,2], Klaasjan J. Raat[1], Marcel Paalman[1], Ate T. Oosterhof[3], Pieter J. Stuyfzand[1,2]

[1]*KWRWatercycle Research Institute, P.O. Box 1072, 3430 BB Nieuwegein, The Netherlands*

[2]*Faculty of Civil Engineering, Technical University Delft, P.O. Box 5048, 2600 GA Delft, The Netherlands*

[3]*Vitens NV, P.O. Box 1205, 8001 BE Zwolle, The Netherlands*

ABSTRACT

Freshwater resources in coastal zones are limited while demands are high, resulting in problems like seasonal water shortage, overexploitation of freshwater aquifers, and seawater intrusion. Three subsurface water technologies (SWT) that can provide robust, effective, and cost-efficient solutions to manage freshwater resources in the subsurface are evaluated using groundwater modelling and validation at field-scale: (1) ASR-coastal to store freshwater surpluses in confined brackish-saline aquifers for recovery in times of demand, (2) the Freshkeeper to counteract salinization of well fields by interception and desalination of upconing brackish groundwater, and (3) the Freshmaker to combine ASR and Freshkeeper to enlarge the volume of natural freshwater lenses for later abstraction. The evaluation indicates that SWT can be used in various hydrogeological settings for various hydrogeological problems like seawater intrusion, upconing, and bubble drift during ASR and have significant economic benefits. Although only sporadically applied to date, we foresee that SWT will stimulate (cost-)efficient and sustainable exploitation of various freshwater sources (like groundwater, rainwater, treated waste water, surface water) in coastal zones. Prolonged SWT testing in the current pilots, replication of SWT in other areas worldwide, and the development of technical and non-technical support tools are required to facilitate potential end-users in investment decision making and SWT implementation.

Keywords: Aquifer storage and recovery, Coastal aquifers, Freshwater, Salinization, Seasonal water shortage, Subsurface water technologies

1 INTRODUCTION

Coastal zones are the most densely populated, productive, and economically dominant regions of the world. About half of the world's population lives within 200 km of a coastline (United Nations 2010), and while this produces many economic benefits, the associated high water demand puts tremendous pressure on the freshwater resources and the coastal ecosystems. This leads to problems like seasonal water shortage (Fig. 1), overexploitation of groundwater resources, saltwater intrusion, and disappearance of wetlands. Further economic growth, population increase, and climate change will intensify these problems, ultimately blocking the sustainable development of coastal zones in industrial, emerging, and developing countries (European Commission 2012). In 2015, water crises were identified as the main global risk in terms of impact (World Economic Forum 2015).

Traditionally, aboveground solutions are sought for these problems, such as construction of reservoirs or saltwater desalination. However, the subsurface may provide more robust, effective, sustainable, and cost-efficient freshwater management solutions due to a better water conservation and limited space requirements aboveground. For instance, the concept of subsurface storage and/or treatment known as managed aquifer recharge (MAR) is increasingly applied worldwide for water storage and treatment (Dillon et al. 2010). In coastal zones, however, the abstraction of (stored) freshwater is generally hampered by saline groundwater, causing early salinization of simple abstraction wells due to buoyancy effects and upconing (Oude Essink 2001; Ward et al. 2009). This makes traditional well configurations vulnerable to salinization and thus application of MAR often inefficient. The same holds for exploitation of fresh groundwater lenses formed by natural recharge, which is difficult due to upconing of deeper saltwater (Fig. 2). The challenge is, therefore, to optimize the management of natural freshwater sources in the subsurface for drinking and irrigation water, thereby creating a valuable ecosystem service in coastal zones.

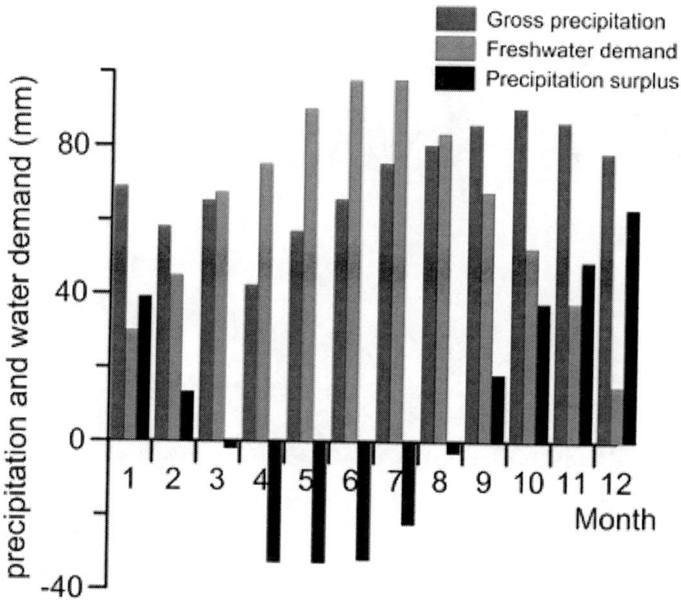

Figure. 1 Illustration of average freshwater availability and demand in a coastal area: mean gross monthly precipitation (1980–2010), estimated monthly water demand of an intensive greenhouse horticulture area (Greenport Westland-Oostland) in the Province of Zuid-Holland in The Netherlands (Paalman et al. 2012), and resulting freshwater surplus/deficit. Source: Zuurbier et al. (2013)

In the past decade, a set of practical tools and concepts that may have the ability to improve freshwater management in coastal-zone aquifers was proposed to solve irrigation and drinking water shortages in The Netherlands. The common feature of these subsurface water technologies (SWT) is to protect, enlarge, and utilize fresh groundwater resources in coastal zones through advanced groundwater management (Fig. 2). Sophisticated and dedicated new well designs, configurations, and management strategies were designed to obtain maximum control over the water resources, and go far beyond the levels of control provided by standard water management techniques. The aim of this study is to report and evaluate the efficiency of three SWT examples (summarized in Table 1) recently applied in the Netherlands, and explore their potential for solving typical freshwater management problems in other coastal zones worldwide. In a first step, the most relevant outcomes of three recently extensively studied SWT examples for freshwater management were presented. Secondly, common freshwater management problems in coastal areas were identified in scientific literature. Finally, the ability of SWT to counteract these typical coastal freshwater management problems was analysed and discussed.

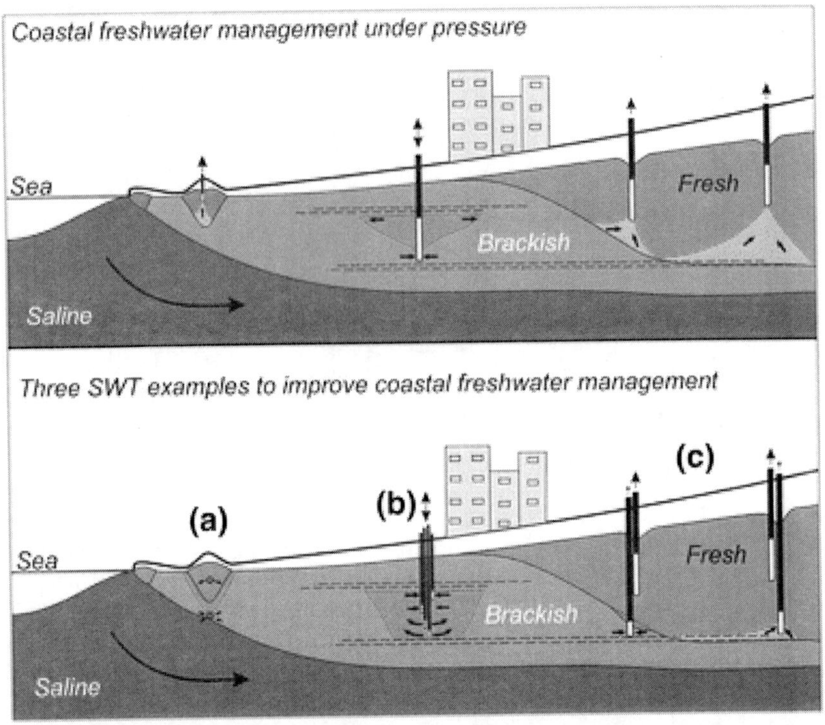

Figure. 2 Three examples of subsurface water technologies to overcome common freshwater issues in coastal aquifers, field-tested in the last 5 years. **a** = Freshmaker (horizontal wells to infiltrate in and recover from shallow freshwater lenses), **b** = ASR-coastal (deep injection, shallow recovery in brackish aquifers), **c** = Freshkeeper (interception of upconing brackish groundwater)

Table 1 Typical characteristics of the different subsurface water technologies (SWT) discussed

SWT	Aim	Target conditions	Artificial recharge	Saltwater interception	Well type	Water treatment
ASR-coastal	Temporal storage	Brackish aquifers	yes	no	MPPW	Pre-treatment
Freshkeeper	Protect wellfields	Stratified groundwater quality	no	yes	PP, MPPW	Optional post-treatment
Freshmaker	Temporal storage	Freshwater lenses	yes	yes	HDDW	Pre-treatment

MPPW multiple partially penetrating wells; *PP* partially penetrating well; *HDDW* horizontal directional drilled well

2 MATERIALS AND METHODS

2.1 Field-Testing of Subsurface Water Technologies (SWT)

2.1.1 ASR-Coastal

Aquifer storage and recovery (ASR) of freshwater using wells in coastal zones may not recover sufficient freshwater to meet the demands due to the mentioned losses by buoyancy effects. With ASR-coastal, multiple partially penetrating wells in a single borehole (MPPW, Fig. 3) are introduced, enabling injection at the base of the aquifer and recovery at the top. An ASR-coastal system with MPPW was successfully applied in a Dutch coastal greenhouse horticulture area (Nootdorp, Province of Zuid-Holland, 12 km from the North Sea coast) and reported by Zuurbier et al. (2014). Here, a brackish (chloride concentration: 150–1100 mg/l) aquifer at 13–41 m below sea level and confined by clay layers was targeted for ASR. The operation of the ASR system, the injected and recovered water quality, and the water quality changes in the aquifer were extensively monitored (Fig. 4) to provide the data to set up a calibrated, density-dependent groundwater transport model (SEAWAT; Langevin et al. 2007) and use this to compare the performance of this advanced configuration with a 'conventional' ASR configuration (Zuurbier et al. 2014). The ASR-system was monitored from January 2012 untill September 2013.

2.1.2 Freshkeeper

The Freshkeeper (Fig. 5) aims to safeguard the water supply from abstraction wells at risk of salinization. The concept follows a three-step approach: (1) intercept upconing brackish groundwater by abstracting freshwater from the top of the aquifer, while pumping intruding brackish water from the lower part of the aquifer; (2) use the intercepted brackish water as an additional water source by desalination through reverse osmosis; and (3) dispose of the RO membrane concentrate by deep-well injection into a confined, more saline aquifer. This 3-way approach was successfully applied in a pilot conducted at the Noardburgum well field (Province of Friesland, The Netherlands) by Vitens Water Supply (Oosterhof et al. 2013), at a well field that was abandoned in 1993 because of salinization. Shallow fresh and deeper brackish groundwater were extracted within a multiple partially penetrating well equipped with two separate well screens, at a rate of 50 m^3/h each. The focus of this pilot was on the management of the freshwater-brackish water interface and the injectivity of the RO-concentrate.

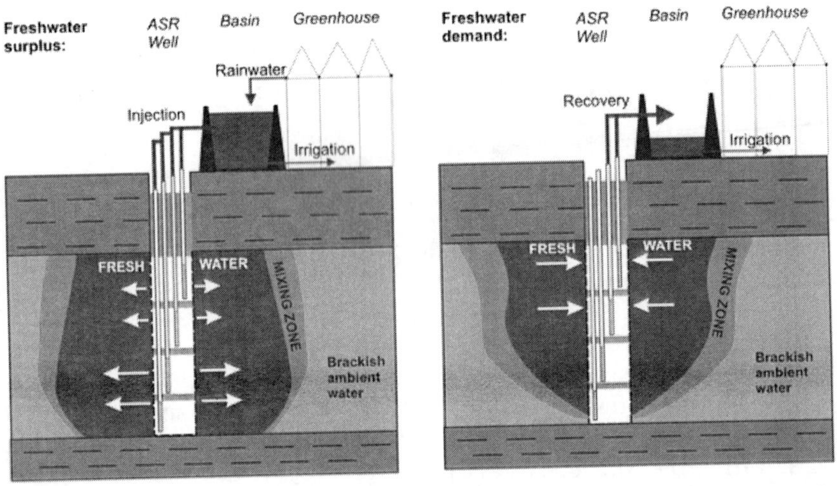

Figure. 3 Use of multiple partially penetrating wells (MPPW) for improvement of freshwater recovery of coastal ASR systems storing rainwater harvested from greenhouse roofs

2.1.3 Freshmaker

It was reasoned that ASR-coastal is inefficient when target aquifers are shallow, saline, and unconfined because of severe buoyancy effects and the risk of unwanted hydrological effects. The Freshmaker technology therefore combines the ASR and Freshkeeper concepts with the use of recently developed horizontal directional drilled wells (HDDWs; Cirkel et al. 2010). With this HDDW-technology, long horizontal wells can be installed at any desired depth in shallow coastal aquifers. The horizontal wells allow for the depth-controlled injection and recovery of large volumes of freshwater and was named the 'Freshmaker'. A shallow HDDW (the actual 'ASR well') is used for artificial recharge and recovery of freshwater surpluses, while buoyancy effects and upconing of saline water are prevented by the use of a deep interception HDDW (Fig. 6). The injected freshwater enlarges the natural freshwater lens along the HDDWs in periods with a freshwater surplus, and this stored water is available for recovery in periods of demand.

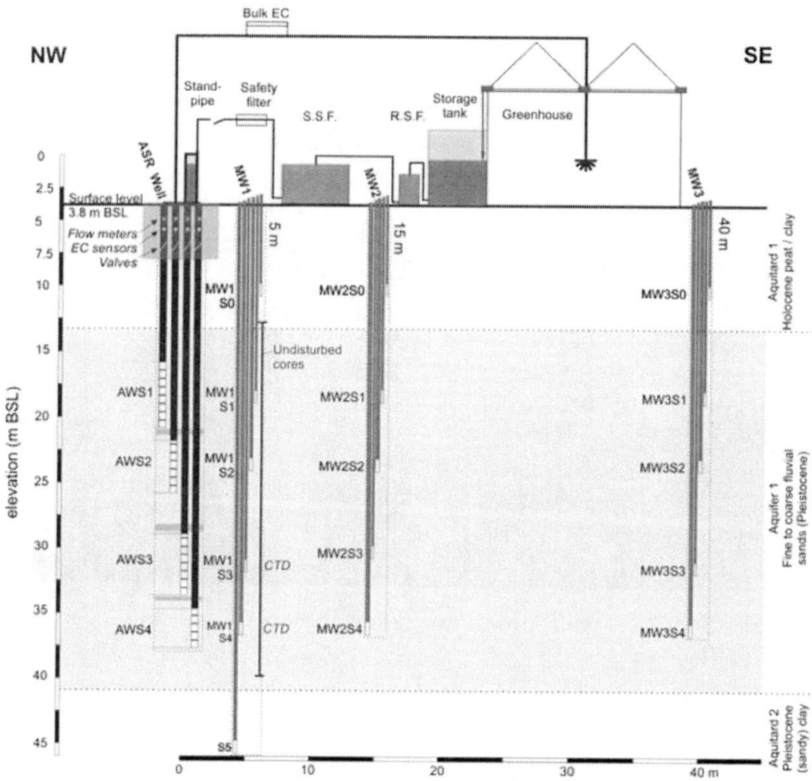

Figure. 4 Cross-section of the Nootdorp ASR system as presented in Zuurbier et al. (2014). Water from the greenhouse is first pre-treated by rapid sand filtration (R.S.F.) and slow sand filtration (S.S.F) and then injected mainly by deeper wells in the aquifer, whereas recovery occurs from the shallow wells

The first Freshmaker was installed in 2013 in the coastal Province of Zeeland to supply a local fruit grower with irrigation water in Ovezande.

Its efficiency was evaluated by 2-D groundwater transport modelling (using SEAWAT) and subsequently using geophysical field measurements (EM-39 borehole logging; McNeill et al. 1990), EC-sensors, and hydrochemical analyses during operation of the Ovezande field pilot. For more information on the modelling approach, the reader is referred to Zuurbier et al. (2015).

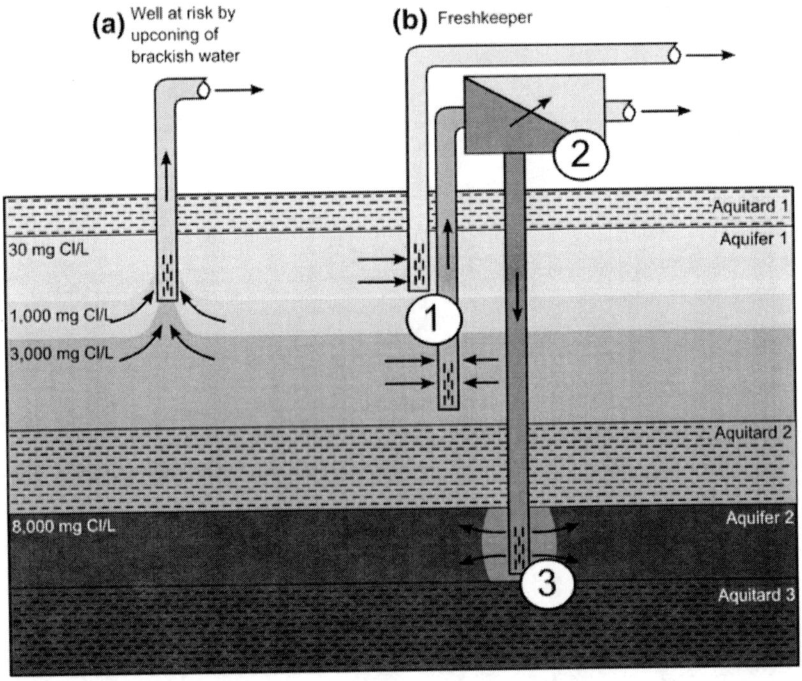

Figure. 5 Water well prone to salinization (**a**) and the Freshkeeper solution (**b**)

2.2 Broader Evaluation of the Efficiency of Subsurface Water Technologies (SWT) to Improve Freshwater Management

In order to explore the (broader) applicability of SWT for freshwater management in coastal zones, its efficiency to solve five common hydrogeological problems in these areas was assessed:

- *Brackish water upconing* (Reilly and Goodman 1987): resorting from shallow abstraction from a stratified aquifer (i.e., freshwater overlying brackish/saline water);

- *Seawater intrusion* (*SWI;* Werner et al. (2013): 'the landward incursion of seawater' via the subsurface;
- *Bubble drift during aquifer storage and recovery* (Ward et al. 2009): 'injected freshwater trying to "float" upwards through the aquifer while the denser native groundwater sinks down and inwards, contaminating the well at the bottom';
- *Thin target aquifer for abstraction / storage*: this may imply a low yield per well, requiring placement of expensive well galleries with many wells and pumps;
- *Saline seepage in deep polder areas* (de Louw et al. 2010): with ongoing land subsidence, sea-level rise, and occasionally (former) peat excavations, saline seepage is a increasing problem in delta areas as it causes salinization of inland surface waters.

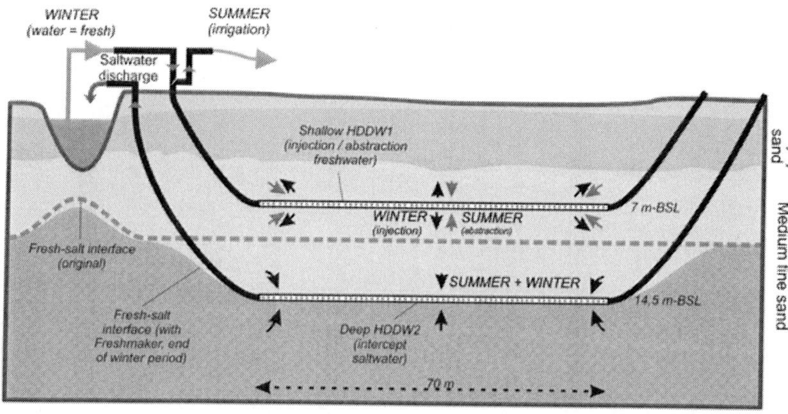

Figure. 6 Use of the Freshmaker principle to enlarge freshwater lenses. Side-view at the HDDWs. Set-up as installed at the Ovezande field-trial. m-BSL = meters below sea-level

3 RESULTS

3.1 Field-Testing of Subsurface Water Technologies (SWT)

3.1.1 ASR-Coastal

The Nootdorp field trial obtained satisfying results, with more than 40 % of the injected freshwater (13,700 m^3) recovered practically unmixed in the first cycle in 2012. A calibrated SEAWAT model was able to reproduce this performance, and predicted even better performance in subsequent cycles, with recovery efficiencies approaching almost 60 % (Fig. 7). This proved to be more than sufficient to guarantee freshwater availability for the greenhouse, even during long periods of drought. Alternatively, without the advanced MPPW set-up, less than 20 % was found recoverable by conventional ASR wells, according to the SEAWAT model (Zuurbier et al. 2014), which would have led to frequent freshwater shortages throughout the summer season. Compared to a situation without buoyancy effects, however, the recovery will always be lower as the formation of a stable, protecting mixing zone is absent in the lower half of the aquifer. The estimated cost price for water supplied by this ASR concept is 0.17 to 1.58 €/m³ (Zuurbier et al. 2012), depending on the scale and recovery efficiency achieved. This makes the concept competitive with less-sustainable, alternative water sources in the area (drinking water, desalinated water: ~1 €/m³).

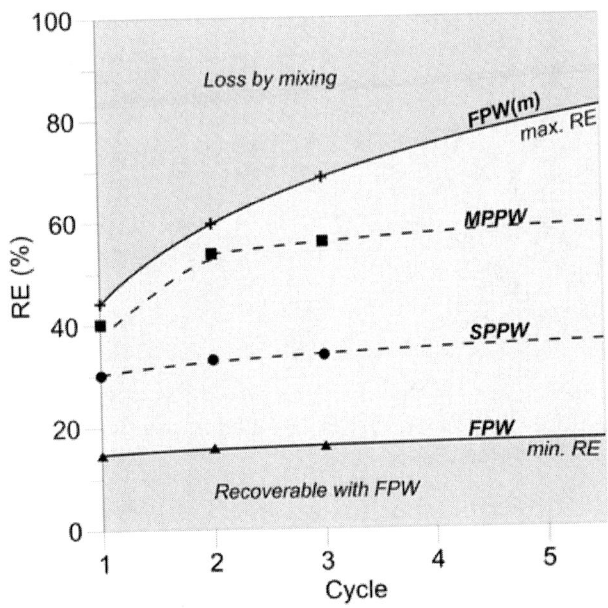

Figure. 7 Modelled RE per cycle versus cycle number for four scenarios. FPW (m) = scenario with only mixing and a fully penetrating ASR well, but no buoyancy. The other scenarios take into account mixing, seepage, and buoyancy for multiple partially penetrating wells (MPPW), a single partially penetrating well in the upper half of the aquifer (SPPW), and a fully penetrating well (FPW). Cycles 1-3 were modelled, cycle 4 and 5 were extrapolated from the modelled cycles. Data from: Zuurbier et al. (2014)

3.1.2 Freshkeeper

At the Noardburgum field site, the abstracted freshwater as well as the brackish abstraction water freshened upon dual zone Freshkeeper abstraction. Chloride concentrations of the abstracted fresh and brackish water decreased in the first months of the pilot, from 45 to 35 and 1000 to 600 mg/L, respectively. Freshening not only occurred in the near surrounding of the abstracting well screens, but also in observation wells at greater distance (Fig. 8). Only the observation well screen just above the confining clay layer did not show any freshening; chloride concentrations remained stable there due to the lateral inflow of brackish groundwater.

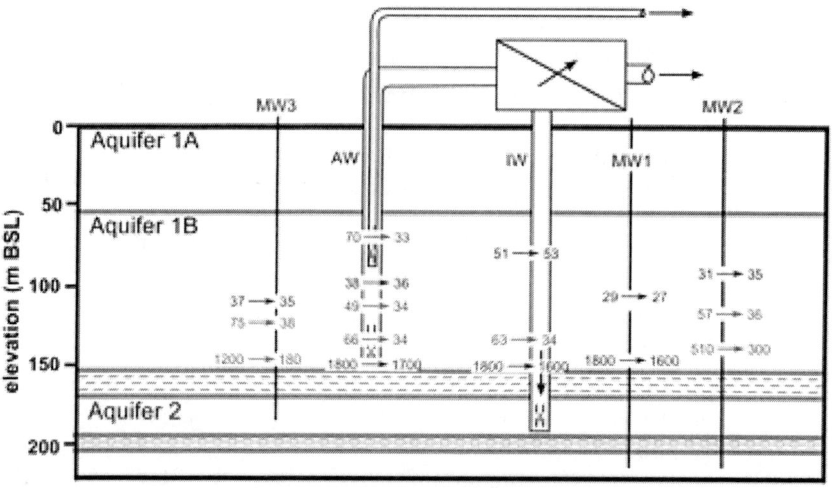

Figure. 8 Changes in chloride concentrations (in mg/l) in the source aquifer, after 8 months of simultaneous abstraction of fresh and brackish groundwater. In blue: freshening; in black: no change in chloride. AW = dual zone Freshkeeper abstraction well; IW = RO concentrate injection well; MW = monitoring well. m BSL = meters below sea-level

While the goal was to stabilize the fresh-brackish water interface, the chosen operation (fresh and brackish water abstraction rates of 50 m^3/h) even provoked downconing of brackish water, which was confirmed by SEAWAT modelling (Van der Valk 2011). In the same study, SEAWAT modelling suggested that the brackish water abstraction rate can be lowered to 16 m^3/h to keep chloride concentrations in the abstracted freshwater well constant at 45 mg/L.

A reverse osmosis recovery level of 70 % was attained, which was accompanied by a waste stream of 30 % containing the dissolved solids from the brackish groundwater and requiring re-injection. Although anti-scalants were not applied in the brackish water desalination process, clogging was not observed during injection of the RO concentrate in a deeper, unconsolidated, sandy aquifer. Production of calcite precipitates was not observed, which was explained by the high iron(II) levels in the feed water, inhibiting calcite precipitation (Wolthek et al. 2012). In a parallel concentrate injection experiment conducted at Zevenbergen, the Netherlands (Brabant Water Water Supply), calcite precipitation caused injection well clogging, which was later prevented by dosing CO_2 to the concentrate.

Following the successful pilot, Vitens Water Supply sees opportunities to apply the Freshkeeper at full-scale and to reopen the abandoned well field. A circular setup of six Freshkeeper wells appears to be the most effective in preventing salinization of the wells and of the central part of the well field (Van der Valk 2011). In this study, the fresh to brackish abstraction ratio was more than 3:1, i.e. for every 3 m^3 of fresh water less than 1 m^3 of brackish water needs to be abstracted. Using a RO recovery of 50 %, this would result in a yearly, sustainable production of 3 Mm^3 of drinking water (abstracted fresh water and RO permeate), while 0.43 Mm^3concentrate is to be disposed of by deep-well injection (Oosterhof et al. 2013).

3.1.3 Freshmaker

Model Results SEAWAT modelling results suggest that the modelled Freshmaker at Ovezande should be able to keep the targeted freshwater volume of 4200 m^3 available for abstraction during at least 6 months (Zuurbier et al. 2015). In the simulation, the deepest HDDW was actively intercepting saline groundwater to lower and stabilize the freshwater-saltwater interface. The modelled chloride concentrations in the abstracted water at the shallowest HDDW (HDDW1), used for injection and recovery of freshwater, indicated that injected surface water will be abstracted in spring. Abstracted water will be mainly native groundwater from the natural freshwater lens in late summers, which is mixed with upconing saltwater at the end of Cycle 1 (Fig. 9; Scenario D). In Cycle 2–5 this simulated upconing was limited and did not impose a risk for the salinity of the abstracted water. The simulation results show the upper part of the aquifer will gradually freshen, which will result in a decrease in saline seepage towards a local water course and eventually even local infiltration via the water course during freshwater recovery.

Significantly less freshwater was found to be attainable when only a shallow single HDDW is used (Scenario A), even when excessive infiltration was applied to form a 'buffer zone' of unrecovered infiltration water (Scenario B).

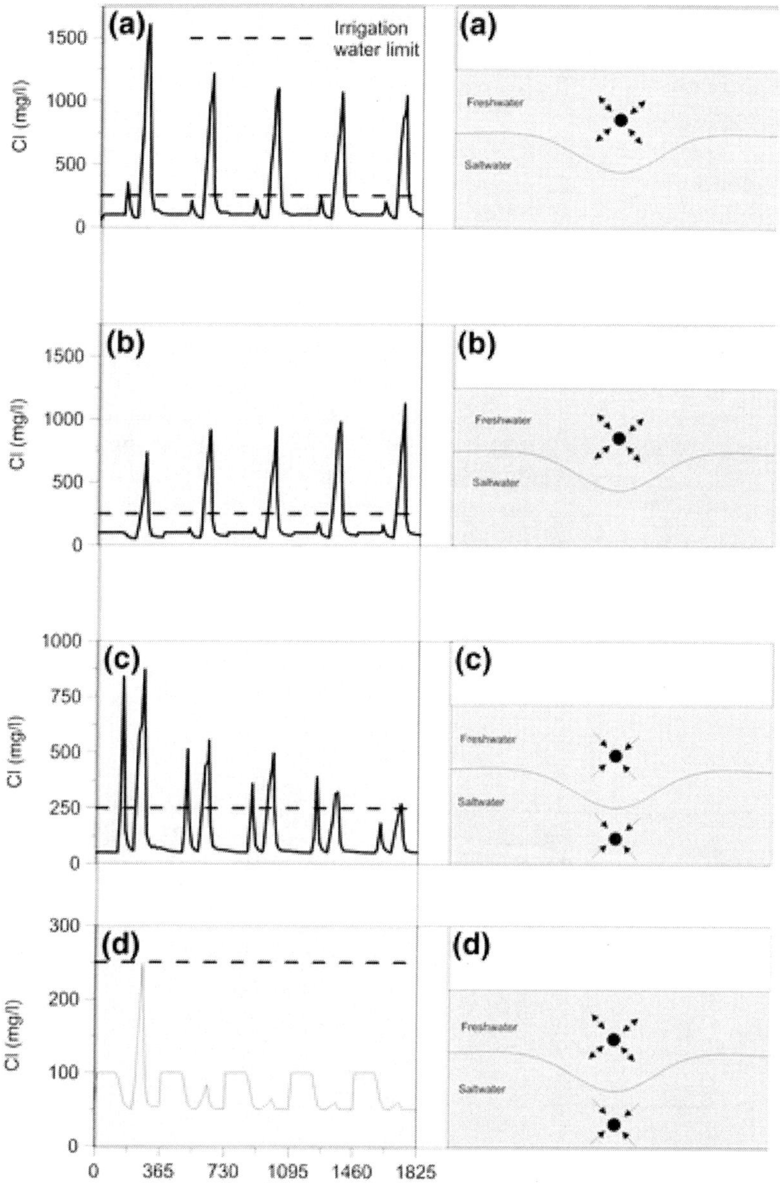

Figure. 9 Modelled chloride concentrations at the upper HDDW for upon yearly injection and abstraction of 4500 m^3 of freshwater without the interception of saline groundwater by a deep HDDW (**a**), same as (**a**), but after 1 year without recovery (**b**), with interception of deep saline groundwater but without injection (**c**), and with a complete Freshmaker (**d**). Source: Zuurbier et al. (2015)

The maximum chloride concentration for irrigation water would be exceeded after abstraction of a volume which was ~50 % of the injected volume. When a Freshmaker was installed, but no water was injected (like a Freshkeeper, Scenario C), a satisfying volume of freshwater could be abstracted from Cycle 5 onwards, due to the almost continuous interception of saltwater by the deep HDDW (HDDW2). This indicates that active injection of freshwater at this location (having a natural recharge of approximately 0.75 mm/d) is not a requirement for the abstraction of a same volume of freshwater, and that continuous interception of saltwater preceding freshwater abstraction can be sufficient.

3.1.4 Field Results

The Freshmaker Ovezande field pilot is being executed since 2013. In Cycle 1 (June – September 2013), 1700 m^3 was injected and a same freshwater volume was successfully abstracted. In Cycle 2 (November 2013 – September 2014) 4450 m^3 was injected, and 4.400 m^{3}of freshwater was abstracted by September 2014. The geophysical EM-39 measurements show that the Freshmaker indeed enlarged the freshwater lens during injection (3 to 4 m, Fig. 10), kept the freshwater at its place during storage (Cycle 1 and 2), and is able to recover a freshwater volume equal to the injected volume. The estimated cost-price for the water supplied by the Freshmaker is 0.35 €/m^3, which is less than the local alternative (piped water: 0.70 €/m^3).

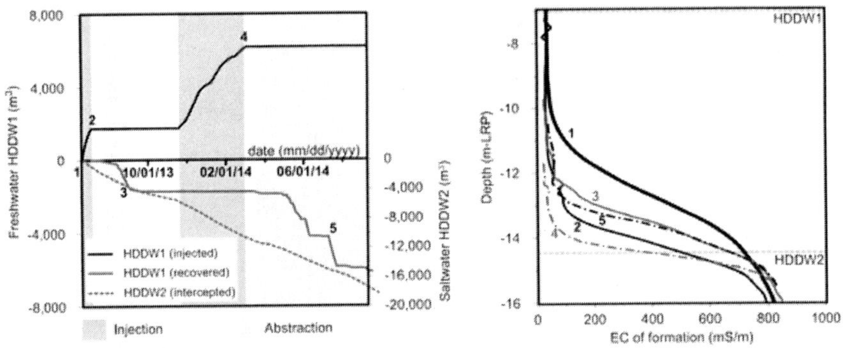

Figure. 10 Pumping by the Freshmaker (positive = injection, negative = abstraction). Changes in the electrical conductivity of the formation in the target aquifer measured by EM-39 demonstrating freshening (EC decrease) and salinization (EC increase) at the centre of the Freshmaker (halfway the HDDW well screens). Black arrows indicate the shift of the freshwater-saltwater interface over time. *m-LRP* = meters below local reference point, which is at approximately 0.4 m above sea-level. *Numbers* 1–5 indicate the moments at which EM-39 recordings were performed

3.2 Broader Evaluation of the Efficiency of Subsurface Water Technologies (SWT) to Improve Freshwater Management

Based on the outcomes and insights of the SWT field and modelling studies, the applicability of SWT to solve common hydrogeological problems and improve freshwater supply in coastal zones was evaluated. The outcomes are summarized in Table 2 and discussed below.

Table 2: Evaluation of the efficiency of SWT concepts to counteract common hydrogeological problems

Hydrological problem	ASR-coastal	Freshkeeper	Freshmaker
Brackish water upconing	+/-	+	+
Seawater intrusion	+/-	+	+
Bubble drift during ASR	+/-	+/-	+
Thin target aquifer for abstraction / storage	-	-	+
Saline seepage	+/-	+	+

+ : *Counteracting the hydrological problem*

+/-: *May counteract when boundary conditions are met*

-: *No positive effect, or even negative effect*

- *Brackish water upconing*: this process can be delayed by ASR-coastal, but full elimination cannot be guaranteed as upconing may still occur during recovery, especially when storage periods are long and deeper water is saline, which is similar to scenario A and B of the Freshmaker in Section 3.1.3. When brackish water is not desalinated or discharged from the groundwater system but only re-injected, upconing may also still threaten shallow abstraction wells in the Freshkeeper case, as recently demonstrated by Alam and Olsthoorn (2014). The Freshkeeper and Freshmaker in their presented form (injecting membrane concentrate in a deeper confined aquifer or discharging abstracted saltwater to sea) can sufficiently eliminate upconing, as demonstrated by field monitoring (Section 3.1.2) and model scenarios C and D (Section 4.1.3). For both the Freshkeeper and Freshmaker it is required that freshwater abstraction rates and saltwater interception rates are coupled: overabstraction of freshwater during limited interception of saltwater will result in upconing;
- *Seawater intrusion*: ASR-coastal may only prevent saltwater intrusion if the net injection exceeds the saltwater intrusion, making it a freshwater hydraulic barrier (e.g. Luyun et al. 2011; Mahesha 1996),

with deep injection at the optimal aquifer interval (Abarca et al. 2006). One may expect, however, that the Freshkeeper and the Freshmaker can even prevent saltwater intrusion, provided that the well placement and their abstraction rates are such that the entire intruding saltwater wedge is intercepted and disposed of or desalinated. This was confirmed using density-dependent transport modelling in combination with an optimization model (Abd-Elhamid and Javadi 2011), which indicated that coupled interception and abstraction (ADR: abstraction, desalination, recharge) can then be considered most (cost-) efficient;

- *Bubble drift during aquifer storage and recovery*: ASR-coastal can reduce the freshwater during ASR, but it was shown by Zuurbier et al. (2014) that it will not lead to 100 % recovery of injected freshwater. Like for conventional ASR: the more saline the aquifer is, the lower the recovery efficiency will be. A Freshkeeper may help to recover a larger part of injected freshwater during ASR, but will also not make the system render 100 % recovery of injected freshwater (Van Ginkel et al. 2014). The Freshmaker concept appeared able to recover a volume equal to the injected freshwater volume in a saline environment (without depleting the natural freshwater lens), given that an existing freshwater lens is enlarged and natural recharge is occurring;
- *Thin target aquifer for abstraction / storage*: ASR-coastal and the Freshkeeper will be hard to apply in thin aquifers. However, the use of HDDWs in the Freshmaker case may make thin aquifers viable for abstraction/storage, since a single, high-capacity well is feasible. Approriate design of the HDDW (length, diameter, pumping rates, amongst others) is however crucial to attain a uniform distribution of the abstraction along the HDDW screen (Sun and Zhan 2006; Wang et al. 2014). At the Freshmaker trial, this was ensured by the relatively limited length of the HDDWs (70 m) and confirmed by the observed lowering of the freshwater-saltwater interface along full length the wells;
- *Saline seepage in deep polders*: ASR-coastal may freshen the diffusive seepage component sourced by shallow groundwater in the upper aquifer, but is less effective in counteracting seepage of deeper, saline groundwater via boils, which can be the largest salt contributor in polder areas (de Louw et al. 2010). The Freshkeeper concept was previously suggested as a suitable technique to counteract saline seepage (Olsthoorn 2008; Stuyfzand and Raat 2010), although it was considered unviable when all abstracted water is directly re-injected in deeper aquifers (De Louw et al. 2007) because of hydrological effects in the surrounding areas and the required high pumping rates. This underlines that disposal or concentration of the abstracted

brackish water is desirable. The Freshmaker can decrease the saline seepage in polder areas based on the modelling performed for the Ovezande field pilot (Section 3.1.3). The current set-up does not contribute to a reduction in salt load to the local surface water system, as the intercepted saltwater is disposed of at a local water course here,. Disposal of intercepted brackish-saline groundwater and membrane concentrate is therefore expected to be a key element in coastal freshwater management.

4 DISCUSSION: CURRENT STATE OF SUBSURFACE WATER TECHNOLOGIES (SWT)

4.1 The Efficiency and Economics of the Demonstrated SWT

The presented subsurface water technologies (SWT) highlight a recent trend in hydrological engineering driven by new drilling techniques, water treatment, and automation of water supply facilities using sophisticated programming and sensors. The SWT-examples (ASR-coastal, Freshkeeper, and Freshmaker) show that despite the increasing complexity, these technologies can realize a significant increase in freshwater availability in coastal areas for a competitive cost price.

SWT will not fully overcome all the mentioned hydrological problems in all coastal zones, but can generally improve freshwater production, or as found in this study: a reduction of freshwater losses during storage of several tens of percents or the complete prevention of freshwater upconing. This has economical relevance. For instance, the greenhouse owner's water demand at the Nootdorp ASR-coastal field site requires an average recovery efficiency of approximately 40 % of the injected water. It was demonstrated that this was not feasible with a conventional ASR well (<20 %) or a shallow partially penetrating well (<35 %). The use of a MPPW at this site (with only minor additional costs for PVC pipelines, standpipes, and valves) boosted the freshwater recovery up to more than the owner's demand (55 %). Instead of investing in more expensive and less sustainable freshwater sources (in this case: desalinated or piped water), there is now a valuable freshwater surplus that can be sold to neighbouring companies with a higher demand.

In the case of the Freshkeeper (Noardburgum) an entire well field was closed and replaced following salinization in 1993. SEAWAT modelling scenarios suggest that installation of only six Freshkeeper wells in a circular set-up is sufficient to prevent salinization of the entire well-field in future (Oosterhof et al. 2013; Van der Valk 2011). Since Vitens Water Supply was looking for additional drinking water in this region, this is a cost-reducing outcome. A recent study has shown the potentials of Freshkeeper to abate salinization problems in Florida (USA), and to guarantee the long-term drinking water supply there (Ross et al. 2014). In the Florida case, a Freshkeeper was found economically much more feasible than alternative water supply options such as full-scale brackish water reverse osmosis. The exact economic benefits of SWT for other cases may vary and likewise for normal MAR-techniques, they are often hard to assess a priori due to feasibility uncertainties and the chance of under-performance (Arshad et al. 2014; Maliva 2014). However, the SWT ability to counteract reductions in freshwater production resulting from unsuitable aquifer conditions will mitigate the increase of operational expenditures, potentially compensating for higher capital expenditures.

4.2 Other SWT Examples

SWT are not limited to the field test examples presented in this paper. For instance, Van Ginkel et al. (2014) proposed an elegant concept to store freshwater in an Egyptian saline aquifer by combining freshwater storage with saltwater abstraction from below the injected freshwater, which has similarities with and can further improve the ASR-coastal concept. Alam and Olsthoorn (2014) proposed to discharge a part of the intercepted brackish water by deep Punjab scavenger wells to achieve a net freshening effect (comparable to elements of both the Freshkeeper and the Freshmaker). In 2013, the Baton Rouge Water Co. (U.S.A.) has installed a brackish water scavenger well that, similar to the Noardburgum Freshkeeper, should prevent brackish water upconing to the overlying freshwater production wells. The pumped brackish water is disposed of to the Mississippi river (Tsai 2011). Olsthoorn (2008) and Stuyfzand and Raat (2010) proposed a Freshkeeper at a polder scale, using the abstracted brackish water for drinking water production and simultaneously solving various environmental problems caused by upward seepage of nutrient-rich brackish groundwater at the same time. However, no Freshkeeper is currently operating for this purpose.

4.3 Wider Scope of Application

The SWT development and studies mentioned above suggest that although the field-tested SWT are all situated in the Netherlands, they potentially have a much wider scope of application. This is underlined by the evaluation of the SWT in this study, which shows SWT can be used to reduce or overcome very common hydrological problems in coastal zones, which are amplified by an expected exacerbation of saltwater intrusion in coastal zones by sea-level rise and changes in both recharge and evaporation due to global climate change (Oude Essink et al. 2010), which will require a more enhanced management of coastal aquifers (Werner et al. 2013). SWT fulfills the demand for more advanced management tools to deal with coastal groundwater salinization and the demand for increased freshwater storage.

Elements of the SWT discussed in this paper may also be combined. For instance, a Freshkeeper was recently added to a new field ASR-coastal system to protect shallow recovery wells and produced additional freshwater via RO-treatment. In this field pilot, clogging of the RO-membranes is monitored with large interest, since these receive a feedwater, which is a mixture of infiltrated fresh, oxic rainwater with saline, anoxic groundwater. In general, abstracted water quality is a relevant aspect when RO-treatment is involved in SWT since the chemical and physical (suspended fines, temperature) quality of water used for RO, which is abstracted close to the freshwater–saltwater transitions may vary significantly over time due to freshening, salinization, and changes in redox conditions, especially upon artificial infiltration of fresh, oxic water. Membrane selection and prevention of membrane clogging are, therefore, critical aspects when desalination via RO is incorporated in the selected SWT.

It should be noted that all current SWT examples are being tested in sandy aquifers in The Netherlands, which are dominated by intergranular flow. However, limestone aquifers are also frequently found in coastal zones, and are targeted for freshwater supply worldwide. Transport processes may differ significantly in such aquifers due to dual-porosity (Bibby 1981). This can lead to underperforming ASR-systems due to early salinization via preferential flow paths (e.g., Maliva and Missimer 2010; Missimer et al. 2002; Pyne 2005). In the same way, this may reduce the effectiveness of SWT, since flow patterns are less predictable and preferential flow paths may hamper for instance the interception of brackish-saline water by the Freshkeeper and the Freshmaker.

Disposal of concentrate produced upon desalination as applied in the Freshkeeper example may be another obstacle, as this is often not allowed on surface waters or sewage systems. Re-injection in deeper aquifers on the other hand may induce (local) groundwater salinization and is therefore under discussion. Important prerequisites for this disposal are often the salinity of the receiving aquifer and the required separation of abstraction and injection well screens by aquitards because local salinization and short-circuiting must be prevented. Since desalination in combination with deep disposal of concentrate does not directly add an additional salt mass to the groundwater system, it may be more relevant to evaluate the regional consequences of this net abstraction of H_2O from the groundwater body. A key question is then if this net abstraction is compensated by either intrusion of more saline groundwater (negative) or by recharge of freshwater (naturally or artificially). The latter is often the case when inland brackish or saline groundwater originates from former transgressions in coastal zones that are currently recharged by freshwater, while seawater intrusion is generally limited to areas close to the shore.

4.4 The Future of SWT

SWT provide a coupled solution of a natural ecosystem service with a technological approach that allows for an enhanced protection and utilization of the freshwater resources in coastal areas. The SWT described in this paper have all been developed within public-private partnerships of innovators in the water market. SWT are gaining more-and-more interest from early adopters in the Netherlands. Following the pilot described in this paper, authorities in western Netherlands now consider ASR-Coastal as an important tool serving their regional water governance, and are stimulating greenhouse owners to increase their water self-supportiveness by applying this technique. Recently, a group of farmers in southwest Netherlands have inquired for a Freshmaker feasibility study to improve the irrigation water supply in their orchards. Vitens Water Supply in the North of the country just started a follow-up Freshkeeper pilot study that should be the final step towards full-scale application in the near future.

Despite the growing interest for SWT, further uptake inside and outside The Netherlands is slowed down by a number of non-technical barriers, including a lack of: (1) demonstration of long-term viability, (2) an analysis of their hydrological effects in their surroundings, (3) knowledge of new technologies and the ability to construct and operate them, (4) capabilities upon making investment decisions, and (5) inherent conservatism due to a lacking *local* track-record of successful implementation of SWT. As a consequence, more expensive and potentially unsustainable but proven technologies are chosen for freshwater management, such as seawater desalination or restrictions on water delivery. We plea for prolonged SWT testing in the current pilots, replication of SWT pilots in other areas worldwide, and the development of technical and non-technical support tools that can facilitate potential end-users in investment decision making and SWT implementation. Such an approach will accelerate acceptance and implementation of subsurface water technologies as robust answers to freshwater resources challenges in coastal areas.

5 CONCLUSIONS

Balancing freshwater availability and demand is a major challenge in especially coastal areas. Subsurface water technologies (SWT) have transformed from idea to proven-technology in the past decade to better manage subsurface freshwater volumes. Both groundwater transport modelling and extensive field operation and monitoring of three SWT examples (ASR-coastal, Freshkeeper, Freshmaker) underline that SWT can be used to protect, enlarge, and utilize fresh groundwater resources in coastal zones for use in times of demand. For this reason, coastal aquifers that have been considered to be or have become unsuitable for freshwater supply or storage have thus become 'instruments' for coastal freshwater management. Local natural freshwater sources such as rainwater can be utilized this way without claiming large areas aboveground and reduce the need for other less sustainable sources of freshwater. SWT can also combine innovations in drilling techniques, information and communications technology (ICT) and online sensoring, and water treatment to counteract very common coastal hydrological problems like saltwater upconing, seawater intrusion, and ASR bubble drift. SWT is not necessarily a freshwater management panacea for every hydrogeological setting, but the required (increase of) water supply may become technically and economically feasibile by SWT. Prolonged SWT testing in the current pilots, replication of SWT in other areas worldwide, and the development of technical and non-technical support tools are required to facilitate potential end-users in investment decision making and SWT implementation. Such an approach will accelerate acceptance of subsurface water technologies as robust answers to freshwater management challenges in coastal zones.

ACKNOWLEDGMENTS

The authors would like to thank the funding agents of the studies discussed in this paper: Knowledge for Climate (ASR-coastal, Freshmaker), Vitens Water Supply (Freshkeeeper), GO-Fresh (Freshmaker), the Joint Water Research Program of the Dutch Water Supply Companies (BTO; all pilots), the EU FP7 project 'Demonstrate Ecosystem Services Enabling Innovation in the Water Sector' (DESSIN, grant agreement no. 619039) and the EU Horizon2020-project 'SUBSOL' (grant agreement no. 642228). Three anonymous reviewers are thanked for their valuable contributions to the earlier version of this manuscript.

REFERENCES

1. Abarca E et al (2006) Optimal design of measures to correct seawater intrusion. Water Resour Res 42(9), W09415
2. Abd-Elhamid H, Javadi A (2011) A cost-effective method to control seawater intrusion in coastal aquifers. Water Resour Manag 25(11):2755–2780
3. Alam N, Olsthoorn TN (2014) Punjab scavenger wells for sustainable additional groundwater irrigation. Agric Water Manag 138:55–67
4. Arshad M, Guillaume J, Ross A (2014) Assessing the feasibility of managed aquifer recharge for irrigation under uncertainty. Water 6(9):2748–2769
5. Bibby R (1981) Mass transport of solutes in dual-porosity media. Water Resour Res 17(4):1075–1081
6. Cirkel DG, Van der Wens P, Rothuizen RD, Kooiman JW, (2010) Water extraction with HDD drillings - One horizontal well for multiple vertical wells. Land+Water, access via www.hddw.nl
7. De Louw PGB, Oude Essink GHP, Maljaars P (2007) Background study: seepage reduction techniques (in Dutch), TNO
8. de Louw PGB, Oude Essink GHP, Stuyfzand PJ, van der Zee SEATM (2010) Upward groundwater flow in boils as the dominant mechanism of salinization in deep polders, The Netherlands. J Hydrol 394(3–4):494–506
9. Dillon P et al (2010) Managed aquifer recharge: rediscovering nature as a leading edge technology. Water Sci Technol 62(10):2338–2345
10. European Commission (2012) A blueprint to safeguard Europe's water resources. Brussel
11. Langevin CD, Thorne DT, Dausman AM, Sukop MC, Guo W (2007) SEAWAT version 4: a computer program for simulation of multi-species solute and heat transport. In: U.S.G.S. (Editor) Techniques and methods, book 6. Reston, Virginia, USA
12. Luyun R, Momii K, Nakagawa K (2011) Effects of recharge wells and flow barriers on seawater intrusion. Ground Water 49(2):239–249
13. Mahesha A (1996) Steady-state effect of freshwater injection on seawater intrusion. J Irrig Drain Eng 122(3):149–154
14. Maliva R (2014) Economics of managed aquifer recharge. Water 6(5):1257–1279
15. Maliva RG, Missimer TM (2010) Aquifer storage and recovery and managed aquifer recharge using wells; planning, hydrogeology, design and operation. Methods in Water Resources Evolution. Schlumberger, Texas, USA, 578 pp
16. McNeill JD, Bosnar M, Snelgrove JB (1990) Technical Note 25: resolution of an electromagnetic Borehole logger for Geotechnical and groundwater applications
17. Missimer TM, Guo W, Walker CW, Maliva RG (2002) Hydraulic and density considerations in the design of aquifer storage and recovery systems. Fla Water Resour J 55(2):30–36
18. Olsthoorn TN (2008) Brackish groundwater as a new resource for drinking water, Specific consequences of density dependent flow, and positive environmental consequences 20th Saltwater Intrusion Meeting (SWIM), Naples, Florida, USA
19. Oosterhof A, Raat KJ, Wolthek N (2013) Reuse of salinized well fields for the production of drinking water by interception and desalination of brackish groundwater, 9th Conference on water reuse. International Water Association (IWA), Windhoek
20. Oude Essink GHP (2001) Improving fresh groundwater supply--problems and solutions. Ocean Coast Manag 44(5–6):429–449
21. Oude Essink GHP, van Baaren ES, de Louw PGB (2010) Effects of climate change on coastal groundwater systems: a modeling study in the Netherlands. Water Resour Res 46, W00F04
22. Paalman M et al (2012). Enlarging selfsufficient horticultural freshwater supply: greenhouse areas Haaglanden (part 1) (in Dutch). KvK105/2013A, Knowledge for Climate, Utrecht, The Netherlands
23. Pyne RDG (2005) Aquifer storage recovery—a guide to groundwater recharge through wells. ASR Systems LLC, Gainesville, 608 pp

24. Reilly TE, Goodman AS (1987) Analysis of saltwater upconing beneath a pumping well. J Hydrol 89(3–4):169–204
25. Ross PS, Raat KJ, Smith DK, Zaadnoordijk WJ (2014) Business case Freshkeeper town of Belleair, ARCADIS / KWR. Rotterdam / Nieuwegein, The Netherlands
26. Stuyfzand P, Raat K (2010) Benefits and hurdles of using brackish groundwater as a drinking water source in the Netherlands. Hydrogeol J 18(1):117–130
27. Sun D, Zhan H (2006) Flow to a horizontal well in an aquitard–aquifer system. J Hydrol 321(1–4):364–376
28. Tsai FTC (2011) Scavenger wells stop saltwater intrusion in Baton Rouge, Louisiana, MODFLOW and More 2011: Integrated Hydrologic Modeling, Golden, Co. (USA), pp. 524–528
29. United Nations (2010) UN atlas of the oceans. In: United Nations (ed) UN atlas of the oceans. United Nations, Rome
30. Van der Valk M (2011) A fresh-keeper for Noard Burgum (Masters Thesis), TU Delft, Delft
31. Van Ginkel M, Olsthoorn TN, Bakker M (2014) A new operational paradigm for small-scale ASR in saline aquifers. Ground Water 52(5):685–693
32. Wang J-Z, Jiang X-W, Wan L, Wang X-S, Li H (2014) An analytical study on groundwater flow in drainage basins with horizontal wells. Hydrogeol J 1–14
33. Ward JD, Simmons CT, Dillon PJ, Pavelic P (2009) Integrated assessment of lateral flow, density effects and dispersion in aquifer storage and recovery. J Hydrol 370(1–4):83–99
34. Werner AD et al (2013) Seawater intrusion processes, investigation and management: recent advances and future challenges. Adv Water Resour 51:3–26
35. Wolthek N, Raat K, de Ruijter JA, Kemperman A, Oosterhof A (2012) Desalination of brackish groundwater and concentrate disposal by deep well injection. Desalin Water Treat 51(4–6):1131–1136
36. World Economic Forum (2015) Global risks 2015. World Economic Forum, Davos
37. Zuurbier K, Bakker M, Zaadnoordijk W, Stuyfzand P (2013) Identification of potential sites for aquifer storage and recovery (ASR) in coastal areas using ASR performance estimation methods. Hydrogeol J 21(6):1373–1383
38. Zuurbier KG, Paalman M, Zwinkels E (2012) Haalbaarheid ondergrondse waterberging glastuinbouw westland. KWR 2012.003. KWR Watercycle Research Institute, Nieuwegein
39. Zuurbier KG, Zaadnoordijk WJ, Stuyfzand PJ (2014) How multiple partially penetrating wells improve the freshwater recovery of coastal aquifer storage and recovery (ASR) systems: a field and modeling study. J Hydrol 509:430–441
40. Zuurbier KG, Kooiman JW, Groen MMA, Maas B, Stuyfzand PJ (2015) Enabling successful aquifer storage and recovery of freshwater using horizontal directional drilled wells in coastal aquifers. J Hydrol Eng 20(3), B4014003

CHAPTER 18

Assessment of the Sustainability of Water Resources Management: A Critical Review of the City Blueprint Approach

Steven H. A. Koop1,2 & Cornelis J. van Leeuwen1,2

1 KWRWatercycle Research Institute, Groningenhaven 7, Nieuwegein 3433 PE, The Netherlands

2 Copernicus Institute for Sustainable Development and Innovation, Utrecht University, Heidelberglaan 2, Utrecht 3584 CS, The Netherlands

ABSTRACT

Climate change, urbanization and water pollution cause adverse effects and rehabilitation costs that may exceed the carrying capacity of cities. Currently, there is no internationally standardized indicator framework for urban Integrated Water Resources Management (IWRM). The City Blueprint® is a first attempt and aims to enhance the transition towards water-wise cities by city-to-city learning. This paper provides a three step revision of the City Blueprint Framework (CBF) based on data of 45 municipalities and regions in 27 countries: (1) A distinction has been made between *trends and pressures* (on which urban IWRM has a negligible influence) and *IWRM performances*. Therefore, a separate trends and pressures framework has been developed; (2) Only the purely performance-oriented indicators have been selected from the CBF. Furthermore, the indicator accuracy and boundaries have been re-assessed, and new indicators have been added; (3) By analyzing correlations and variances, the performance-oriented indicators have been rearranged in order to establish a proportional contribution of all indicators and categories to the overall score, i.e., the Blue City Index®. In conclusion, six indicators have been removed because of insufficient accuracy, overlap or lack of focus on IWRM. Seven indicators have been added, i.e., secondary and tertiary wastewater treatment, operation cost recovery, green space and three indicators concerning solid waste treatment. The geometric aggregation method has been selected because it emphasizes the need to improve the lowest scoring indicators. In conclusion, the improved CBF is more performance-oriented and therefore more suitable to assist cities in their transition towards water-wise cities.

Keywords: Urban water management, Climate change, Sustainability, indicators, Blue City, Index®, EIP Water

1 URBAN WATER CHALLENGES

Rapid urbanization and climate change pose increasing pressures on Integrated Water Resources Management (IWRM), especially in cities (OECD 2015). In 2014, about 4 billion people lived in cities, mostly situated along coasts and major rivers (accounting for 54 % of the world's population). IWRM becomes even more challenging as the global urban population is estimated to increase with 2.5 billion people by 2050 (UN 2014).

Climate change amplifies urban water vulnerabilities such as flooding, heat stress, water scarcity and water pollution. Sea level rise and increased river discharges pose a projected 15 % of the global population at risk of flooding. This is mainly in cities including almost all worlds' megacities (Ligtvoet et al. 2014). Also, extreme rainfall and heat waves will become more severe due to global warming (Jongman et al. 2014). As urban surfaces are often largely sealed and lack green areas, the impact of drainage flooding and Urban Heat Islands (UHI) are even more exacerbated (Shuster et al. 2005; Gill et al. 2007). Water withdrawals are estimated to increase by 50 % in developing countries and by 18 % in developed countries by 2025 (WWDR 2006). This increased water demand will lead to an estimated 40 % fresh water shortage by 2030 (2030 WRG 2009). Climate change will exacerbate these fresh water shortages (Iglesias et al. 2007) and amplify the spread of water-borne diseases (IPCC

2013). Pollution from combined sewer overflows and stormwater runoff will rise due to climate induced increase in extreme rainfall events (Nilsen et al. 2011). Rapid urbanization poses extra stresses, for example, wastewater treatment in Asia and Africa is already sparse, while nutrient emissions are projected to double or triple within 40 years. This will strongly enhance eutrophication, biodiversity loss, threaten drinking water, fisheries, aquaculture, and tourism (Ligtvoet et al. 2014). Moreover, cities produce massive amounts of solid waste, in particular plastics. Worldwide 280 million tons of plastics are produced annually (Sigler 2014). These plastics photodegrade into small particles that affect marine ecosystems (Derraik 2002).

The prospect of increased urban flooding, heat stress, water scarcity and pollution emphasizes the need for adaptive and reliable urban water infrastructures (Short et al. 2012). However, water infrastructures are often old and require refurbishment to meet current standards, whereas standards to withstand future conditions of increased storm events and urbanization are often not accounted for (OECD 2015). An estimated US$ 41 trillion (41×10^{12}) is needed to refurbish the urban infrastructure in the period 2005–2030. Over 50 % will be needed to refurbish the water systems (UNEP 2013). This is roughly 60 % more than is spent on infrastructure in the same period until now (McKinsey 2013). In developed countries water infrastructure investments amount to 1 % of the GDP every year. For developing countries this is even more substantial, i.e., about 3.5 % with extremes up to 6 % or more (Cashman and Ashley 2008). The costs of climate change related damages are expected to be large. In Europe this is estimated to be € 190 billion by 2080 (JRC 2014), especially flood damage is predicted to increase five-fold by 2050 (Jongman et al. 2014). Recent extreme weather events resulted in much damage. For example, in 2011 in Copenhagen a storm event caused a lot of damage of nearly € 1 billion (Leonardsen 2012). In the USA, hurricane Sandy (2012) caused 117 deaths and costs US$ 19 billion including US$ 2.6 billion to repair New York's drinking and wastewater infrastructure (Johnson 2013). These alarming events demonstrate the climate vulnerability of cities and the urgency to improve urban IWRM.

2 THE CITY BLUEPRINT

Approximately 80 % of the world's GDP is produced, and 75 % of the global energy and material flows are consumed in cities (UNEP 2013). This makes cities the major contributors of economic growth but also the major sources of environmental pressures. It also implies that cities have the highest potential to reduce these pressures. Urban water management is often locked-in to the large-scale, centralized infrastructure approaches limiting the adoption of more flexible and resilient technologies and approaches such as fit-for-purpose water use, nutrient and energy recovery from wastewater, and blue-green infrastructures (Brown et al. 2011). Hence, the main objective of the City Blueprint® action is to create awareness among decision makers and resource managers. It may help them envisioning, developing and implementing stepwise measures to transform towards water-wise or water sensitive cities (EC 2015). Improving the implementation capacities of cities by sharing information can be described as city–to-city learning. The first step in the strategic planning process is that stakeholders are identified and information is provided for a baseline assessment. Hereafter, long-term goals and priorities are set resulting in follow-up actions leading to measures that promote sustainable IWRM (Philip et al. 2011; Van Leeuwen and Sjerps 2015). The City Blueprint Framework (CBF) aims to be the first step in strategic planning and consists of 24 indicators divided over eight broad categories, i.e., 1) Water security, 2) Water quality, 3) Drinking water, 4) Sanitation, 5) Infrastructure, 6) Climate robustness, 7) Biodiversity and attractiveness, and 8) Governance. The output is a spider diagram and a Blue City Index® (BCI). The BCI is the arithmetic mean of the 24 indicators. The methodology and its application have been published in this journal (Van Leeuwen et al. 2012; Van Leeuwen 2013) and a detailed description of the calculation methods is provided in the City Blueprint questionnaire available on the website of the European Innovation Partnership on Water (EC 2015).

This paper provides a critical revision of the CBF, based on the learning experiences obtained during the assessments in 45 cities in 27 countries. The aim of this revision is to:

- Provide a clear separation between indicators describing urban *trends and pressures* (on which local water managers have a negligible influence, e.g., current climatic conditions, demographic chances etc.) and indicators measuring *IWRM performance* in cities. The improved performance framework (CBF) will be more adequate in showing the potential for improvements and enhances city-to-city learning because emphasizes the city's own IWRM performance.

- Develop a separate framework describing the most important trends and pressures that may limit IWRM or, on the contrary, poses opportunity windows. This supplementary framework is key in providing context and may assist in the setting of priorities.

- Include solid waste indicators as important urban performers to decrease water pollution.

- Update existing indicators by including new developments in data accuracy and availability.
- Design a coherent framework in which indicators and categories make a balanced contribution to the overall score, i.e., the BCI.
- Select an aggregation method that penalizes unbalanced indicator scores in order to express the urgency to improve the lowest scoring indicators.
- Make sure that the indicator results are easy to understand, timely and relevant, and useful for the end-users, i.e., policy makers, decision makers, water managers and citizens in general.

3 METHODOLOGY

The application of the CBF followed a learning by doing approach. Based on constructive feedback from stakeholders, we have performed a detailed revision of the indicator choice and framework, together with a revision of the indicator scaling and aggregation method (Koop and Van Leeuwen 2015b). The process is summarized in Fig. 1 and is based on data of 45 municipalities and regions, mainly in Europe (Koop and Van Leeuwen 2015a). In the rest of this paper these municipalities, regions and Malta will be referred to as cities. This review consists of three consecutive steps:

1) Step 1 Development of the Trends and Pressure Framework (TPF).
2) Step 2 Improvements of the City Blueprint indicators.
3) Step 3 Indicator rearrangements and aggregation methods.

The application of the revised CBF and new TPF, the clustering and categorization of cities' IWRM performances, as well as comparisons with other city descriptors, have been published separately in this journal (Koop and Van Leeuwen 2015a).

3.1 Step 1: Development of the Trends and Pressures Framework

Every city has its own social, financial and environmental setting in which water managers have to operate. In order to promote city-to-city learning, it is essential to solely measure urban water management performances. For example, a city situated in an arid area may not necessarily experience water stress due to overconsumption, but simply due to the low natural availability of fresh water. In this case, water consumption or the use of water saving techniques are performance indicators, whereas the natural availability of fresh water is a descriptive indicator belonging to the TPF. A more performance-oriented set of indicators (CBF) is more adequate in showing the potential for improvements and sharing of knowledge, experiences and best practices between cities.

A separate TPF may provide the context to obtain insight in the limitations and windows of opportunities for urban IWRM. Therefore, the first step in this review was the development of the TPF with the aim to provide an overview of the most important social, environmental and financial characteristics affecting urban IWRM.

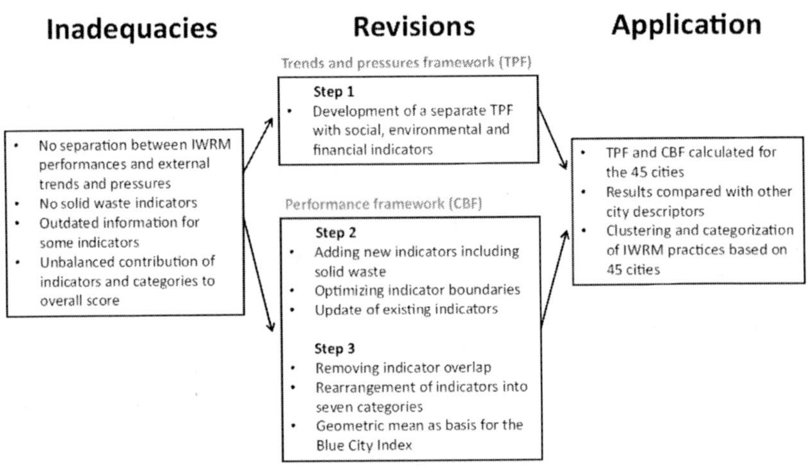

Figure. 1 Flow chart of the City Blueprint revision described in this paper using data of 45 cities divided over 27 countries. The applications of the Trends and Pressure Framework (TPF) and Performance framework (CBF) have been published in this journal (Koop and Van Leeuwen 2015a)

3.2 Step 2: Improvements of the City Blueprint Indicators

All indicators of the CBF have been critically assessed on data reliability, scoring method and whether the used data are time-series in order to ensure that the indicators are up-to-date. At the same time new indicators are proposed to replace indicators with data problems.

Also boundary issues are reviewed in order to adjust extreme indicator variances. Alternative scaling methods for a few indicators with sufficiently large data sets (n > 100) are proposed as well. Only performance-oriented indicators have been selected in this step.

3.3 Step 3: Indicator Rearrangements and Aggregation Method

Finally, the revised performance-oriented indicators have been rearranged to arrive at a more coherent framework. The aim was to obtain intra-category correlations (correlations between indicators of the same category) that were higher than the inter-category correlations (correlations of the indicators belonging to different categories). Hence, high correlations between indicators suggest that these indicators should be united into one category. Finally, it is aimed to equalize the number of indicators per category, to make sure that all categories equally contribute to the BCI. Furthermore, an aggregation method has been selected that penalizes unbalanced scores in order to emphasize the need to improve the lowest indicator scores.

4 RESULTS

4.1 Step 1: Development of the Trends and Pressures Framework

Because the regenerative capacity and renewable resources on earth are limited (Hoekstra and Wiedman 2014), environmental pressures of cities need to be reduced provided that adequate living standards are maintained (Mori and Yamashita 2015). Hence, an equal number and weighing of indicators for social, environmental and financial classes has been strived for. Each indicator has been scaled from 0 to 4 points, where a higher score represents a higher urban pressure or concern. The following ordinal classes, expressed as 'degree of concern', have been used: 0–0.5 points (no concern), 0.5–1.5 (little concern), 1.5–2.5 (medium concern), 2.5–3.5 (concern), and 3.5–4 (great concern). In this way, a TPF is provided that depicts the most relevant topics that either hamper sustainable IWRM or, on the contrary, pose opportunity windows. Examples for three cities are shown in Section 4.4. For the social and financial categories, four well-known descriptive indicators have been selected, that are widely used to describe urban pressures. The environmental category is focussed on urban water and climate issues, i.e., flooding, water scarcity and heat risk (EEA 2012). Water quality is also included because cities are often situated at rivers and in deltas that receive pollution from upstream activities.

Table 1 describes the scaling method and data sources of the indicators and sub-indicators that are proposed. Most of these indicators are scaled according to existing scaling methods. Furthermore, sub-indicator 5.4 (*Land subsidence*) and 6.3 (*Salinization and seawater intrusion*) are scaled according to an ordinal self-assessment.

Table 1 Summary of calculation methods for the indicators of the Trends and Pressures Framework developed in step 1 of this revision (Fig. 1). Three indicator categories are distinguished: (a) social indicators (nrs. 1–4), (b) environmental indicators (nrs. 5–8), and (c) financial indicators (nrs. 9–12). All indicators are categorized into five concern scores varying from 0 to 4, i.e., 0–0.5 points (no concern), 0.5–1.5 (little concern), 1.5–2.5 (medium concern), 2.5–3.5 (concern), and 3.5–4 (great concern). The equations are based on publicly available data and the derived relations are all significant ($p \max < 10^{-6}$). For a more detailed description of the data source and calculation method reference is made to the EIP City Blueprint website (EC 2015)

Indicator	**Unit X**	**Method/Equation**	**Source**
1. Urbanization rate	% per year	Score = $-0.114X^2 + 1.3275X + 0.1611$ r = 0.999	CIA 2014
2. Burden of disease	DALY's	Classification World Health Organization	WHO 2014
3. Education rate	% primary education	Score = $-10^{-5}X^3 + 0.0012X^2 - 0.0426X + 4.3057$ r = 0.974	World 2012
4. Political instability	World Bank standardized	Classification of the World Bank	World 2013a
5.1 Urban drainage flood	%flooded	Score = $6.10^{-8}X - 2.10^{-5}X^{-4} + 0.0014X^3 - 0.0526X^2 + 0.8302X - 3.8745$ r = 0.992	EEA 2015b
5.2. River peak discharges	% flooded	Classification of the EEA (2012)	EEA 2015b
5.3. Sea level rise	% flooded	Classification of the EEA (2012)	EEA 2015b
5.4. Land subsidence	Self-assessment	0 = No infrastructure damage, no flood risk. 1 = Low infrastructure damage expected, no increase in flood risk expected.	Case studies Local reports

		2 = Infrastructure damage or <0.50 m subsidence. by 2100 in substantial urban area. 3 = Serious infrastructural damage or <1 m subsidence substantial urban area by 2100. 4 = As 3 with Imminent flood risk.	
6.1. Freshwater scarcity	% use of renewable resource	Classification of the OECD (2004) and WRI (2013)	FAO 2015
6.2. Groundwater scarcity	% use of renewable resource	Classification of the IGRAC (2010)	FAO 2015
6.3. Salinization and seawater intrusion	Self-assessment	0 = Both not reported and city not vulnerable 1 = Both not reported but city is vulnerable in the coming century and please use capitals in One/both 2 = one/both not reported but imminent threat 3 = one/both reported 4 = one/both reported seriously affecting the city	Case studies Local report s
7.1. Surface water quality	WQI (Water Quality Index)	Score = [100 – WQI] / 25	EPI 2010
7.2. Biodiversity	% Water (effect on ecosystems) index	For EU: % water bodies less than good quality[a] For other: Score = 100 – [Water (effects on ecosystems)/25]	EEA 2015a EPI 2010
8. Heat island effect		Average of the following two scores: (1) Number of hot days >35 °C and nights >20 °C[b] (2) Green space coverage (%)[c]	EEA 2015b EEA 2012
9. Economic pressure	GDP $capita^{-1}day^{-1}$	Score = −0.783 ln(X) + 4.115 r = 0.995	IMF 2013
10. Unemploymen	%	Score = $0.0002X^2 + 0.5077X - 0.8356$	World

t rate		r = 0.989 If <2 % score is 0 points and if x > 20 score is 4	2014
11. Poverty rate	% <2US$ day^{-1}	Score = $-0.0001X^2 + 0.0404X + 1.168$ 6 r = 0.994	World 2015
12. Inflation rate	% year^{-1}	Score = $0.0025X^3 - 0.0744X^2 + 0.866X + 0.0389$ r = 0.996	World 2013b

[a] The scores for the classification of water bodies (EEA 2015a) are as follows: 0–10 % (0), 10–30 % (1), 30–50 % (2), 50–70 % (3), >70 % (4)

[b] Annual number of hot days >35 °C and nights >20 °C standardized in a 0–4 range (boundaries: min. 0; max. 50)

[c] Green space coverage (%) standardized in a 0–4 range (boundaries have been set at the upper and lower 10 % of approximately 600 cities according to EEA (2012))

For seven indicators and sub-indicators we have proposed a scoring method as no international standards are available. The scores are determined using the ranking of the city amongst all available country scores (Fig. 2). The rankings are linearly standardized on a scale from 0 to 4 points (y-axis Fig. 2). Subsequently, an equation has been selected that best fits this ranking (lowest correlation coefficient: r = 0.97; Table 1). Next, these equations are applied in order to determine the indicator scores for each city. The numbers are rounded and scores of 3 or 4 points are marked in red and communicated to the stakeholders because they are considered as a concern or great concern (see Table 3 in Section 4.4). These scores are not normative but only provide an indication of the urban pressures. Finally, the Trends and Pressure Index (TPI; the arithmetic mean of all twelve TPF indicators) can be calculated for each city.

4.2 Step 2: Improvements of the City Blueprint Indicators

4.2.1 The Use of Wastewater Treatment as Indicator for Surface Water Quality

Baseline assessments of IWRM in cities should preferably include surface water quality and biodiversity. Unfortunately, this appears problematic as a result of a lack of local data and questionable data reliability. In the current CBF, the indicator *surface water quality* uses data from the water quality index (WQI) of the environmental performance index (EPI 2010). The WQI is the only global database of water quality for inland waters. However, insufficient spatial and temporal coverage of measurements, poor reporting and inconsistent sampling design, all posed data reliability problems (Srebotnjak et al. 2012). Therefore, the EPI stopped using the WQI after the year 2010 (EPI 2012).

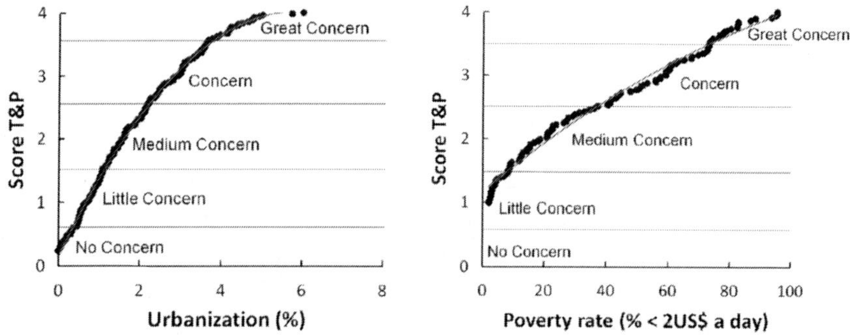

Figure. 2 Two examples of trends and pressure indicators developed in step 1 of this revision (Fig. 1). Left: urbanization rate (CIA 2014). Right: poverty rate (World Bank 2014). The urbanization and poverty rates of all available countries are ranked and subsequently linearly converted into scores of 0–4 (y-axis). Based on this score, ordinal classes, expressed as 'degree of concern', are used. Table 1 provides the calculation method

A similar concern is the current indicator *biodiversity*, for which information is provided for member states of the European Union (EU), but this information is not available for non-EU countries (EEA 2014a). As an estimate for non-EU countries, the EPI water (effects on ecosystems) index has been used (EPI 2010). However, data reliability and the use of different databases appeared to be problematic (Emerson et al. 2010; Koop and Van Leeuwen 2015b). Moreover, both indices are national averages which lead to overestimations of urban water quality and biodiversity because cities are often large emitters of pollutants (Van Leeuwen 2013; Gessner et al. 2014).

Cities may have hardly any direct influence on the scores for surface water quality and biodiversity because of upstream pollution. Hence, the indicators *water quality* and *biodiversity* should focus on the city's own IWRM performance to prevent pollution. The coverage of secondary Waste Water Treatment (WWT) greatly determines the quality and biodiversity in urban waters because detrimental effects on aquatic species due to oxygen depletion are avoided. WWT removes at least 70 % of the biological and 75 % of the chemical oxygen demand (OECD 2013). Furthermore, the coverage of tertiary WWT avoids eutrophication by removing nitrogen and phosphorous, as well as other (microbiological) pollutants (OECD 2013). Secondary and tertiary WWT data are often available at the level of cities, frequently updated and widely used by international organizations (ISO 2014; UN Water 2014; OECD 2013; Siemens 2015; SOPAC 2004).

The scores for indicators 10 (*nutrient recovery*), 11 (*energy recovery*) and 12 (*sewage sludge recycling*) have been updated. These indicators reflect the reuse of wastewater as a fraction of the water that is passing the city's wastewater treatment plants; whereas the city's WWT service coverage is disregarded. Hence, a city that only treats a small fraction of its wastewater, but fully recycles the nutrients, sludge and applies full energy recovery from this small fraction, would receive a high score for these indicators while loads of potentially abstractable nutrients, sludge and energy actually flush away. In order to take all urban wastewater into account, the scores are therefore multiplied by the fraction of wastewater (F) that is actually treated at the WWT facilities (Eqs. 1, 2 and 3).

$$\text{Score for nutrient recovery} = (\text{nutrient recovery } (\%) * F)/10 \quad (1)$$

$$\text{Score for energy recovery} = (\text{energy recovery}(\%) * F)/10 \quad (2)$$

$$\text{Score for sewage sludge recycling} = (\text{sewage sludge recycling } (\%) * F)/10 \quad (3)$$

4.2.2 Public Participation

The Voluntary Participation Index (VPI) is used for indicator 23 (*public participation*) and represents the average number of memberships in voluntary organizations as measured for EU-countries in 2003 (EFILWC 2006). We have updated this indicator by using the percentage of people involved in unpaid work (EFILWC 2012). For non-EU countries a best estimate is provided (Eq. 4) based on the high correlation with the World Bank indicator *Rule of Law* ($n = 27$; $r = 0.84$; $p < 0.0000001$).

$$\text{Public participationscore} = 0.0657 * \text{score for Rule of Law} - 2.2278 \quad (4)$$

4.2.3 Indicator Boundaries

The indicators of the CBF are standardized according to the min-max method, by using percentages or by a self-assessment scoring method as explained in the City Blueprint questionnaire (EC 2015). The min-max standardization method is sensitive for outliers that may result in unrealistic scores and therefore indicator boundaries have been reviewed. For instance, indicator 16 (*water system leakages*) measures the leakage rate of the drinking water distribution system.

The indicator boundaries for leakage have been set at 0 and 100 %, respectively. Consequently, 0 % is equivalent to a completely leaking water distribution system which results in a low variance of this indicator. Therefore, the maximum boundary of water leakage has been arbitrarily reset at 50 %.

The same holds for indicator 15 (*average age sewer system*), where the maximum age of the urban sewers has been lowered from 100 to 60 year. For large datasets (n > 100), the average of the lower and upper 10 % have been used as minimum and maximum. This is done for indicator 4 (*solid waste collected*), 17 (*operation cost recovery*) and 18 (*green space*).

4.2.4 Water Footprint Indicators

Based on constructive criticism from local water authorities concerning the use of the Water Footprint (WF) concept (used in the former category *water security*), it has been decided to discontinue their use. The WF describes the total volume of water over the full supply chain that is needed to produce a product or that is used by a country (Hoekstra et al. 2009). Water imports and exports of a country are highly dependent on many social-economic processes and national and global trends on which the local water authorities have a negligible influence.

Moreover, the indicator *water scarcity* and *water self-sufficiency* are largely determined by the total renewable water resource which is abundant or not, regardless of the interventions by local water authorities. In fact, temperature, rainfall, soil type, river course, etc., largely determine the total renewable water resources.

4.2.5 Solid Waste

Cities are prone to water pollution due to their highly efficient drainage systems that quickly collect and discharge polluted water. Cities are the largest source of plastic waste that enters the oceans via rivers and canals. About 280 million tonnes of plastics ends up in the oceans annually, where it photodegrades into small particles and affects marine ecosystems (Derraik 2002; Sigler 2014). The degradation of plastic waste in landfills is approximately 1 to 5 % during 100 years (Bez et al. 1998).

Urban solid waste treatment is therefore a key performance-oriented aspect that determines the water quality in cities, rivers and oceans. According to the European Commission (EC 2008), the priority order for waste reduction is waste prevention, reuse, recycling, recovery and disposal. Hence, indicator 4 (*solid waste collected*), 5 (*solid waste recycled*) and 6 (*solid waste energy recovered*) have been included.

4.3 Step 3: Indicator Rearrangements and Aggregation Method

The contribution of the indicators and categories to the BCI are currently highly variable. Indicators with low variances contribute less to the BCI. Moreover, the variances differ strongly and the number of indicators per category are not equal (Koop and Van Leeuwen 2015b). However, an equal contribution of each category and indicator is strived for. The final rearrangement of the indicators and categories is shown in Table 2. Currently, no particular penalty for unbalanced scores are given, and therefore the need to improve the achievements for low scoring indicators needs to be addressed (JRC 2014).

Table 2 Overview of the performance indicators of the improved City Blueprint Framework (CBF) as described in step 2 and 3 of the revision (Fig. 1). Indicators in bold are new and indicators with adjusted scaling or new data bases are in italics. The other indicators have not been adjusted

I. Water quality	**1. Secondary WWT**
	2. Tertiary WWT
	3. Groundwater quality
II. Solid waste treatment	**4. Solid waste collected**
	5. Solid waste recycled
	6. Solid waste energy recovered
III. Basic water services	7. Access to drinking water
	8. Access to sanitation
	9. Drinking water quality
IV. Wastewater treatment	*10. Nutrient recovery*
	11. Energy recovery
	12. Sewage sludge recycling
	13. WWT energy efficiency
V. Infrastructure	14. Stormwater separation
	15. Average age sewer
	16. Water system leakages

	17. Operation cost recovery
VI. Climate robustness	**18. Green space**
	19. Climate adaptation
	20. Drinking water consumption
	21. Climate-robust buildings
VII. Governance	22. Management and action plans
	23. Public participation
	24. Water efficiency measures
	25. Attractiveness

4.3.1 Basic Water Services

An indicator intra-correlation analysis revealed that the indicators *safe sanitation, sufficient to drink* and *drinking water quality* strongly correlate (Koop and Van Leeuwen 2015b). Access to proper drinking water and improved sanitation are basic water services which are united into a new category, i.e., category III (*basic water services*). The remaining indicators in the former category *sanitation*, all deal with wastewater treatment and the category has been renamed to '*wastewater treatment*' (category IV; Table 2). This is in accordance with the ISO37120, which is a much broader urban indicator framework that also includes a '*water and sanitation*' and '*wastewater*' category (ISO 2014).

4.3.2 Climate Robustness

The correlation coefficient between the indicators *climate commitment* and *adaptation measures* is high ($r = 0.93$). Given the similar ordinal assessment method, it is most likely that these indicators overlap. Furthermore, given the high correlations with indicators 19, 20, 22 and 23 ($r > 0.71$), it is presumable that the indicator *climate commitment* is redundant and has therefore been removed (Koop and Van Leeuwen 2015b). Although climate change impacts are complex and ubiquitous, three climate vulnerabilities with particular reference to urban areas have been identified, i.e., UHI, water scarcity and flooding (EEA 2012). The UHI exacerbates heat waves in cities because the cooling effect of vegetation is replaced by surfaces sealed with concrete, asphalt and stone (EEA 2012). Green and blue areas store rainwater and evaporate this water, thereby mitigating heat waves and storm events.

In addition, green and blue areas alleviate air pollution, increase the city's recreational value and reduce water polluted (Li et al. 2014; Jia et al. 2013; Jonker et al. 2014; Czemiel Berndtsson 2014). Therefore, the share of blue and green area (%) is added as a new indicator in category VI (*climate robustness*), shown in Table 2. Climate change will also increase the urban vulnerability to water scarcity (EEA 2012). The domestic and industrial use of drinking water is an adequate indication of how cities combat water scarcity. Therefore indicator 20 (*drinking water consumption*) is also included in category VI (*climate robustness*) (Table 2).

4.3.3 Governance

At this stage of the framework rearrangement, the former category *biodiversity and attractiveness* only includes indicator 25 (*attractiveness*), because the indicator *biodiversity* has previously been removed. Indicator 25 (*attractiveness*) measures the incorporation of blue-green areas which is mainly the result of urban planning by local authorities, the local community and private companies (shops, factories, restaurants, bars etc.) who want to shape an attractive place to live.

The relation with governance is also indicated by the correlations with the indicator 22 (*management and action plans*) ($r = 0.53$), 23 (*public participation*) ($r = 0.61$) and 24 (*water efficiency measures*) ($r = 0.74$). *Water efficiency measures* assesses to what extent water efficiency measures are applied. Because correlated indicators 22, 23, 24 and 25 all assess how local authorities manage water issues, these indicators have been united into category VII (*governance*).

4.3.4 Infrastructure

The former category *infrastructure* is only poorly accounted for given the large maintenance cost and large global investment deficit in water infrastructure (AWWA 2001; Cashman and Ashley 2008). Therefore indicator 16 (*water system leakages*) has been added because it is suitable to represent the maintenance of the drinking water infrastructure. In contrast with data on total investments in water infrastructure, data of operation costs and revenues of drinking water and sanitary services is often available (IBNET 2015; OECD 2010).

The ratio of the total yearly operating revenues divided by the operating costs of drinking water and sanitation services is an important indicator of the financial state of the local water authorities and their ability to make the necessary investments in infrastructure (OECD 2015). Hence, indicator 17 (*operation cost recovery*) is added to strengthen category V (*infrastructure*), given in Table 2.

4.3.5 Aggregation Method

For composite indices, issues of weighting and aggregation are sensitive and subjective (EPI 2010). It should be recognized that assigning explicit weightings, by definition, represents only one viewpoint. Therefore, no indicator or category weightings are applied. Moreover, the implicit weighting, due to correlations and/or differences in variances are addressed before the indicators are aggregated.

Most frequently used aggregation methods are the arithmetic and geometric mean. The arithmetic mean gives no particular penalty for unbalanced scores, and consequently does not address the urgent need to improve achievements for the lowest scores (JRC 2014).

However, it is essential to regard water management in an integrative way, e.g., increasing access to sanitation greatly improves human hygiene but without adequate investments in WWT, this leads to a strong emission increase in hazardous pollutants and nutrients. Hence, a high score for access to sanitation should not fully compensate a low score for WWT coverage (Ligtvoet et al. 2014).

Since a geometric mean can be defined as the nth root of the product of n numbers or as the anti-log of the sum of logs divided by the number of samples and the log zero (0) is not defined, the calculation of the geometric mean method requires strictly positive values. It is therefore chosen to re-standardize the indicators to a 1–11 score, aggregate the indicators with the geometric mean and finally subtract 1 point from this score. In this way, balanced indicator performances are rewarded.

4.4 Examples

For all 45 cities data have been gathered for the analyses of their TPF and improved CBF. In this section examples of only three cities are presented, i.e., Dar es Salaam, Melbourne and Amsterdam. The TPF is shown in Table 3.

The overall Trends and Pressure Index (TPI) for Dar es Salaam, Melbourne and Amsterdam are 2.7, 1.1 and 1.2 respectively. The improved City Blueprints for these cities are shown in Fig. 3.

The new BCI for Dar es Salaam, Melbourne and Amsterdam are 1.3, 5.4 and 8.3 respectively. The performance ranking of these cities is shown for each category (Fig. 4).

Table 3 Trends and Pressures Framework (TPF) of Dar es Salaam, Melbourne and Amsterdam with their concern scores: 0 (no concern), 1 (little concern), 2 (medium concern), 3 (concern), and 4 (great concern)

		Dar es Salaam	Melbourne	Amsterdam
Social	1. Urbanization rate	4	1	1
	2. Burden of disease	3	1	0
	3. Education rate	3	0	1
	4. Political instability	2	1	1
Environmental	5. Water scarcity	2	1	1
	6. Flood risk	3	2	3
	7. Water quality	1	2	2
	8. Heat risk	3	4	1
Financial	9. Economic pressure	4	0	1
	10. Unemployment rate	1	1	1
	11. Poverty rate	4	0	0
	12. Inflation rate	3	2	1

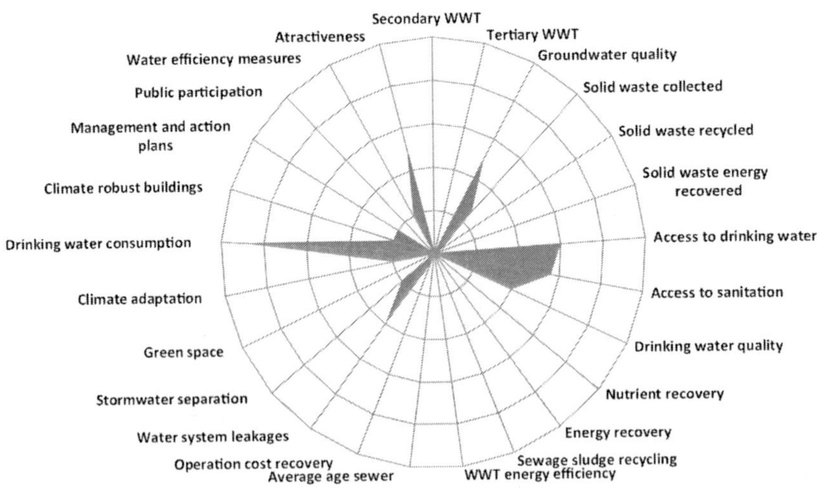

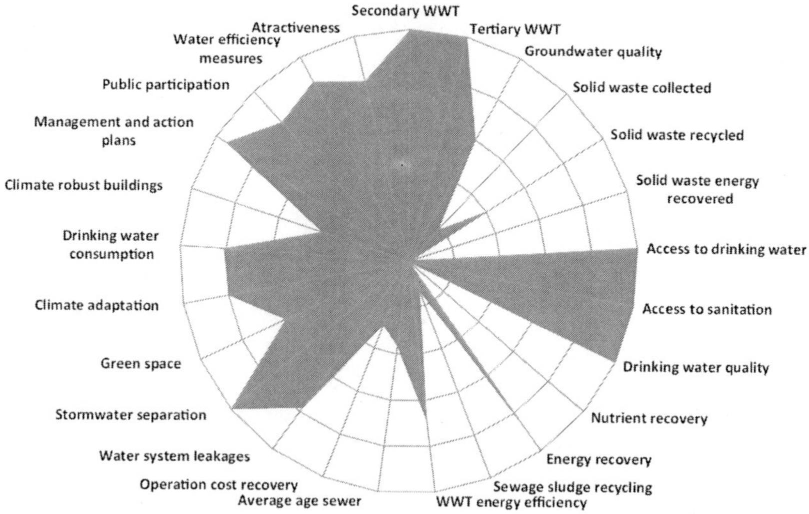

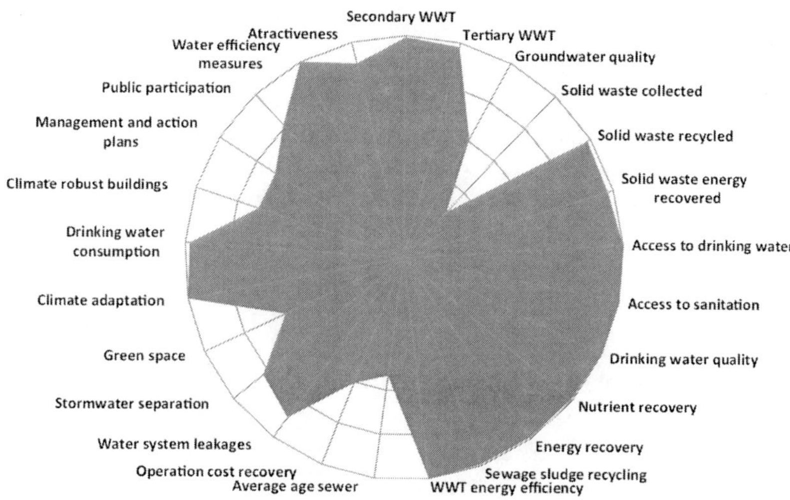

Figure. 3 Spider diagram application of the improved performance framework (Fig. 1). Examples of three spider diagrams based on 25 performance indicators for Dar es Salaam (top), Melbourne (centre) and Amsterdam (bottom). The geometric mean of the indicators, i.e., the BCI, are 1.3, 5.4 and 8.3, respectively

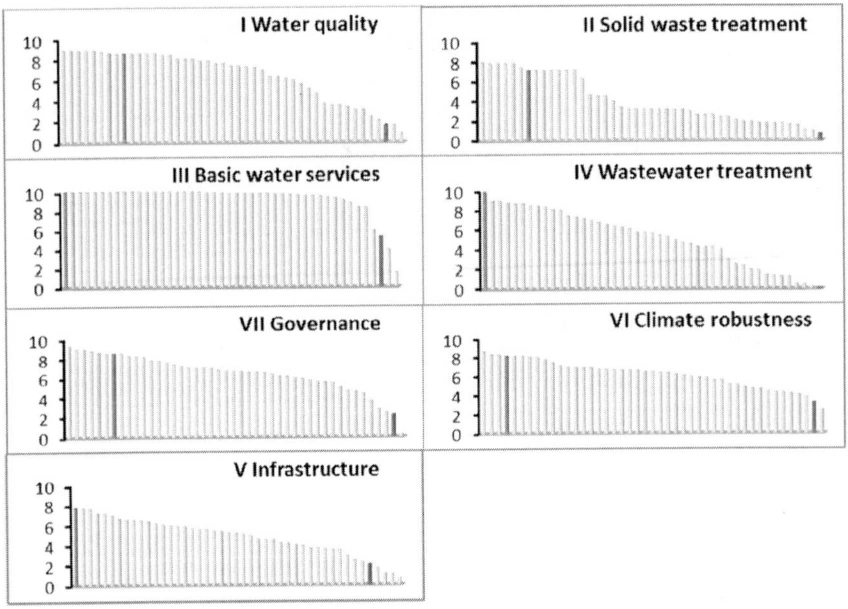

Figure. 4 Category ranking of the City Blueprints of Dar es Salaam (*red*), Melbourne (*orange*) and Amsterdam (*green*). The arithmetic average of the indicator scores for each of the seven categories (Table 2) are ranked from high to low for each of the 45 municipalities and regions. This ranking is part of the application of the performance framework (Fig. 1)

5 DISCUSSION

5.1 City Blueprint Improvements

We have developed two separated indicator frameworks that embody the distinction between trends and pressures (TPF) and IWRM performance of a city (CBF). In the TPF only indicators that are of concern or great concern (3 or 4 points) are explicitly communicated to the stakeholders (Table 3). The CBF is essential to show the potential gain possible by sharing knowledge, experiences and best practices amongst cities. Moreover, it more accurately depicts the own activities and efforts of cities to improve the sustainability of their IWRM and how they can alleviate environmental pressures such as emissions of poorly treated wastewater and solid waste. The TPF provides a wider context, which is key in obtaining insight into the limitations and windows of opportunities to improve urban IWRM. Finally, the categories of the 45 cities are averaged and ranked in order to provide a quick overview of potential improvements compared to other cities (Fig. 4). These simple graphical representations allow for a quick overview of cities on their path to become water-wise and climate prove.

5.2 Limitations

Only publicly available data is added to this improved CBF in order to promote transparency, and to reduce time and costs. City scale data concerning urban IWRM appeared to be particularly scarce. In fact, this information is often available, but publicly inaccessible which hampers city-to-city learning (EEA 2014b). Moreover, existing urban water-related indicators are often not standardized, consistent, or comparable over time and between cities (ISO 2014). For a considerable number of indicators national data are used because local data is not publicly available for all cities. This limits the accuracy of city assessments, especially in large countries with high regional variety.

Basic statistics obtained from the 45 cities have been used to reconstruct the CBF and to arrive at an approximately proportional contribution of all indicators and categories. However, the 45 cities that are used for this statistical analysis have a distribution bias as most cities are located in North-West Europe and are therefore not representative for the global urban challenges. This is a limitation of our work. Hence, it should be taken into account that the actual variance of the indicators may differ from the output of cities that have been analysed. For instance, the variance of category 3 (*basic water services*) and 7 (*governance*) is expected to be larger if the cities would really represent a global sample, i.e., if cities from developing countries would not be underrepresented.

5.2.1 Indicator Limitations

The data for the baseline assessment proposed in this paper is gathered by means of a questionnaire (EC 2015) and by using publicly available data. In order to include a large number of cities of different size and geography, the indicator framework is limited by the information that was available. Hence, the set of indicators can and should be extended for some cities. Public data concerning flood risks appeared to be particularly sparse. Information on urban flood vulnerability is available, i.e., vulnerability to river peaks and sea level rise (Jongman et al. 2014; EEA 2012). However, these data do not represent flood protection performances. Hence, these indicators are included in the TPF. Ideally, a flood return interval is used to quantify the performance of flood defence. However, flood return intervals are not consistently reported worldwide, let alone, calculated in comparable manner.

The percentage of uncollected solid waste is inconsistently reported. Only for a very few cities, it was more than zero, preventing it to become a suitable comparative indicator. However, uncollected solid waste can strongly contribute to the release of a variety of pollutants such as heavy metals, persistent bio-accumulative chemicals, pesticides, pharmaceuticals and plastic (Katsanevakis 2011). However, efforts to reduce water polluting activities such as landfilling and solid waste production are explicitly included.

Another major cause of urban water pollution is stormwater runoff. This pollution pathway is related to traffic intensity and uncollected solid waste (Czemiel Berndtsson 2014; Revitt et al. 2014). Urban soil permeability or the use of best practices to reduce or filter stormwater runoff, e.g., by biofilters, infiltration ponds or bioswales, may be insufficiently addressed in the CBF due to limited data availability on soil permeability. Only indicator 18 (*green space*) and 19 (*climate adaptation*) implicitly address these issues. Hence, a supplementary indicator that assesses the application of measures to decrease and filter stormwater runoff could be added.

The min-max method can be sensitive for extreme outliers that may disrupt the scoring. Therefore the average of the lower and upper 10 % from a large data set ($n > 100$) is taken as the minimum and maximum for indicators 4 (*solid waste collection*), 17 (*operation cost recovery*) and 18 (*green space*).

However, for drinking water consumption this was hindered because there was no large dataset that included residential, commercial, industrial and public purposes. Often only domestic water consumption data is provided which represents only 10 % of the total drinking water consumption in the EU (ISO 2014). Likewise, indicator 15 (*average age sewer*) is prone for outliers because minimum and maximum numbers are taken from a limited dataset.

Data coverage of the selected set of indicators is high. Only the operation cost recovery (ratio) of Helsinki and data according to the EEA (2012) assessment method for indicator 18 (*green space*) were not available for non-EU cities. In most cases a realistic default value of 20 % green space is taken (e.g., for Ho Chi Minh City and Istanbul) and only for Melbourne we arbitrarily set this score equal to that of Amsterdam. Finally, the operation cost recovery for Helsinki has been arbitrarily set at 1, as data have not been provided.

5.2.2 Water Infrastructure Investment Deficit

Upgrading of water infrastructure is crucial for water security in cities (OECD 2015). Water infrastructure investment requirements are already high in developed countries (yearly 0.35–1.2 % of GDP) and even higher for developing countries (yearly 0.71–6.30 % of GDP) (Cashman and

Ashley 2008). Hence, the state of the water infrastructure network (indicator 15 *average age sewer*) is decisive for the city's financial performance.

As an indication of the maintenance of the sewer system in a city, we have divided the average by an assumed sewer lifespan. This is an inaccurate approximation since local circumstances that determine the sewer lifespan are not incorporated. A first improvement could be to determine a site-specific maximum sewer lifespan and compare this with the current site-specific average age. A more advanced approach would be to calculate the yearly sewer maintenance investment requirement (Equation 5) as proposed by Prof. Dr. Bosseler (personal communication). The infrastructure lifespan and system asset value should be calculated for each city. By using equation (5), an annual investment can be calculated that should be reserved for long-term infrastructure maintenance. The actual investment in infrastructure can be scored as a fraction of this annual investment requirement.

$$\text{Investment requirements } (€/\text{year}) = \frac{System\ Asset\ Value(€)}{Lifespan(years)} \tag{5}$$

6 CONCLUSIONS

The goal of this study is a methodological review of the City Blueprint indicator framework to improve the assessment of the sustainability of IWRM in cities (Fig. 1). The results of the application of this improved method has been reported in this journal (Koop and Van Leeuwen 2015a). This revision was necessary to better emphasize cities' opportunities to envision, develop and implement stepwise measures to transform towards water-wise or water sensitive cities. The following changes have been made:

1) A distinction between descriptive and performance-oriented indicators has been made by developing a Trends and Pressures Framework (TPF; on which the city's IWRM has a negligible influence) and improving the city-level IWRM performance framework, i.e., the City Blueprint Framework (CBF).

2) The TPF includes the following social, environmental and financial indicators: urbanization rate, burden of disease, education rate, political instability, water scarcity, flood risk, water quality, heat risk, economic pressure, unemployment rate, poverty rate and inflation rate. The data are publicly available (Table 1).

3) We distinguish 5 ordinal classes, varying from no concern to great concern, for each of these 12 TPF indicators.

4) In the revised CBF, seven indicators have been added, i.e., secondary and tertiary wastewater treatment, operation cost recovery, green space and three indicators concerning *solid waste treatment* (Table 2).

5) Adjusted scaling or new databases are used for the following CBF indicators: nutrient recovery, energy recovery, sewage sludge recycling, average age sewer, water system leakages and public participation (Table 2).

6) The geometric aggregation method has been chosen to calculate the Blue City Index for the 25 indicators of the CBF.

7) We have applied this new CBF, TPF and BCI to 45 cities in 27 countries (Koop and Van Leeuwen 2015a).

8) The focus on performance will enhance city-to-city learning, i.e., sharing of knowledge, experiences and best practices between cities. This is the ultimate goal of our EIP Water Action Group (EC 2015) as the need to improve IWRM in cities is crucial and the time window to do this, is closing rapidly (OECD 2015; Van Leeuwen 2013).

ACKNOWLEDGMENTS

We would like to thank Prof. Dr. Bosseler (Institute for Underground Infrastructure, Gelsenkirchen, Germany) for his advice on water infrastructure investment deficits. The City Blueprint Action Group is part of the governance activity of the European Innovation Partnership on Water of the European Commission, coordinated by both Dr. Richard Elelman of Fundació CTM Centre Tecnològic and NETWERC H2O (Manresa, Spain) and Prof. Dr. Kees Van Leeuwen of KWR Watercycle Research Institute. The European Commission is acknowledged for the support of our City Blueprint action and for the BlueSCities project in H2020-Water under Grant Agreement No. 642354.

REFERENCES

1. AWWA (2001) American water works association. Reinvesting in drinking water infrastructure. Government Affair Office, New York
2. Bez J, Heyde M, Goldhan G (1998) Waste treatment in product specific life cycle inventories. An approach of material-related modelling. Part II: Sanitary landfill. Int J Life Cycle Ass 3:100–105

3. Brown R, Ashley R, Farrelly M (2011) Political and professional agency entrapment: an agenda for urban water research. Water Resour Manag 25:4037–4050
4. Cashman A, Ashley R (2008) Costing the long-term demand for water sector infrastructure. Foresight 10:9–26
5. CIA (2014) Central Intelligence Agency: The World Factbook. Urbanization https://www.cia.gov/library/publications/the-world-factbook/fields/2212.html Accessed 18 March 2015
6. Czemiel Berndtsson J (2014) Storm water quality of first flush urban runoff in relation to different traffic characteristics. Urban Water J 11:284–296
7. Derraik JGB (2002) The pollution of the marine environment by plastic debris: a review. Mar Pollut Bull 44:842–852
8. EC (2008) European Commission. Directive 2008/98/EC of the European parliament and of the council on waste and repealing certain directives – Appendix II, 19 November 2008
9. EC (2015) European Commission: European Innovation Partnership on Water. Brussels, Belgium. http://www.eip-water.eu/City_Blueprints Accessed 12 March 2015
10. EEA (2012) European environment agency. Urban adaptation to climate change in Europe. Challenges and opportunities for cities together with supportive national and European policies. Synthesis, Copenhagen, Denmark
11. EEA (2014a) European Environmental Agency: Water Framework Directive and monitoring http://www.eea.europa.eu/themes/water/status-and-monitoring/water-framework-directive-and-monitoring Accessed 29 January 2015
12. EEA (2014b) European environmental agency: performance of water utilities beyond compliance. Sharing knowledge bases to support environmental and resource-efficiency policies and technical improvements. Technical report No 5/2014, Copenhagen, Denmark
13. EEA (2015a) European environmental agency: Percent of classified water bodies in less than good ecological status of potential. http://www.eea.europa.eu/data-and-maps/figures/proportion-of-classified-surface-water/proportion-of-classified-surface-water/image_original Accessed 8 January 2015
14. EEA (2015b) European environmental agency: Urban adaptation to climate change. Annex II. City data sensitivity. http://www.eea.europa.eu/data-and-maps/figures/mean-soil-sealing-in-european Accessed 19 January 2015
15. EFILWC (2006) European foundation for the improvement of living and working conditions. First European Quality of life survey: Participation in civil society, Dublin
16. EFILWC (2012) European foundation for the improvement of living and working conditions. Quality of life in Europe: Impact of the crisis, Luxembourg
17. Emerson JDC, Esty MA, Levy CH, Kim V, Mara ADE, Sherbinin E, Srebotnjak T (2010) 2010 Environmental performance index. Yale Center for Environmental Law and Policy, New Haven
18. EPI (2010) Environmental performance index: http://www.ciesin.columbia.edu/repository/epi/data/2010EPI_country_profiles.pdf Accessed 19 January 2015
19. EPI (2012) Environmental Performance Index: Environmental Performance Index and Pilot Trend Environmental Performance Index. http://epi.yale.edu/files/ 2012_ epi_report.pdf Accessed 21 February 2015
20. FAO (2015) Food and Agriculture Organization of the United Nations: AQUASTAT database. http://www.fao.org/nr/water/aquastat/data/query/index.html; jsessionid=B022D1C2732DF571D2A384B57E0128D6 Accessed 19 January 2015
21. Gessner MO, Hinkelmann R, Nützmann G, Jekel M, Singer G, Lewandowski J, Nehls T, Barjenbruch M (2014) Urban water interfaces. J Hydrol 514:226–232
22. Gill S, Handley J, Ennos A, Pauleit S (2007) Adapting cities for climate change: the role of the green infrastructure. Build Environ 1:115–133
23. Hoekstra AY, Wiedman TO (2014) Humanity's unsustainable environmental footprint. Science 344:1114–1117
24. Hoekstra AY, Chapagain AK, Aldaya MM, Mekonnen MM (2009) The water footprint manual. Setting the global standard, 2009. Water Footprint Network, Enschede

25. IBNET (2015) The International Benchmarking Network for Water and Sanitation Utilities: Search by country http://www.ib-net.org/en/production/?action=country Accessed 10 March 2015
26. Iglesias A, Garrote L, Flores F, Moneo M (2007) Challenges to manage the risk of water scarcity and climate change in the Mediterranean. Water Resour Manag 21:775–788
27. Igrac (2010) International Groundwater Resources Assessment Center: Groundwater Development stress http://www.unesco.org/new/fileadmin/MULTIMEDIA/ HQ/SC/temp/wwap_pdf/Groundwater_development_stress_GDS.pdf Accessed 21 January 2015
28. IMF (2013) International Monetary Fund: World economic outlook database, April 2013 http://www.imf.org/external/pubs/ft/weo/2013/01/weodata/index.aspx Accessed 19 January 2015
29. IPCC (2013) International panel on Climate Change: Climate change 2013. The Physical science basis. Contribution of Working Group I to the Fifth Assessment Report of the Intergovernmental Panel on Climate Change [Stocker TF, Qin D, Plattner GK, Tignor M, Allen SK, Boschung J, Nauels A, Xia Y, Bex V and Midgley PM (eds.)]. Cambridge University Press, Cambridge, United Kingdom and New York, NY, USA
30. ISO (2014) International standards organization. Sustainable development of communities. Indicators for city services and quality of life. ISO, Geneva
31. Jia B, Wang C, Qiu E, Qie G (2013) The status and trend on the urban tree canopy research. Acta Ecol Sin 33:23–32
32. Johnson T (2013) Hurricane Sandy leaves state with $2.6B tab for water infrastructure. http://www.njspotlight.com/stories/13/04/09/hurricane-sandy-leaves-state-with-2-6b-tab-for-water-infrastructure/ Accessed 27 July 2015
33. Jongman B, Hochrainer-Stigler S, Feyen L, Aerts JCJH, Mechler R, Botzen WJW, Bouwer LM, Pflug G, Rojas R, Ward PJ (2014) Increasing stress on disaster-risk finance due to large floods. Nat Clim Change 4:264–268
34. Jonker MFJ, Van Lenthe F, Donkers B, Mackenbach JP, Burdorf A (2014) The effect of urban green on small-area (healthy) life expectancy. J Epidemiol Community Health 68:999–1002
35. JRC (2014) Joint Research Center: Climate Impacts in Europe The JRC PESETA II Project. Report EUR 26586EN, Seville http://publications.jrc.ec.europa.eu/repository/ handle/111111111/31757 Accessed 17 March 2015
36. Katsanevakis S (2011) Marine debris, a growing problem: sources, distribution, composition, and impacts. Nova, New York
37. Koop SHA, and Van Leeuwen CJ (2015a) Application of the improved City Blueprint Framework in 45 municipalities and regions. Water Resour Manag 29:4629–4647
38. Koop SHA, and Van Leeuwen CJ (2015b) Towards sustainable water resources management: Improving the city blueprint framework. Report KWR 2015.025. KWR Watercycle Research Institute, Nieuwegein, the Netherlands. http://www.eip-water.eu/City_Blueprints Accessed 4 August 2015
39. Leonardsen L (2012) Financing adaptation in Copenhagen http://resilient-cities.iclei.org/fileadmin/sites/resilient cities/files/Webinar_Series/Webinar_Presentations/Leonardsen__financing_adaptation_in_Copenhagen_ICLEI_sept_2012.pdf Accessed 12 May 2015
40. Li F, Wang RS, Zhao D (2014) Urban ecological infrastructure based on ecosystem services: status, problems and perspectives. Acta Ecol Sin 34:190–200
41. Ligtvoet W, Hilderink H, Bouwman A, Puijenbroek P, Lucas P, Witmer M (2014) Towards a world of cities in 2050. An outlook on water-related challenges. Background report to the UN-Habitat Global Report. PBL Netherlands Environmental Assessment Agency
42. McKinsey (2013) Infrastructure productivity: How to save $1 trillion a year. McKinsey & Company
43. Mori K, Yamashita T (2015) Methodological framework of sustainability assessment in city sustainability index (CSI): a concept of constraint and maximisation indicators. Habitat Int 45:10–14
44. Nilsen V, Lier JA, Bjerkholt J, Lindholm OG (2011) Analysing urban floods and combined sewer overflows in a changing climate. J WateR Clim Change 2:260–271

45. OECD (2004) Organization for economic communication and development: OECD key environmental indicators, Paris, France http://www.oecd.org/environment/indicators-modelling-outlooks/31558547.pdf Accessed 9 March 2015
46. OECD (2010) Organization for Economic Co-operation and Development: Pricing water resources and water and sanitation services http://browse.oecdbookshop.org/ oecd/pdfs/product/9710041e.pdf Accessed 10 March 2015
47. OECD (2013) Organization for Economic Co-operation and Development: Environment at a glance 2013. OECD indicators. OECD Publishing. DOI: 10.1787/9789264185715-enhttp://www.oecd-ilibrary.org/environment/environment-at-a-glance-2013_9789264185715-en Accessed 7 March 2015
48. OECD (2015) Organization for economic co-operation and development: water and cities. Ensuring sustainable futures. OECD studies on water. OECD Publishing. doi:10.1787/9789264230149-en
49. Philip R, Anton B, van der Steen P (2011) SWITCH training kit. Integrated urban water management in the city of the future. Module 1. Strategic planning, ICLEI, Freiburg, Germany. http://www.switchtraining.eu/ Accessed 10 March 2015
50. Revitt DM, Lundy L, Coulon F, Fairley M (2014) The sources, impact and management of car park runoff pollution: a review. J Environ Manage 146:552–567
51. Short MD, Peirson WL, Peters GM, Cox RJ (2012) Managing adaptation of urban water systems in a changing climate. Water Resour Manag 26:1953–1981
52. Shuster WD, Bonta J, Thurston H, Warnemuende E, Smith DR (2005) Impacts of impervious surface on watershed hydrology: a review. Urban Water J 2:263–275
53. Siemens (2015) Green city index. http://www.siemens.com/entry/ cc/en/greencityindex.htm Accessed 2 April 2015
54. Sigler M (2014) The effects of plastic pollution on aquatic wildlife: Current situations and future solutions. Water Air Soil Poll 225
55. SOPAC (2004) Applied Geoscience and Technology division of Secretariat of the Pacific Community: Environmental Vulnerability Index http://www.sopac.org/index.php/environmental- vulnerability-index Accessed 25 March 2015
56. Srebotnjak T, Carr GDE, Sherbinin A, Rickwood C (2012) A global water quality index and hot-deck imputation of missing data. Ecol Indic 17:108–119
57. UN (2014) United Nations. World urbanization prospect. The revision 2014. Highlights, New York
58. UNEP (2013) United Nations Environment programme. City-level decoupling. Urban resource flows and the governance of infrastructure transitions. A Report of the Working Group on Cities of the International Resource Panel. Swilling M, Robinson B, Marvin S, Hodson M. Nairobi, Kenia
59. UN Water (2014) United Nations Water: Set of Key Indicators for the Water Sector. http://www.unwater.org/activities/task-forces/indicators/key-indicators/en/ Accessed 19 March 2015
60. Van Leeuwen CJ (2013) City blueprints: baseline assessments of sustainable water management in 11 cities of the future. Water Resour Manag 27:5191–5206
61. Van Leeuwen CJ, Sjerps R (2015) Istanbul: the challenges of integrated water resources management in Europa's megacity. Environ Dev Sustain. doi:10.1007/s10668-015-9636-z
62. Van Leeuwen CJ, Frijns J, Van Wezel A, Van De Ven FHM (2012) City blueprints: 24 indicators to assess the sustainability of the urban water cycle. Water Resour Manag 26:2177–2197
63. WHO (2014) World Health Organization: Burden of disease. Age-standardized disability-adjusted life year –DALY- rates per 100 000 (population): 2012 http://gamapserver.who.int/gho/interactive_charts/mbd/as_daly_rates/atlas.html Accessed 9 January 2015
64. World Bank (2012) Primary completion rate, total (% of relevant age group) http://data.worldbank.org/indicator/SE.PRM.CMPT.ZS/countries/1W-AO?display=default Accessed 13 January 2015
65. World Bank (2013a) Worldwide Governance Indicators http://info.worldbank.org/governance/ wgi/index.aspx#faq. Accessed 12 January 2015

66. World Bank (2013b) Inflation, costumers price (annual %) http:// data. worldbank.org/indicator/FP.CPI.TOTL.ZG. Accessed 19 January 2015
67. World Bank (2014) Unemployment total (% of total labor force) (modelled ILO estimate) http://data.worldbank.org/indicator/SL.UEM.TOTL.ZS Accessed 19 January 2015
68. World Bank (2015) Poverty gap at $2 a day (PPP) (%) http://data. worldbank. org/indicator/SI.POV.GAP2/countries/1W?display=default Accessed 12 February 2015
69. WRG (2009) 2030 Water resources group. Charting our water future. Economic frameworks to inform decision-making, West Perth
70. WRI (2013) World Resources Institute. Aqueduct metadata document. Aqueduct global maps 2.0. http://www.wri.org/sites/default/files/pdf/aqueduct_metadata_global.pdf Accessed 15 May 2015
71. WWDR (2006) World Water Development Report: Water a shared responsibility. Section 2: Changing Natural Systems. The state of the Resource http://www.unesco.org/bpi/wwap/press/pdf/wwdr2_chapter_4.pdf Accessed 25 March 2015

INDEX